Bayesian Predictive Inference and Related Asymptotics—Festschrift for Eugenio Regazzini's 75th Birthday

Bayesian Predictive Inference and Related Asymptotics—Festschrift for Eugenio Regazzini's 75th Birthday

Editors

Emanuele Dolera
Federico Bassetti

MDPI • Basel • Beijing • Wuhan • Barcelona • Belgrade • Manchester • Tokyo • Cluj • Tianjin

Editors
Emanuele Dolera
University of Pavia
Pavia, Italy

Federico Bassetti
Politecnico of Milano
Milan, Italy

Editorial Office
MDPI
St. Alban-Anlage 66
4052 Basel, Switzerland

This is a reprint of articles from the Special Issue published online in the open access journal *Mathematics* (ISSN 2227-7390) (available at: https://www.mdpi.com/si/mathematics/bayesian_predictive_inference_asymptotics).

For citation purposes, cite each article independently as indicated on the article page online and as indicated below:

LastName, A.A.; LastName, B.B.; LastName, C.C. Article Title. *Journal Name* **Year**, *Volume Number*, Page Range.

ISBN 978-3-0365-5113-5 (Hbk)
ISBN 978-3-0365-5114-2 (PDF)

© 2022 by the authors. Articles in this book are Open Access and distributed under the Creative Commons Attribution (CC BY) license, which allows users to download, copy and build upon published articles, as long as the author and publisher are properly credited, which ensures maximum dissemination and a wider impact of our publications.

The book as a whole is distributed by MDPI under the terms and conditions of the Creative Commons license CC BY-NC-ND.

Contents

Emanuele Dolera
Preface to the Special Issue on "Bayesian Predictive Inference and Related Asymptotics—Festschrift for Eugenio Regazzini's 75th Birthday"
Reprinted from: *Mathematics* **2022**, *10*, 2567, doi:10.3390/math10152567 1

Emanuele Dolera and Stefano Favaro
A Compound Poisson Perspective of Ewens–Pitman Sampling Model
Reprinted from: *Mathematics* **2021**, *9*, 2820, doi:10.3390/math9212820 5

Sandra Fortini, Sonia Petrone and Hristo Sariev
Predictive Constructions Based on Measure-Valued Pólya Urn Processes
Reprinted from: *Mathematics* **2021**, *9*, 2845, doi:10.3390/math9222845 17

Federico Camerlenghi and Stefano Favaro
On Johnson's "Sufficientness" Postulates for Feature-Sampling Models
Reprinted from: *Mathematics* **2021**, *9*, 2891, doi:10.3390/math9222891 37

Giacomo Aletti and Irene Crimaldi
The Rescaled Pólya Urn and the Wright—Fisher Process with Mutation
Reprinted from: *Mathematics* **2021**, *9*, 2909, doi:10.3390/math9222909 53

Federico Bassetti and Lucia Ladelli
Mixture of Species Sampling Models
Reprinted from: *Mathematics* **2021**, *9*, 3127, doi:10.3390/math9233127 65

Patrizia Berti, Luca Pratelli and Pietro Rigo
A Central Limit Theorem for Predictive Distributions
Reprinted from: *Mathematics* **2021**, *9*, 3211, doi:10.3390/math9243211 93

Alexander Gnedin and Zakaria Derbazi
Trapping the Ultimate Success
Reprinted from: *Mathematics* **2022**, *10*, 158, doi:10.3390/math10010158 105

Persi Diaconis
Partial Exchangeability for Contingency Tables
Reprinted from: *Mathematics* **2022**, *10*, 442, doi:10.3390/math10030442 125

Lancelot F. James
Single-Block Recursive Poisson–Dirichlet Fragmentations of Normalized Generalized Gamma Processes
Reprinted from: *Mathematics* **2022**, *10*, 561, doi:10.3390/math10040561 137

Emanuele Dolera
Asymptotic Efficiency of Point Estimators in Bayesian Predictive Inference
Reprinted from: *Mathematics* **2022**, *10*, 1136, doi:10.3390/math10071136 147

Sandy Zabell
Fisher, Bayes, and Predictive Inference [†]
Reprinted from: *Mathematics* **2022**, *10*, 1634, doi:10.3390/math10101634 175

Editorial

Preface to the Special Issue on "Bayesian Predictive Inference and Related Asymptotics—Festschrift for Eugenio Regazzini's 75th Birthday"

Emanuele Dolera

Department of Mathematics, University of Pavia, Via Adolfo Ferrata 5, 27100 Pavia, Italy; emanuele.dolera@unipv.it

Citation: Dolera, E. Preface to the Special Issue on "Bayesian Predictive Inference and Related Asymptotics—Festschrift for Eugenio Regazzini's 75th Birthday". *Mathematics* 2022, *10*, 2567. https://doi.org/10.3390/math10152567

Received: 21 July 2022
Accepted: 21 July 2022
Published: 23 July 2022

Publisher's Note: MDPI stays neutral with regard to jurisdictional claims in published maps and institutional affiliations.

Copyright: © 2022 by the authors. Licensee MDPI, Basel, Switzerland. This article is an open access article distributed under the terms and conditions of the Creative Commons Attribution (CC BY) license (https://creativecommons.org/licenses/by/4.0/).

It is my pleasure to write this Preface to the Special Issue of Mathematics entitled "Bayesian Predictive Inference and Related Asymptotics—Festschrift for Eugenio Regazzini's 75th Birthday". As the title suggests, this Special Issue is dedicated to Professor Eugenio Regazzini to honor his more than quinquagenary career (which is still ongoing!). For many years, Eugenio served as both a lecturer and scholar of various Italian Universities (Torino, Bologna, Milano Statale, Bocconi, Pavia), visited a number of foreign academic institutions (including Stanford University in the US, where he lectured for a brief period) and organized various summer schools to promote advanced studies in probability and statistics around Italy. Indeed, more than one generation of Italian scholars has learned and consolidated the study of probability and mathematical statistics under his supervision. It is evident that, besides transmitting enthusiasm and expertise to his students, Eugenio created a solid bridge between the actual academic generation, working in probability and mathematical statistics, and the great Italian masters of the first half of the twentieth century, such as de Finetti, Cantelli and Gini. As a scholar, Eugenio's activity has received—and still receives—appreciation from both colleagues and academic institutions worldwide. Apropos of this, it would be remiss not to mention the prestigious IMS fellowship he received in July 2007.

To briefly outline his scientific contributions, MathSciNet includes 84 of his publications: of these, 41 are concerned with Mathematical Statistics, 9 with Pure Probability, 16 with Mathematical Physics and Economics, and 18 with historical issues. His most significant works can be thematically grouped as follows:

(a) *Bayesian Nonparametrics*: means of the Dirichlet process [1–4], means of normalized completely random measures [5], approximations of posterior distributions by mixtures of Dirichlet probability laws [6];
(b) *General Bayesian Inference*: Bayesian sufficiency [7–9], asymptotics for Bayesian predictive inference [10–12];
(c) *Classical Inference*: minimum distance estimation [13–16], classical point estimation, and testing theory [17,18];
(d) *Descriptive Statistics*: theory of concentration [19,20], theory of monotone dependence [21];
(e) *Abstract Probability Theory*: finitely additive probability [22–27], mixtures of distributions of Markov chains [28], CLT for exchangeable summands [29,30];
(f) *Mathematical Physics and Economics*: analysis of some kinetic Boltzmann-type equations [31–39].

Returning to the Special Issue, we present 11 papers, which are briefly summarized below.

In [40], the author reviews the historical position of Sir R.A. Fisher towards Bayesian inference, particularly regarding the classical Bayes–Laplace paradigm. The main focus of the paper is on Fisher's fiducial argument.

In [41], the author considers point estimation problems concerned with random quantities which depend on both observable and non-observable variables, starting from decision-theoretical principles. A two-phase strategy is proposed, the former relying on estimation of the random parameter of the model, the latter concerning estimation of the original quantity sampled from the distinguished element of the statistical model after plug-in of the estimated parameter in the place of the random parameter. The asymptotic efficiency of the entire procedure is finally discussed.

In [42] the authors obtains explicit descriptions of properties of some Markov chains, called Mittag-Leffler Markov chains, conditioned with a mixed Poisson process when it equates to an integer n, which has interpretations in a species sampling context. This is equivalent to obtaining properties of the fragmentation operations when applied to mass partitions formed by the normalized jumps of a generalized gamma subordinator and its generalizations.

The author of [43] develops a parameter-free version of classical models for contingency tables, along the lines of de Finetti's notions of partial exchangeability.

In [44], the authors introduce a betting game where the gambler aims to guess the last success epoch in a series of inhomogeneous Bernoulli trials paced at random. At a given stage, the gambler may bet on either the event that no further successes occur, or the event that exactly one success is yet to occur, or may choose any proper range of future times (a trap). When a trap is chosen, the gambler wins if the final success epoch is the only one that falls in the trap. Then, the authors use this tool to analyse the best-choice problem, with random arrivals generated via a Pólya–Lundberg process.

In [45], the authors consider a sequence $\{X_n\}_{n \geq 1}$ of conditionally identically distributed random variables. They show that, under suitable conditions, the finite dimensional distributions of the empirical process stably converge to a Gaussian kernel with a known covariance structure.

In [46], the authors introduce mixtures of species sampling sequences and discuss how these sequences are related to various types of Bayesian models. They prove that mixtures of species sampling sequences are obtained by assigning the values of an exchangeable sequence to the classes of a latent exchangeable random partition. Using this representation, they give an explicit expression of the Exchangeable Partition Probability Function of the partition generated by a mixture of species sampling sequences. Finally, they discuss some special cases.

The authors of [47] pursue a project in which the authors introduce, study, and apply a variant of the Eggenberger–Pólya urn, called the "rescaled" Pólya urn. This variant exhibits a reinforcement mechanism based on the most recent observations, a random persistent fluctuation of the predictive mean, and the almost certain convergence of the empirical mean to a deterministic limit. Then, the authors show that the multidimensional Wright–Fisher diffusion with mutation can be obtained as a suitable limit of the predictive means associated with a family of rescaled Pólya urns.

In [48], the authors review "sufficientness" postulates for species-sampling models, and investigate analogous predictive characterizations for the more general feature-sampling models. In particular, they present a "sufficientness" postulate for a class of feature-sampling models referred to as Scaled Processes, and then discuss analogous characterizations in the general setup of feature-sampling models.

In [49], the authors study the asymptotic properties of the predictive distributions and the empirical frequencies of certain sequences $\{X_n\}_{n \geq 1}$ of random variables that are connected to the so-called measure-valued Pólya urn processes, under different assumptions on the weights. They also investigate a generalization of the above models via a randomization of the law of reinforcement.

Finally, in [50], the authors consider a generalization of the log-series compound Poisson sampling model, and they show that it leads to an extension of the compound Poisson perspective of the Ewens sampling model to the more general Ewens–Pitman sampling model. The interplay between the negative Binomial compound Poisson sampling model and the Ewens–Pitman sampling model is then applied to the study of the large n asymptotic behavior of the number of blocks in the corresponding random partitions, leading to new proof of Pitman's α diversity.

Funding: This research received no external funding

Conflicts of Interest: The authors declare no conflict of interest.

References

1. Lijoi, A.; Regazzini, E. Means of a Dirichlet process and multiple hypergeometric functions. *Ann. Probab.* **2004**, *32*, 1469–1495. [CrossRef]
2. Regazzini, E.; Guglielmi, A.; Di Nunno, G. Theory and numerical analysis for exact distributions of functionals of a Dirichlet process. *Ann. Statist.* **2002**, *30*, 1376–1411. [CrossRef]
3. Cifarelli, D.M.; Regazzini, E. Distribution functions of means of a Dirichlet process. *Ann. Statist.* **1990**, *18*, 429–442. [CrossRef]
4. Cifarelli, D.M.; Regazzini, E. A general approach to Bayesian analysis of nonparametric problems. The associative mean values within the framework of the Dirichlet process. II. *Riv. Mat. Sci. Econom. Social.* **1979**, *2*, 95–111. (In Italian)
5. Regazzini, E.; Lijoi, A.; Prünster, I. Distributional results for means of normalized random measures with independent increments. *Ann. Statist.* **2003**, *31*, 560–585. [CrossRef]
6. Regazzini, E.; Sazonov, V.V. Approximation of distributions of random probabilities by mixtures of Dirichlet distributions with applications to nonparametric Bayesian statistical inferences. *Theory Probab. Appl.* **2000**, *45*, 93–110. [CrossRef]
7. Fortini, S.; Ladelli, L.; Regazzini, E. Exchangeability, predictive distributions and parametric models. *Sankhya* **2000**, *62*, 86–109.
8. Cifarelli, D.M.; Regazzini, E. On the role of predictive sufficient statistics in a Bayesian context. II. *Riv. Mat. Sci. Econom. Social.* **1981**, *4*, 3–11. (In Italian)
9. Cifarelli, D.M.; Regazzini, E. On the role of prediction sufficiency in prediction in the Bayesian context. I. *Riv. Mat. Sci. Econom. Social.* **1980**, *3*, 109–125. (In Italian)
10. Dolera, E.; Regazzini, E. Uniform rates of the Glivenko-Cantelli convergence and their use in approximating Bayesian inferences. *Bernoulli* **2019**, *25*, 2982–3015. [CrossRef]
11. Cifarelli, D.M.; Dolera, E.; Regazzini, E. Note on "Frequentistic approximations to Bayesian prevision of exchangeable random elements". *Internat. J. Approx. Reason.* **2017**, *86*, 26–27. [CrossRef]
12. Cifarelli, D.M.; Dolera, E.; Regazzini, E. Frequentistic approximations to Bayesian prevision of exchangeable random elements. *Internat. J. Approx. Reason.* **2016**, *78*, 138–152. [CrossRef]
13. Bassetti, F.; Bodini, A.; Regazzini, E. Consistency of minimum divergence estimators based on grouped data. *Statist. Probab. Lett.* **2007**, *77*, 937–941. [CrossRef]
14. Bassetti, F.; Bodini, A.; Regazzini, E. On minimum Kantorovich distance estimators. *Statist. Probab. Lett.* **2006**, *76*, 1298–1302. [CrossRef]
15. Bassetti, F.; Regazzini, E. Asymptotic properties and robustness of minimum dissimilarity estimators of location-scale parameters. *Teor. Veroyatn. Primen.* **2005**, *50*, 312–330; reprinted in *Theory Probab. Appl. 50*, 171–186. [CrossRef]
16. Bassetti, F.; Regazzini, E. Asymptotic distribution and robustness of minimum total variation distance estimators. *Metron* **2005**, *63*, 55–80.
17. Cifarelli, D.M.; Regazzini, E. On a distribution-free test of independence based on Gini's rank association coefficient. In *Recent Developments in Statistics*; European Meeting of Statisticians, Grenoble: Amsterdam, The Netherland, 1977.
18. Cifarelli, D.M.; Regazzini, E. Sugli stimatori non distorti a varianza minima di una classe di funzioni di probabilità troncate. *Giorn. Econom. Ann. Econom.* **1973**, *32*, 492–501. (In Italian)
19. Regazzini, E. Concentration comparison between probability measures. *Sankhya* **1992**, *54*, 129–149.
20. Cifarelli, D.M.; Regazzini, E. On a general definition of concentration function. *Sankhya* **1987**, *49*, 307–319.
21. Cifarelli, D.M.; Conti, P.L.; Regazzini, E. On the asymptotic distribution of a general measure of monotone dependence. *Ann. Statist.* **1996**, *24*, 1386–1399. [CrossRef]
22. Berti, P.; Regazzini, E.; Rigo, P. Modes of convergence in the coherent framework. *Sankhya* **2007**, *69*, 314–329.
23. Berti, P.; Regazzini, E.; Rigo, P. Well-calibrated, coherent forecasting systems. *Theory Probab. Appl.* **1997**, *42*, 82–102. [CrossRef]
24. Berti, P.; Regazzini, E.; Rigo, P. Finitely additive Radon-Nikodým theorem and concentration function of a probability with respect to a probability. *Proc. Am. Math. Soc.* **1992**, *114*, 1069–1078.
25. Regazzini, E. Finitely additive conditional probabilities. *Rend. Sem. Mat. Fis. Milano* **1988**, *55*, 69–89. [CrossRef]
26. de Regazzini, E. Finetti's coherence and statistical inference. *Ann. Statist.* **1987**, *15*, 845–864. [CrossRef]
27. Regazzini, E. Strengthening the conglomerability property of finitely additive probability laws. *Rend. Mat.* **1984**, *4*, 169–178. (In Italian)
28. Fortini, S.; Ladelli, L.; Petris, G.; Regazzini, E. On mixtures of distributions of Markov chains. *Stochastic Process. Appl.* **2002**, *100*, 147–165. [CrossRef]
29. Regazzini, E.; Sazonov, V.V. On the central limit problem for partially exchangeable random variables with values in a Hilbert space. *Theory Probab. Appl.* **1997**, *42*, 656–670. [CrossRef]
30. Fortini, S.; Ladelli, L.; Regazzini, E. A central limit problem for partially exchangeable random variables. *Theory Probab. Appl.* **1996**, *41*, 224–246. [CrossRef]
31. Perversi, E.; Regazzini, E. Inequality and risk aversion in economies open to altruistic attitudes. *Math. Models Methods Appl. Sci.* **2016**, *26*, 1735–1760. [CrossRef]

32. Dolera, E.; Regazzini, E. Proof of a McKean conjecture on the rate of convergence of Boltzmann–equation solutions. *Probab. Theory Related Fields* **2014**, *160*, 315–389. [CrossRef]
33. Gabetta, E.; Regazzini, E. Complete characterization of convergence to equilibrium for an inelastic Kac model. *J. Stat. Phys.* **2012**, *147*, 1007–1019. [CrossRef]
34. Dolera, E.; Regazzini, E. The role of the central limit theorem in discovering sharp rates of convergence to equilibrium for the solution of the Kac equation. *Ann. Appl. Probab.* **2010**, *20*, 430–461. [CrossRef]
35. Gabetta, E.; Regazzini, E. Central limit theorems for solutions of the Kac equation: speed of approach to equilibrium in weak metrics. *Probab. Theory Related Fields* **2010**, *146*, 451–480. [CrossRef]
36. Dolera, E.; Gabetta, E.; Regazzini, E. Reaching the best possible rate of convergence to equilibrium for solutions of Kac's equation via central limit theorem. *Ann. Appl. Probab.* **2009**, *19*, 186–209. [CrossRef]
37. Gabetta, E.; Regazzini, E. Central limit theorem for the solution of the Kac equation. *Ann. Appl. Probab.* **2008**, *18*, 2320–2336. [CrossRef]
38. Carlen, E.; Gabetta, E.; Regazzini, E. Probabilistic investigations on the explosion of solutions of the Kac equation with infinite energy initial distribution. *J. Appl. Probab.* **2008**, *45*, 95–106. [CrossRef]
39. Bassetti, F.; Gabetta, E.; Regazzini, E. On the depth of the trees in the McKean representation of Wild's sums. *Transport Theory Statist. Phys.* **2007**, *36*, 421–438. [CrossRef]
40. Zabell, S. Fisher, Bayes, and predictive inference. *Mathematics* **2022**, *10*, 1634. [CrossRef]
41. Dolera, E. Asymptotic efficiency of point estimators in Bayesian predictive inference. *Mathematics* **2022**, *10*, 1136. [CrossRef]
42. James, L.F. Single-block recursive Poisson–Dirichlet fragmentations of normalized generalized Gamma processes. *Mathematics* **2022**, *10*, 561. [CrossRef]
43. Diaconis, P. Partial exchangeability for contingency tables. *Mathematics* **2022**, *10*, 442. [CrossRef]
44. Gnedin, A.; Derbazi, Z. Trapping the ultimate success. *Mathematics* **2022**, *10*, 158. [CrossRef]
45. Berti, P.; Pratelli, L.; Rigo, P. A central limit theorem for predictive distributions. *Mathematics* **2021**, *9*, 3211. [CrossRef]
46. Bassetti, F.; Ladelli, L. Mixture of species sampling models. *Mathematics* **2021**, *9*, 3127. [CrossRef]
47. Aletti, G.; Crimaldi, I. The rescaled Pólya urn and the Wright–Fisher process with mutation. *Mathematics* **2021**, *9*, 2909. [CrossRef]
48. Camerlenghi, F.; Favaro, S. On Johnson's "Sufficientness" postulates for feature–sampling models. *Mathematics* **2021**, *9*, 2891. [CrossRef]
49. Fortini, S.; Petrone, S.; Sariev, H. Predictive constructions based on measure–valued Pólya urn processes. *Mathematics* **2021**, *9*, 2845. [CrossRef]
50. Dolera, E.; Favaro, S. A compound Poisson perspective of Ewens–Pitman sampling model. *Mathematics* **2021**, *9*, 2820. [CrossRef]

Article

A Compound Poisson Perspective of Ewens–Pitman Sampling Model

Emanuele Dolera [1,2,3] and Stefano Favaro [2,3,4,*]

1. Department of Mathematics, University of Pavia, Via Adolfo Ferrata 5, 27100 Pavia, Italy; emanuele.dolera@unipv.it
2. Collegio Carlo Alberto, Piazza V. Arbarello 8, 10122 Torino, Italy
3. IMATI-CNR "Enrico Magenes", 27100 Pavia, Italy
4. Department of Economic and Social Sciences, Mathematics and Statistics, University of Torino, Corso Unione Sovietica 218/bis, 10134 Torino, Italy
* Correspondence: stefano.favaro@unito.it

Abstract: The Ewens–Pitman sampling model (EP-SM) is a distribution for random partitions of the set $\{1,\ldots,n\}$, with $n \in \mathbb{N}$, which is indexed by real parameters α and θ such that either $\alpha \in [0,1)$ and $\theta > -\alpha$, or $\alpha < 0$ and $\theta = -m\alpha$ for some $m \in \mathbb{N}$. For $\alpha = 0$, the EP-SM is reduced to the Ewens sampling model (E-SM), which admits a well-known compound Poisson perspective in terms of the log-series compound Poisson sampling model (LS-CPSM). In this paper, we consider a generalisation of the LS-CPSM, referred to as the negative Binomial compound Poisson sampling model (NB-CPSM), and we show that it leads to an extension of the compound Poisson perspective of the E-SM to the more general EP-SM for either $\alpha \in (0,1)$, or $\alpha < 0$. The interplay between the NB-CPSM and the EP-SM is then applied to the study of the large n asymptotic behaviour of the number of blocks in the corresponding random partitions—leading to a new proof of Pitman's α diversity. We discuss the proposed results and conjecture that analogous compound Poisson representations may hold for the class of α-stable Poisson–Kingman sampling models—of which the EP-SM is a noteworthy special case.

Keywords: Berry–Esseen type theorem; Ewens–Pitman sampling model; exchangeable random partitions; log-series compound poisson sampling model; Mittag–Leffler distribution function; negative binomial compound poisson sampling model; Pitman's α-diversity; wright distribution function

1. Introduction

The Pitman–Yor process is a discrete random probability measure indexed by real parameters α and θ such that either $\alpha \in [0,1)$ and $\theta > -\alpha$, or $\alpha < 0$ and $\theta = -m\alpha$ for some $m \in \mathbb{N}$—as can be seen in, e.g., Perman et al. [1], Pitman [2] and Pitman and Yor [3]. Let $\{V_i\}_{i \geq 1}$ be independent random variables such that V_i is distributed as a Beta distribution with parameter $(1-\alpha, \theta+i\alpha)$, for $i \geq 1$, with the convention for $\alpha < 0$ that $V_m = 1$ and V_i is undefined for $i > m$. If $P_1 := V_1$ and $P_i := V_i \prod_{1 \leq j \leq i-1}(1-V_j)$ for $i \geq 2$, such that almost definitely $\sum_{i \geq 1} P_i = 1$, then the Pitman–Yor process is the random probability measure $\tilde{p}_{\alpha,\theta}$ on $(\mathbb{N}, 2^{\mathbb{N}})$ such that $\tilde{p}_{\alpha,\theta}(\{i\}) = P_i$ for $i \geq 1$. The Dirichlet process (Ferguson [4]) arises for $\alpha = 0$. Because of the discreteness of $\tilde{p}_{\alpha,\theta}$, a random sample $(X_1, \ldots, X_n)$ induces a random partition Π_n of $\{1, \ldots, n\}$ by means of the equivalence $i \sim j \iff X_i = X_j$ (Pitman [5]). Let $K_n(\alpha, \theta) := K_n(X_1, \ldots, X_n) \leq n$ be the number of blocks of Π_n and let $M_{r,n}(\alpha, \theta) := M_{r,n}(X_1, \ldots, X_n)$, for $r = 1, \ldots, n$, be the number of blocks with frequency r of Π_n with $\sum_{1 \leq r \leq n} M_{r,n} = K_n$ and $\sum_{1 \leq r \leq n} r M_{r,n} = n$. Pitman [2] showed that:

$$\Pr[(M_{1,n}(\alpha,\theta),\ldots,M_{n,n}(\alpha,\theta)) = (x_1,\ldots,x_n)] = n! \frac{\left(\frac{\theta}{\alpha}\right)_{(\sum_{i=1}^n x_i)}}{(\theta)_{(n)}} \prod_{i=1}^n \frac{\left(\frac{\alpha(1-\alpha)_{(i-1)}}{i!}\right)^{x_i}}{x_i!}, \quad (1)$$

with $(x)_{(n)}$ being the ascending factorial of x of order n, i.e., $(x)_{(n)} := \prod_{0 \leq i \leq n-1}(x+i)$. The distribution (1) is referred to as the Ewens–Pitman sampling model (EP-SM), and for $\alpha = 0$, it reduces to the Ewens sampling model (E-SM) in Ewens [6]. The Pitman–Yor process plays a critical role in a variety of research areas, such as mathematical population genetics, Bayesian nonparametrics, machine learning, excursion theory, combinatorics and statistical physics. See Pitman [5] and Crane [7] for a comprehensive treatment of this subject.

The E-SM admits a well-known compound Poisson perspective in terms of the log-series compound Poisson sampling model (LS-CPSM). See Charalambides [8] and the references therein for an overview of compound Poisson models. We consider a population of individuals with a random number K of distinct types, and let K be distributed as a Poisson distribution with parameter $\lambda = -z \log(1-q)$ for $q \in (0,1)$ and $z > 0$. For $i \in \mathbb{N}$, let N_i denote the random number of individuals of type i in the population, and let the N_i's be independent of K and independent from each other, with the same distribution:

$$\Pr[N_1 = x] = -\frac{1}{x \log(1-q)} q^x \tag{2}$$

for $x \in \mathbb{N}$. Let $S = \sum_{1 \leq i \leq K} N_i$ and let $M_r = \sum_{1 \leq i \leq K} \mathbb{1}_{\{N_i = r\}}$ for $r = 1, \ldots, S$, that is, M_r is the random number of N_i equal to r such that $\sum_{r \geq 1} M_r = K$ and $\sum_{r \geq 1} r M_r = S$. If $(M_1(z,n), \ldots, M_n(z,n))$ denotes a random variable whose distribution coincides with the conditional distribution of $(M_1, \ldots, M_S)$ given $S = n$, then (Section 3, Charalambides [8]) it holds:

$$\Pr[(M_1(z,n), \ldots, M_n(z,n)) = (x_1, \ldots, x_n)] = \frac{n!}{(z)_{(n)}} \prod_{i=1}^{n} \frac{\left(\frac{z}{i}\right)^{x_i}}{x_i!}. \tag{3}$$

The distribution (3) is referred to as the LS-CPSM, and it is equivalent to the E-SM. That is, the distribution (3) coincides with the distribution (1) with $\alpha = 0$. Therefore, the distributions of $K(z,n) = \sum_{1 \leq r \leq n} M_r(z,n)$ and $M_r(z,n)$ coincide with the distributions of $K_n(0,z)$ and $M_{r,n}(0,z)$, respectively. Let $\xrightarrow{w}$ denote the weak convergence. From Korwar and Hollander [9], $K(z,n)/\log n \xrightarrow{w} z$ as $n \to +\infty$, whereas from Ewens [6], it follows that $M_r(z,n) \xrightarrow{w} P_{z/r}$ as $n \to +\infty$, where P_z is a Poisson random variable with parameter z.

In this paper, we consider a generalisation of the LS-CPSM referred to as the negative binomial compound Poisson sampling model (NB-CPSM). The NB-CPSM is indexed by real parameters α and z such that either $\alpha \in (0,1)$ and $z > 0$ or $\alpha < 0$ and $z < 0$. The LS-CPSM is recovered by letting $\alpha \to 0$ and $z > 0$. We show that the NB-CPSM leads to extend the compound Poisson perspective of the E-SM to the more general EP-SM for either $\alpha \in (0,1)$, or $\alpha < 0$. That is, we show that: (i) for $\alpha \in (0,1)$, the EP-SM (1) admits a representation as a randomised NB-CPSM with $\alpha \in (0,1)$ and $z > 0$, where the randomisation acts on z with respect a scale mixture between a Gamma and a scaled Mittag–Leffler distribution (Pitman [5]); (ii) for $\alpha < 0$ the NB-CPSM admits a representation in terms of a randomised EP-SM with $\alpha < 0$ and $\theta = -m\alpha$ for some $m \in \mathbb{N}$, where the randomisation acts on m with respect to a tilted Poisson distribution arising from the Wright function (Wright [10]). The interplay between the NB-CPSM and the EP-SM is then applied to the large n asymptotic behaviour of the number of distinct blocks in the corresponding random partitions. In particular, by combining the randomised representation in (i) with the large n asymptotic behaviour or the number of distinct blocks under the NB-CPSM, we present a new proof of Pitman's α-diversity (Pitman [5]), namely the large n asymptotic behaviour of $K_n(\alpha, \theta)$ under the EP-SM for $\alpha \in (0,1)$.

2. A Compound Poisson Perspective of EP-SM

To introduce the NB-CPSM, we considered a population of individuals with a random number K of types and let K be distributed as a Poisson distribution with parameter $\lambda = z[1 - (1-q)^\alpha]$ such that either $q \in (0,1)$, $\alpha \in (0,1)$ and $z > 0$, or $q \in (0,1)$, $\alpha < 0$ and

$z < 0$. For $i \in \mathbb{N}$, let N_i be the random number of individuals of type i in the population, and let the N_i be independent of K and independent from each other with the same distribution:

$$\Pr[N_1 = x] = -\frac{1}{[1 - (1-q)^\alpha]} \binom{\alpha}{x} (-q)^x \tag{4}$$

for $x \in \mathbb{N}$. Let $S = \sum_{1 \leq i \leq K} N_i$ and $M_r = \sum_{1 \leq i \leq K} \mathbb{1}_{\{N_i = r\}}$ for $r = 1, \ldots, S$, that is, M_r is the random number of N_i equal to r such that $\sum_{r \geq 1} M_r = K$ and $\sum_{r \geq 1} r M_r = S$. If $(M_1(\alpha, z, n), \ldots, M_n(\alpha, z, n))$ is a random variable whose distribution coincides with the conditional distribution of $(M_1, \ldots, M_S)$, given $S = n$, then it holds (Section 3, Charalambides [8]):

$$\Pr[(M_1(\alpha, z, n), \ldots, M_n(\alpha, z, n)) = (x_1, \ldots, x_n)] = \frac{n!}{\sum_{j=0}^n \mathscr{C}(n, j; \alpha) z^j} \prod_{i=1}^n \frac{\left[z \frac{\alpha(1-\alpha)_{(i-1)}}{i!}\right]^{x_i}}{x_i!}, \tag{5}$$

where $\mathscr{C}(n, j; \alpha) = \frac{1}{j!} \sum_{0 \leq i \leq j} \binom{j}{i} (-1)^i (-i\alpha)_{(n)}$ is the generalised factorial coefficient (Charalambides [11]), with the proviso $\mathscr{C}(n, 0, \alpha) = 0$ for all $n \in \mathbb{N}$, $\mathscr{C}(n, j, \alpha) = 0$ for all $j > n$ and $\mathscr{C}(0, 0, \alpha) = 1$. The distribution (5) is referred to as the NB-CPSM. As $\alpha \to 0$, the distribution (4) reduces to the distribution (2), and hence the NB-CPSM (5) is reduced to the LS-CPSM (3). The next theorem states the large n asymptotic behaviour of the counting statistics $K(\alpha, z, n) = \sum_{1 \leq r \leq n} M_r(\alpha, z, n)$ and $M_r(\alpha, z, n)$ arising from the NB-CPSM.

Theorem 1. *Let P_λ denote a Poisson random variable with the parameter $\lambda > 0$. As $n \to +\infty$,*

(i) *for $\alpha \in (0, 1)$ and $z > 0$:*

$$K(\alpha, z, n) \xrightarrow{w} 1 + P_z \tag{6}$$

and:

$$M_r(\alpha, z, n) \xrightarrow{w} P_{\frac{\alpha(1-\alpha)_{(r-1)}}{r!} z}; \tag{7}$$

(ii) *for $\alpha < 0$ and $z < 0$:*

$$\frac{K(\alpha, z, n)}{n^{\frac{-\alpha}{1-\alpha}}} \xrightarrow{w} \frac{(\alpha z)^{\frac{1}{1-\alpha}}}{-\alpha} \tag{8}$$

and:

$$M_r(\alpha, z, n) \xrightarrow{w} P_{\frac{\alpha(1-\alpha)_{(r-1)}}{r!} z}. \tag{9}$$

Proof. As regards the proof of (6), we start by recalling that the probability generating function $G(\cdot; \lambda)$ of P_λ is $G(s; \lambda) = \exp\{-\lambda(s-1)\}$ for any $s > 0$. Now, let $G(\cdot; \alpha, z, n)$ be the probability generating function of $K(\alpha, z, n)$. The distribution of $K(\alpha, z, n)$ follows by combining the NB-CPSM (5) with Theorem 2.15 of Charalambides [11]. In particular, it follows that:

$$G(s; \alpha, z, n) = \frac{\sum_{j=1}^n \mathscr{C}(n, j; \alpha)(sz)^j}{\sum_{j=1}^n \mathscr{C}(n, j; \alpha) z^j}.$$

Hereafter, we show that $G(s; \alpha, z, n) \to s \exp\{z(s-1)\}$ as $n \to +\infty$, for any $s > 0$, which implies (6). In particular, by the direct application of the definition of $\mathscr{C}(n, k; \alpha)$, we write the following:

$$\sum_{j=1}^n \mathscr{C}(n, j; \alpha) z^j = \sum_{i=1}^n (-1)^i (-i\alpha)_{(n)} \sum_{k=i}^n \frac{1}{k!} \binom{k}{i} z^k = \sum_{i=1}^n (-1)^i (-i\alpha)_{(n)} e^z z^i \frac{\Gamma(n-i+1, z)}{i! \Gamma(n-i+1)},$$

where $\Gamma(a, x) := \int_x^{+\infty} t^{a-1} e^{-t} dt$ denotes the incomplete gamma function for $a, x > 0$ and $\Gamma(a) := \int_0^{+\infty} t^{a-1} e^{-t} dt$ denotes the Gamma function for $a > 0$. Accordingly, we write the identity:

$$G(s; \alpha, z, n) = e^{z(s-1)} \frac{-zs \frac{\Gamma(n,zs)}{\Gamma(n)} + \sum_{i=2}^n (-1)^i \frac{(-i\alpha)_{(n)}}{(-\alpha)_{(n)}} (zs)^i \frac{\Gamma(n-i+1,zs)}{i! \Gamma(n-i+1)}}{-z \frac{\Gamma(n,z)}{\Gamma(n)} + \sum_{i=2}^n (-1)^i \frac{(-i\alpha)_{(n)}}{(-\alpha)_{(n)}} z^i \frac{\Gamma(n-i+1,z)}{i! \Gamma(n-i+1)}}.$$

Since $\lim_{n \to +\infty} \frac{\Gamma(n,x)}{\Gamma(n)} = 1$ for any $x > 0$, the proof (6) is completed by showing that, for any $t > 0$:

$$\lim_{n \to +\infty} \sum_{i=2}^n (-1)^i \frac{(-i\alpha)_{(n)}}{(-\alpha)_{(n)}} \frac{\Gamma(n-i+1,t)}{\Gamma(n-i+1)} \frac{t^i}{i!} = 0. \tag{10}$$

By the definition of ascending factorials and the reflection formula of the Gamma function, it holds:

$$\frac{(-i\alpha)_{(n)}}{(-\alpha)_{(n)}} = \frac{\Gamma(n - i\alpha)}{\Gamma(n - \alpha)} \frac{\sin i\pi\alpha}{\pi} \Gamma(i\alpha + 1)\Gamma(-\alpha).$$

In particular, by means of the monotonicity of the function $[1, +\infty) \ni z \mapsto \Gamma(z)$, we can write:

$$\frac{1}{i!} \left| \frac{(-i\alpha)_{(n)}}{(-\alpha)_{(n)}} \right| \leq \frac{|\Gamma(-\alpha)|}{\pi} \frac{\Gamma(n - 2\alpha)}{\Gamma(n - \alpha)} \frac{\Gamma(i\alpha + 1)}{i!} \tag{11}$$

for any $n \in \mathbb{N}$ such that $n > 1/(1 - \alpha)$, and $i \in \{2, \ldots, n\}$. Note that $\frac{\Gamma(n,x)}{\Gamma(n)} \leq 1$. Then, we apply (11) to obtain:

$$\left| \sum_{i=2}^n (-1)^i \frac{(-i\alpha)_{(n)}}{(-\alpha)_{(n)}} \frac{\Gamma(n-i+1,t)}{\Gamma(n-i+1)} \frac{t^i}{i!} \right| \leq \sum_{i=2}^n \frac{t^i}{i!} \left| \frac{(-i\alpha)_{(n)}}{(-\alpha)_{(n)}} \right|$$

$$\leq \frac{|\Gamma(-\alpha)|}{\pi} \frac{\Gamma(n - 2\alpha)}{\Gamma(n - \alpha)} \sum_{i \geq 0} t^i \frac{\Gamma(i\alpha + 1)}{i!}.$$

Now, by means of Stirling approximation, it holds $\frac{\Gamma(n-2\alpha)}{\Gamma(n-\alpha)} \sim \frac{1}{n^\alpha}$ as $n \to +\infty$. Moreover, we have:

$$\sum_{i \geq 0} t^i \frac{\Gamma(i\alpha + 1)}{i!} = \int_0^{+\infty} e^{tz^\alpha - z} dz < +\infty$$

where the finiteness of the integral follows, for any fixed $t > 0$, from the fact that $tz^\alpha < \frac{1}{2} z$ if $z > (2t)^{\frac{1}{1-\alpha}}$. This completes the proof of (10) and hence the proof of (6). As regards the proof of (7), we make use of the falling factorial moments of $M_r(\alpha, z, n)$, which follows by combining the NB-CPSM (5) with Theorem 2.15 of Charalambides [11]. Let $(a)_{[n]}$ be the falling factorial of a of order n, i.e., $(a)_{[n]} = \prod_{0 \leq i \leq n-1} (a - i)$, for any $a \in \mathbb{R}^+$ and $n \in \mathbb{N}_0$ with the proviso $(a)_{[0]} = 1$. Then, we write:

$$\mathbb{E}[(M_r(\alpha, z, n))_{[s]}]$$

$$= (-1)^{rs} (n)_{[rs]} \binom{\alpha}{r}^s (-z)^s \frac{\sum_{j=0}^{n-rs} \mathscr{C}(n - rs, j; \alpha) z^j}{\sum_{j=0}^n \mathscr{C}(n, j; \alpha) z^j}$$

$$= (-1)^{rs} (n)_{[rs]} \binom{\alpha}{r}^s (-z)^s \frac{(-z) \frac{\Gamma(n-rs,z)}{\Gamma(n-rs)} + \sum_{i=2}^{n-rs} (-1)^i \frac{(-i\alpha)_{(n-rs)}}{(-\alpha)_{(n-rs)}} (z)^i \frac{\Gamma(n-rs-i+1,z)}{i! \Gamma(n-rs-i+1)}}{(-z) \frac{\Gamma(n,z)}{\Gamma(n)} + \sum_{i=2}^n (-1)^i \frac{(-i\alpha)_{(n)}}{(-\alpha)_{(n)}} (z)^i \frac{\Gamma(n-i+1,z)}{\Gamma(n-i+1)}}$$

$$= (-1)^{rs} (n)_{[rs]} \binom{\alpha}{r}^s (-z)^s$$

$$\times \frac{(-\alpha)_{(n-rs)}}{(-\alpha)_{(n)}} \frac{(-z)\frac{\Gamma(n-rs,z)}{\Gamma(n-rs)} + \sum_{i=2}^{n-rs}(-1)^i \frac{(-i\alpha)_{(n-rs)}}{(-\alpha)_{(n-lr)}}(z)^i \frac{\Gamma(n-rs-i+1,z)}{i!\Gamma(n-rs-i+1)}}{(-z)\frac{\Gamma(n,z)}{\Gamma(n)} + \sum_{i=2}^{n}(-1)^i \frac{(-i\alpha)_{(n)}}{(-\alpha)_{(n)}}(z)^i \frac{\Gamma(n-i+1,z)}{\Gamma(n-i+1)}}.$$

Now, by means of the same argument applied in the proof of statement (6), it holds true that:

$$\lim_{n \to +\infty} \frac{(-z)\frac{\Gamma(n-rs,z)}{\Gamma(n-rs)} + \sum_{i=2}^{n-rs}(-1)^i \frac{(-i\alpha)_{(n-rs)}}{(-\alpha)_{(n-lr)}}(z)^i \frac{\Gamma(n-rs-i+1,z)}{i!\Gamma(n-rs-i+1)}}{(-z)\frac{\Gamma(n,z)}{\Gamma(n)} + \sum_{i=2}^{n}(-1)^i \frac{(-i\alpha)_{(n)}}{(-\alpha)_{(n)}}(z)^i \frac{\Gamma(n-i+1,z)}{\Gamma(n-i+1)}} = 1.$$

Then:

$$\lim_{n \to +\infty} \mathbb{E}[(M_r(\alpha,z,n))_{[s]}] = (-1)^{rs}\binom{\alpha}{r}^s (-z)^s = \left[\frac{\alpha(1-\alpha)_{(r-1)}}{r!}z\right]^s$$

follows from the fact that $(n)_{[rs]} \sim \frac{(-\alpha)_{(n-rs)}}{(-\alpha)_{(n)}}$ as $n \to +\infty$. The proof of the large n asymptotics (7) is completed by recalling that the falling factorial moment of order s of P_λ is $\mathbb{E}[(P_\lambda)_{[s]}] = \lambda^s$.

As regards the proof of statement (8), let $\alpha = -\sigma$ for any $\sigma > 0$ and let $z = -\zeta$ for any $\zeta > 0$. Then, by direct application of Equation (2.27) of Charalambides [11], we write the following identity:

$$\sum_{j=0}^{n} \mathscr{C}(n,j;-\sigma)(-\zeta)^j = (-1)^n \sum_{v=0}^{n} s(n,v)(-\sigma)^v \sum_{j=0}^{v} \zeta^j S(v,j),$$

where $S(v,j)$ is the Stirling number of that second type. Now, note that $\sum_{0 \le j \le v} \zeta^j S(v,j)$ is the moment of order v of a Poisson random variable with parameter $\zeta > 0$. Then, we write:

$$\sum_{j=0}^{n} \mathscr{C}(n,j;-\sigma)(-\zeta)^j = \sum_{v=0}^{n} |s(n,v)|\sigma^v \sum_{j \ge 0} j^v e^{-\zeta}\frac{\zeta^j}{j!} = \sum_{j \ge 0} e^{-\zeta}\frac{\zeta^j}{j!} \int_{0}^{+\infty} x^n f_{G_{\sigma j,1}}(x)\mathrm{d}x. \quad (12)$$

That is:
$$B_n(w) = \mathbb{E}[(G_{\sigma P_w,1})^n], \quad (13)$$

where $G_{a,1}$ and P_w are independent random variables such that $G_{a,1}$ is a Gamma random variable with a shape parameter $a > 0$ and a scale parameter 1, and P_w is a Poisson random variable with a parameter w. Accordingly, the distribution of $G_{\sigma P_w,1}$, say $\mu_{\sigma,w}$, is the following:

$$\mu_{\sigma,w}(\mathrm{d}t) = e^{-w}\delta_0(\mathrm{d}t) + \left(\sum_{j \ge 1} \frac{e^{-w}w^j}{j!}\frac{1}{\Gamma(j\sigma)}e^{-t}t^{j\sigma-1}\right)\mathrm{d}t$$

for $t > 0$. The discrete component of $\mu_{\sigma,w}$ does not contribute to the expectation (13) so that we focus on the absolutely continuous component, whose density can be written as follows:

$$\sum_{j \ge 1} \frac{e^{-w}w^j}{j!}\frac{1}{\Gamma(j\sigma)}e^{-t}t^{j\sigma-1} = \frac{e^{-(w+t)}}{t}W_{\sigma,0}(wt^\sigma),$$

where $W_{\sigma,\tau}(y) := \sum_{j \ge 0} \frac{y^j}{j!\Gamma(j\sigma+\tau)}$ is the Wright function (Wright [10]). In particular, for $\tau = 0$:

$$B_n(w) = \int_{0}^{+\infty} t^n \frac{e^{-(w+t)}}{t}W_{\sigma,0}(wt^\sigma)\mathrm{d}t. \quad (14)$$

If we split the integral as $\int_0^M + \int_M^{+\infty}$ for any $M > 0$, the contribution of the latter integral is overwhelming with respect to the contribution of the former. Then, $W_{\sigma,0}$ can be equivalently replaced by the asymptotics $W_{\sigma,0}(y) \sim c(\sigma) y^{\frac{1}{2(1+\sigma)}} \exp\{\sigma^{-1}(\sigma+1)(\sigma y)^{\frac{1}{1+\sigma}}\}$, as $y \to +\infty$, for some constant $c(\sigma)$ solely depending on σ. See Theorem 2 in Wright [10]. Hence:

$$B_n(w) \sim c(\sigma) \int_0^{+\infty} t^{n-1} e^{-(w+t)} (wt^\sigma)^{\frac{1}{2(1+\sigma)}} \exp\left\{\frac{\sigma+1}{\sigma}(\sigma w t^\sigma)^{\frac{1}{1+\sigma}}\right\} dt$$
$$= c(\sigma) e^{-w} w^{\frac{1}{2(1+\sigma)}} \int_0^{+\infty} t^{n+\frac{\sigma}{2(1+\sigma)}-1} \exp\{A(w,\sigma) t^{\frac{\sigma}{1+\sigma}} - t\} dt,$$

where $A(w,\sigma) := \frac{\sigma+1}{\sigma}(\sigma w)^{\frac{1}{1+\sigma}}$. Then, the problem is reduced to an integral whose asymptotic behaviour is described in Berg [12]. From Equation (31) of the Berg [12] and Stirling approximation, we have:

$$B_n(w) \sim c(\sigma) e^{-w} w^{\frac{1}{2(1+\sigma)}} \Gamma(n) \exp\left\{A(w,\sigma) n^{\frac{\sigma}{1+\sigma}}\right\}. \tag{15}$$

This last asymptotic expansion leads directly to (8). Indeed, let $G(\cdot; -\sigma, -\zeta, n)$ be the probability generating function of the random variable $K(-\sigma, -\zeta, n)$, which reads as $G(s; -\sigma, -\zeta, n) = B_n(s\zeta)/B_n(\zeta)$ for $s > 0$. Then, by means of (15), for any fixed $s > 0$ we write:

$$G(s; -\sigma, -\zeta, n) \sim e^{-w(s-1)} s^{\frac{1}{2(1+\sigma)}} \exp\left\{n^{\frac{\sigma}{1+\sigma}} \frac{\sigma+1}{\sigma}(\sigma\zeta)^{\frac{1}{1+\sigma}} [s^{\frac{1}{1+\sigma}} - 1]\right\}. \tag{16}$$

Since (15) holds uniformly in w in a compact set, we consider the function $G(s; -\sigma, -\zeta, n)$ evaluated at some point s_n and extend the validity of (16) with s_n in the place of s, as long as $\{s_n\}_{n\geq 1}$ varies in a compact subset of $[0, +\infty)$. Thus, we can choose $s_n = s^{\beta(n)}$ and $\beta(n) = \frac{1}{n^{\frac{\sigma}{1+\sigma}}}$ and notice that $\beta(n) \to 0$ as $n \to +\infty$. Thus, $s_n \simeq 1 + \beta(n) \log s \to 1$ and we have:

$$n^{\frac{\sigma}{1+\sigma}} \frac{\sigma+1}{\sigma}(\sigma w)^{\frac{1}{1+\sigma}} [s_n^{\frac{1}{1+\sigma}} - 1] \to \frac{(\sigma\zeta)^{\frac{1}{1+\sigma}}}{\sigma} \log s,$$

which implies that $K(-\sigma, -\zeta, n) \to \frac{(\sigma\zeta)^{\frac{1}{1+\sigma}}}{\sigma}$ as $n \to +\infty$. This completes the proof of (8). As regards the proof (9), let $\alpha = -\sigma$ for any $\sigma > 0$ and let $z = -\zeta$ for any $\zeta > 0$. Similarly to the proof of (7), here we make use of the falling factorial moments of $M_r(-\sigma, -\zeta, n)$, that is:

$$\mathbb{E}[(M_r(-\sigma, \zeta, n))_{[s]}] = (-1)^{rs} (n)_{[rs]} \binom{-\sigma}{r}^s \zeta^s \frac{\sum_{j=0}^{n-rs} \mathscr{C}(n-rs, j; -\sigma)(-\zeta)^j}{\sum_{j=0}^{n} \mathscr{C}(n, j; -\sigma)(-\zeta)^j}.$$

At this point, we make use of the same large n arguments applied in the proof of statement (7). In particular, by means of the large n asymptotic (15), as $n \to +\infty$, it holds true that:

$$\frac{\sum_{j=0}^{n-rs} \mathscr{C}(n-rs, j; -\sigma)(-\zeta)^j}{\sum_{j=0}^{n} \mathscr{C}(n, j; -\sigma)(-\zeta)^j} \sim n^{-rs}.$$

Then:

$$\lim_{n \to +\infty} \mathbb{E}[(M_r(-\sigma, -\zeta, n))_{[s]}] = (-1)^{rs} \binom{-\sigma}{r}^s \zeta^s = \left[\frac{-\sigma(1+\sigma)_{(r-1)}}{r!}(-\zeta)\right]^s$$

it follows from the fact that $(n)_{[rs]} \sim n^{rs}$ as $n \to +\infty$. The proof of the large n asymptotics (9) is completed by recalling that the falling factorial moment of order s of P_λ is $\mathbb{E}[(P_\lambda)_{[s]}] = \lambda^s$. □

In the rest of the section, we make use of the NB-CPSM (5) to introduce a compound Poisson perspective of the EP-SM. In particular, our result extends the well-known compound Poisson perspective of the E-SM to the EP-SM for either $\alpha \in (0,1)$, or $\alpha < 0$. For $\alpha \in (0,1)$ let f_α denote the density function of a positive α-stable random variable X_α, that is X_α is a random variable for which $\mathbb{E}[\exp\{-tX_\alpha\}] = \exp\{-t^\alpha\}$ for any $t > 0$. For $\alpha \in (0,1)$ and $\theta > -\alpha$, let $S_{\alpha,\theta}$ be a positive random variable with the density function:

$$f_{S_{\alpha,\theta}}(s) = \frac{\Gamma(\theta+1)}{\alpha \Gamma(\theta/\alpha+1)} s^{\frac{\theta-1}{\alpha}-1} f_\alpha(s^{-\frac{1}{\alpha}}).$$

That is, $S_{\alpha,\theta}$ is a scaled Mittag–Leffler random variable (Chapter 1, Pitman [5]). Let $G_{a,b}$ be a Gamma random variable with the scale parameter $b > 0$ and shape parameter $a > 0$, and let us assume that $G_{a,b}$ is independent of $S_{\alpha,\theta}$. Then, for $\alpha \in (0,1)$, $\theta > -\alpha$ and $n \in \mathbb{N}$ let:

$$\bar{X}_{\alpha,\theta,n} \stackrel{d}{=} G^\alpha_{\theta+n,1} S_{\alpha,\theta}. \tag{17}$$

Finally, for $\alpha < 0$, $z < 0$ and $n \in \mathbb{N}$, let $\tilde{X}_{\alpha,z,n}$ be a random variable on $\mathbb{N}$ whose distribution is a tilted Poisson distribution arising from the identity (12). Precisely, for any $x \in \mathbb{N}$:

$$\Pr[\tilde{X}_{\alpha,z,n} = x] = \frac{1}{\sum_{j=1}^n \mathscr{C}(n,j;\alpha)z^j} \frac{e^z(-z)^x \Gamma(-x\alpha+n)}{x! \Gamma(-x\alpha)}. \tag{18}$$

The next theorem makes use of $\bar{X}_{\alpha,\theta,n}$ and $\tilde{X}_{\alpha,z,n}$ to set an interplay between NB-CPSM (5) and EP-SM (1). This extends the compound Poisson perspective of the E-SM.

Theorem 2. *Let $(M_{1,n}(\alpha,\theta),\ldots,M_{n,n}(\alpha,\theta))$ be distributed as the EP-SM (1) and let $\bar{X}_{\alpha,\theta,n}$ be the random variable defined in (17), which is independent of $(M_{1,n}(\alpha,\theta),\ldots,M_{n,n}(\alpha,\theta))$. Moreover, let $(M_1(\alpha,z,n),\ldots,M_n(\alpha,z,n))$ be distributed as the NB-CPSM (5), and let $\tilde{X}_{\alpha,z,n}$ be the random variable defined in (18), which is independent of $(M_1(\alpha,z,n),\ldots,M_n(\alpha,z,n))$. Then:*

(i) *for $\alpha \in (0,1)$ and $\theta > -\alpha$:*

$$(M_{1,n}(\alpha,\theta),\ldots,M_{n,n}(\alpha,\theta)) \stackrel{d}{=} (M_1(\alpha,\bar{X}_{\alpha,\theta,n},n),\ldots,M_n(\alpha,\bar{X}_{\alpha,\theta,n},n));$$

(ii) *for $\alpha < 0$ and $z < 0$:*

$$(M_1(\alpha,z,n),\ldots,M_n(\alpha,z,n)) \stackrel{d}{=} (M_{1,n}(\alpha,-\tilde{X}_{\alpha,z,n}\alpha),\ldots,M_{n,n}(\alpha,-\tilde{X}_{\alpha,z,n}\alpha)).$$

Proof. As regards the proof of statement (i), it relies on the classical integral representation of the Gamma function. That is, by applying the integral representation of $\Gamma(\theta/\alpha+k)$ to the EP-SM (1), for $x_1,\ldots,x_n \in \{0,\ldots,n\}$ with $\sum_{i=1}^n x_i = k$ and $\sum_{i=1}^n i x_i = n$, we can write that:

$$\Pr[(M_{1,n}(\alpha,\theta),\ldots,M_{n,n}(\alpha,\theta)) = (x_1,\ldots,x_n)]$$

$$= n! \frac{\alpha^k}{\Gamma(\theta+n)} \prod_{i=1}^n \frac{\left(\frac{(1-\alpha)_{(i-1)}}{i!}\right)^{x_i}}{x_i!} \frac{\Gamma(\theta+1)}{\alpha\Gamma(\theta/\alpha+1)}$$

$$\times \int_0^{+\infty} z^{\theta/\alpha-1} e^{-z} \frac{z^k}{\sum_{j=1}^n \mathscr{C}(n,j;\alpha)z^j} \left(\sum_{j=1}^n \mathscr{C}(n,j;\alpha)z^j\right) dz$$

By Equation (13) of Favaro et al. [13]:

$$= n! \frac{\alpha^k}{\Gamma(\theta+n)} \prod_{i=1}^n \frac{\left(\frac{(1-\alpha)_{(i-1)}}{i!}\right)^{x_i}}{x_i!} \frac{\Gamma(\theta+1)}{\alpha\Gamma(\theta/\alpha+1)}$$

$$\times \int_0^{+\infty} z^{\theta/\alpha-1} e^{-z} \frac{z^k}{\sum_{j=1}^n \mathscr{C}(n,j;\alpha)z^j} \left(e^z z^{n/\alpha} \int_0^{+\infty} y^n e^{-yz^{1/\alpha}} f_\alpha(y) dy \right) dz$$

$$= \int_0^{+\infty} \frac{n!}{\sum_{j=0}^n \mathscr{C}(n,j;\alpha)z^j} \prod_{i=1}^n \frac{\left(z \frac{\alpha(1-\alpha)_{(i-1)}}{i!} \right)^{x_i}}{x_i!}$$

$$\times \frac{\Gamma(\theta+1)}{\alpha \Gamma(\theta+n)\Gamma(\theta/\alpha+1)} z^{\theta/\alpha+n/\alpha-1} \int_0^{+\infty} y^n e^{-yz^{1/\alpha}} f_\alpha(y) dy dz$$

$$= \int_0^{+\infty} \Pr[(M_1(\alpha,x,n),\ldots,M_n(\alpha,x,n)) = (x_1,\ldots,x_n)]$$

$$\times \frac{\Gamma(\theta+1)}{\alpha \Gamma(\theta+n)\Gamma(\theta/\alpha+1)} z^{\theta/\alpha+n/\alpha-1} \int_0^{+\infty} y^n e^{-yz^{1/\alpha}} f_\alpha(y) dy dz$$

By the distribution of $\tilde{X}_{\alpha,\theta,n}$:

$$= \int_0^{+\infty} \Pr[(M_1(\alpha,z,n),\ldots,M_n(\alpha,z,n)) = (x_1,\ldots,x_n)] f_{\tilde{X}_{\alpha,\theta,n}}(z) dz,$$

where $f_{\tilde{X}_{\alpha,\theta,n}}$ is the density function of the random variable $\tilde{X}_{\alpha,\theta,n}$. This completes the proof of (i).

As regards the proof of statement (ii), for any $\alpha < 0$, $m \in \mathbb{N}$, $k \leq m$ and $n \in \mathbb{N}$, we define the function $m \mapsto A(m;k,\alpha,n) = \frac{m!}{(m-k)!} \frac{\Gamma(-m\alpha)}{\Gamma(-m\alpha+n)}$, and then consider the following identity:

$$\frac{(-z)^k}{\sum_{j=1}^n \mathscr{C}(n,j;\alpha)z^j} = \sum_{m \geq k} A(m;k,\alpha,n) \Pr[\tilde{X}_{\alpha,z,n} = m]. \quad (19)$$

By applying (19) to the NB-CPSM (5), for $x_1,\ldots,x_n \in \{0,\ldots,n\}$ with $\sum_{i=1}^n x_i = k$ and $\sum_{i=1}^n i x_i = n$, we write:

$$\Pr[(M_1(\alpha,z,n),\ldots,M_n(\alpha,z,n)) = (x_1,\ldots,x_n)]$$

$$= \sum_{m \geq k} n!(-1)^k A(m;k,\alpha,n) \Pr[\tilde{X}_{\alpha,z,n} = m] \prod_{i=1}^n \frac{\left(\frac{\alpha(1-\alpha)_{(i-1)}}{i!} \right)^{x_i}}{x_i!}$$

$$= \sum_{m \geq k} n!(-1)^k \frac{m!}{(m-k)!} \frac{\Gamma(-m\alpha)}{\Gamma(-m\alpha+n)} \Pr[\tilde{X}_{\alpha,z,n} = m] \prod_{i=1}^n \frac{\left(\frac{\alpha(1-\alpha)_{(i-1)}}{i!} \right)^{x_i}}{x_i!}$$

$$= \sum_{m \geq k} n! \frac{\left(\frac{-m\alpha}{\alpha} \right)_{(k)}}{(-m\alpha)_{(n)}} \prod_{i=1}^n \frac{\left(\frac{\alpha(1-\alpha)_{(i-1)}}{i!} \right)^{x_i}}{x_i!} \Pr[\tilde{X}_{\alpha,z,n} = m]$$

$$= \sum_{m \geq k} \Pr[(M_1(\alpha,-m\alpha),\ldots,M_n(\alpha,-m\alpha)) = (x_1,\ldots,x_n)] \Pr[\tilde{X}_{\alpha,z,n} = m].$$

This completes the proof of (ii). □

Theorem 2 presents a compound Poisson perspective of the EP-SM in terms of the NB-CPSM, thus extending the well-known compound Poisson perspective of the E-SM in terms of the LS-CPSM. Statement (i) of Theorem 2 shows that for $\alpha \in (0,1)$ and $\theta > -\alpha$, the EP-SM admits a representation in terms of the NB-CPSM with $\alpha \in (0,1)$ and $z > 0$, where the randomisation acts on the parameter z with respect to the distribution (17). Precisely, this is a compound mixed Poisson sampling model. That is, a compound sampling model in which the distribution of the random number K of distinct types in the population is a mixture of Poisson distributions with respect to the law of $\tilde{X}_{\alpha,\theta,n}$. Statement (ii) of Theorem 2 shows that for $\alpha < 0$ and $z < 0$, the NB-CPSM admits a representation in terms of a randomised EP-SM with $\alpha < 0$ and $\theta = -m\alpha$ for some $m \in \mathbb{N}$, where the randomisation acts on the parameter m with respect to the distribution (17).

Remark 1. *The randomisation procedure introduced in Theorem 2 is somehow reminiscent of a class of Gibbs-type sampling models introduced in Gnedin and Pitman [14]. This class is defined from the EP-SM with $\alpha < 0$ and $\theta = -m\alpha$, for some $m \in \mathbb{N}$, and then it assumes that the parameter m is distributed according to an arbitrary distribution on $\mathbb{N}$. This can be seen in Theorem 12 of Gnedin and Pitman [14] and Gnedin [15] for example. However, differently from the definition of Gnedin and Pitman [14], in our context, the distribution of m depends on the sample size n.*

For $\alpha \in (0,1)$ and $\theta > -\alpha$, Pitman [5] first studied the large n asymptotic behaviour of $K_n(\alpha, \theta)$. This can also be seen in Gnedin and Pitman [14] and the references therein. Let $\xrightarrow{a.s.}$ denote the almost sure convergence, and let $S_{\alpha,\theta}$ be the scaled Mittag–Leffler random variable defined above. Theorem 3.8 of Pitman [5] exploited a martingale convergence argument to show that:

$$\frac{K_n(\alpha,\theta)}{n^\alpha} \xrightarrow{a.s.} S_{\alpha,\theta} \tag{20}$$

as $n \to +\infty$. The random variable $S_{\alpha,\theta}$ is referred to as Pitman's α-diversity. For $\alpha < 0$ and $\theta = -m\alpha$ for some $m \in \mathbb{N}$, the large n asymptotic behaviour of $K_n(\alpha, \theta)$ is trivial, that is:

$$K_n(\alpha,\theta) \xrightarrow{w} m \tag{21}$$

as $n \to +\infty$. We refer to Dolera and Favaro [16,17] for Berry–Esseen type refinements of (20) and to Favaro et al. [18,19] and Favaro and James [13] for generalisations of (20) with applications to Bayesian nonparametrics. This can also be seen in Pitman [5] (Chapter 4) for a general treatment of (20). According to Theorem 2, it is natural to ask whether there exists an interplay between Theorem 1 and the large n asymptotic behaviours (20) and (21). Hereafter, we show that: (i) (20), with the almost sure convergence replaced by the convergence in distribution, arises by combining (6) with (i) of Theorem 2; (ii) (8) arises by combining (21) with (ii) of Theorem 2. This provides an alternative proof of Pitman's α-diversity.

Theorem 3. *Let $K_n(\alpha, \theta)$ and $K(\alpha, z, n)$ under the EP-SM and the NB-CPSM, respectively. As $n \to +\infty$:*

(i) *For $\alpha \in (0,1)$ and $\theta > -\alpha$:*

$$\frac{K_n(\alpha,\theta)}{n^\alpha} \xrightarrow{w} S_{\alpha,\theta}. \tag{22}$$

(ii) *For $\alpha < 0$ and $z < 0$:*

$$\frac{K(\alpha,z,n)}{n^{\frac{-\alpha}{1-\alpha}}} \xrightarrow{w} \frac{(\alpha z)^{\frac{1}{1-\alpha}}}{-\alpha}. \tag{23}$$

Proof. We show that (22) arises by combining (6) with statement (i) of Theorem 2. For any pair of $\mathbb{N}$-valued random variables U and V, let $d_{TV}(U;V)$ be the total variation distance between the distribution of U and the distribution of V. Furthermore, let P_c denote a Poisson random variable with parameter $c > 0$. For any $\alpha \in (0,1)$ and $t > 0$, we show that as $n \to +\infty$:

$$d_{TV}(K(\alpha, tn^\alpha, n); 1 + P_{tn^\alpha}) \to 0. \tag{24}$$

This implies (22). The proof of (24) requires a careful analysis of the probability generating function of $K(\alpha, tn^\alpha, n)$. In particular, let us define $\omega(t;n,\alpha) := tn^\alpha + \frac{tM'_\alpha(t)}{M_\alpha(t)}$, where $M_\alpha(t) := \frac{1}{\pi} \sum_{m=1}^{\infty} \frac{(-t)^{m-1}}{(m-1)!} \Gamma(\alpha m) \sin(\pi \alpha m)$ is the Wright–Mainardi function (Mainardi et al. [20]). Then, we apply Corollary 2 of Dolera and Favaro [16] to conclude that $d_{TV}(K(\alpha, tn^\alpha, n); 1 + P_{\omega(t;n,\alpha)}) \to 0$ as $n \to +\infty$. Finally, we applied inequality (2.2) in Adell and Jodrá [21] to obtain:

$$d_{TV}(1 + P_{tn^\alpha}; 1 + P_{\omega(t;n,\alpha)}) = d_{TV}(P_{tn^\alpha}; P_{\omega(t;n,\alpha)}) \leq \frac{tM'_\alpha(t)}{M_\alpha(t)} \min\left\{1, \frac{\sqrt{(2/e)}}{\sqrt{\omega(t;n,\alpha)} + \sqrt{tn^\alpha}}\right\}$$

So that $d_{TV}(1 + P_{tn^\alpha}; 1 + P_{\omega(t;n,\alpha)}) \to 0$ as $n \to +\infty$, and (24) follows. Now, keeping α and t fixed as above, we show that (24) entails (22). To this aim, we introduced the Kolmogorov distance d_K which, for any pair of $\mathbb{R}_+$-valued random variables U and V, is defined by $d_K(U; V) := \sup_{x \geq 0} |\Pr[U \leq x] - \Pr[V \leq x]|$. The claim to be proven is equivalent to:

$$d_K(K_n(\alpha, \theta)/n^\alpha; S_{\alpha,\theta}) \to 0$$

as $n \to +\infty$. We exploit statement (i) of Theorem 2. This leads to the distributional identity $K_n(\alpha, \theta) \stackrel{d}{=} K(\alpha, \bar{X}_{\alpha,\theta,n}, n)$. Thus, in view of the basic properties of the Kolmogorov distance:

$$d_K(K_n(\alpha, \theta)/n^\alpha; S_{\alpha,\theta}) \leq d_K(K_n(\alpha, \theta); K(\alpha, n^\alpha S_{\alpha,\theta}, n)) \tag{25}$$
$$+ d_K(K(\alpha, n^\alpha S_{\alpha,\theta}, n); 1 + P_{n^\alpha S_{\alpha,\theta}})$$
$$+ d_K([1 + P_{n^\alpha S_{\alpha,\theta}}]/n^\alpha; S_{\alpha,\theta}),$$

where the $\{P_\lambda\}_{\lambda \geq 0}$ is thought of here as a homogeneous Poisson process with a rate of 1, independent of $S_{\alpha,\theta}$. The desired conclusion will be reached as soon as we will prove that all the three summands on the right-hand side of (25) go to zero as $n \to +\infty$. Before proceeding, we recall that $d_K(U; V) \leq d_{TV}(U; V)$. Therefore, for the first of these terms, we write:

$$d_K(K_n(\alpha, \theta); K(\alpha, n^\alpha S_{\alpha,\theta}, n))$$
$$\leq \frac{1}{2} \sum_{k=1}^n \left| \mathscr{C}(n, k; \alpha) \frac{\Gamma(k + \theta/\alpha)}{\alpha \Gamma(\theta/\alpha + 1)} \frac{\Gamma(\theta + 1)}{\Gamma(n + \theta)} - \int_0^{+\infty} \frac{\mathscr{C}(n, k; \alpha)(tn^\alpha)^k}{d_n(t)} f_{S_{\alpha,\theta}}(t) dt \right|$$

with $d_n(t) := \sum_{j=1}^n \mathscr{C}(n, j; \alpha)(tn^\alpha)^j$. Now, let us define $d_n^*(t) := e^{tn^\alpha}(n - 1)! \frac{1}{t^{1/\alpha}} f_\alpha(\frac{1}{t^{1/\alpha}})$. Accordingly, we can make the above right-hand side major by means of the following quantity:

$$\frac{1}{2} \sum_{k=1}^n \left| \mathscr{C}(n, k; \alpha) \frac{\Gamma(k + \theta/\alpha)}{\alpha \Gamma(\theta/\alpha + 1)} \frac{\Gamma(\theta + 1)}{\Gamma(n + \theta)} - \int_0^{+\infty} \frac{\mathscr{C}(n, k; \alpha)(tn^\alpha)^k}{d_n^*(t)} f_{S_{\alpha,\theta}}(t) dt \right|$$
$$+ \frac{1}{2} \int_0^{+\infty} \frac{|d_n^*(t) - d_n(t)|}{d_n^*(t)} f_{S_{\alpha,\theta}}(t) dt .$$

Then, by exploiting the identity $\int_0^{+\infty} \frac{(tn^\alpha)^k}{d_n^*(t)} f_{S_{\alpha,\theta}}(t) dt = \frac{1}{(n-1)!} \frac{\Gamma(k+\theta/\alpha)}{n^\theta} \frac{\Gamma(\theta+1)}{\alpha \Gamma(\theta/\alpha+1)}$, we can write:

$$\sum_{k=1}^n \left| \mathscr{C}(n, k; \alpha) \frac{\Gamma(k + \theta/\alpha)}{\alpha \Gamma(\theta/\alpha + 1)} \frac{\Gamma(\theta + 1)}{\Gamma(n + \theta)} - \int_0^{+\infty} \frac{\mathscr{C}(n, k; \alpha)(tn^\alpha)^k}{d_n^*(t)} f_{S_{\alpha,\theta}}(t) dt \right| = \left| 1 - \frac{\Gamma(n + \theta)}{\Gamma(n) n^\theta} \right|$$

which goes to zero as $n \to +\infty$ for any $\theta > -\alpha$, by Stirling's approximation. To show that the integral $\int_0^{+\infty} \frac{|d_n^*(t) - d_n(t)|}{d_n^*(t)} f_{S_{\alpha,\theta}}(t) dt$ also goes to zero as $n \to +\infty$, we may resort to identities (13)–(14) of Dolera and Favaro [16], as well as Lemma 3 therein. In particular, let $\Delta : (0, +\infty) \to (0, +\infty)$ denote a suitable continuous function independent of n, and such that $\Delta(z) = O(1)$ as $z \to 0$ and $\Delta(z) f_\alpha(1/z) = O(z^{-\infty})$ as $z \to +\infty$. Then, we write that:

$$\int_0^{+\infty} \frac{|d_n^*(t) - d_n(t)|}{d_n^*(t)} f_{S_{\alpha,\theta}}(t) dt$$
$$\leq \left| \frac{(n/e)^n \sqrt{2\pi n}}{n!} - 1 \right| + \left(\frac{(n/e)^n \sqrt{2\pi n}}{n!} \right) \frac{1}{n} \int_0^{+\infty} \Delta(t^{1/\alpha}) f_{S_{\alpha,\theta}}(t) dt .$$

Since $\int_0^{+\infty} \Delta(t^{1/\alpha}) f_{S_{\alpha,\theta}}(t) dt < +\infty$ by Lemma 3 of Dolera and Favaro [16], both the summands on the above right-hand side go to zero as $n \to +\infty$, again by Stirling's approximation. Thus, the first summand on the right-hand side of (25) goes to zero as $n \to +\infty$. As for the second summand on the right-hand side of (25), it can be bounded by

$$\int_0^{+\infty} d_{TV}(K(\alpha, tn^\alpha, n); 1 + P_{tn^\alpha}) f_{S_{\alpha,\theta}}(t) dt.$$

By a dominated convergence argument, this quantity goes to zero as $n \to +\infty$ as a consequence of (24). Finally, for the third summand on the right-hand side of (25), we can resort to a conditioning argument in order to reduce the problem to a direct application of the law of large numbers for renewal processes (Section 10.2, Grimmett and Stirzaker [22]). In particular, this leads to $n^{-\alpha} P_{tn^\alpha} \xrightarrow{a.s.} t$ for any $t > 0$, which entails that $n^{-\alpha} P_{n^\alpha S_{\alpha,\theta}} \xrightarrow{a.s.} S_{\alpha,\theta}$ as $n \to +\infty$. Thus, this third term also goes to zero as $n \to +\infty$ and (22) follows.

Now, we consider (23), showing that it arises by combining (21) with statement (ii) of Theorem 2. In particular, by an obvious conditioning argument, we can write that as $n \to +\infty$:

$$\frac{K_n(\alpha, \tilde{X}_{\alpha,z,n}|\alpha|)}{\tilde{X}_{\alpha,z,n}} \xrightarrow{a.s.} 1.$$

At this stage, we consider the probability generating function of $\tilde{X}_{\alpha,z,n}$ and we immediately obtain $\mathbb{E}[s^{\tilde{X}_{\alpha,z,n}}] := B_n(-sz)/B_n(-z)$ for $n \in \mathbb{N}$ and $s \in [0,1]$ with the same B_n as in (13) and (14). Therefore, the asymptotic expansion we already provided in (15) entails:

$$\frac{\tilde{X}_{\alpha,z,n}}{n^{\frac{-\alpha}{1-\alpha}}} \xrightarrow{w} \frac{(\alpha z)^{\frac{1}{1-\alpha}}}{-\alpha} \quad (26)$$

as $n \to +\infty$. In particular, (26) follows by applying exactly the same arguments used to prove (8). Now, since:

$$\frac{K_n(\alpha, \tilde{X}_{\alpha,z,n}|\alpha|)}{n^{\frac{-\alpha}{1-\alpha}}} \stackrel{d}{=} \frac{K_n(\alpha, \tilde{X}_{\alpha,z,n}|\alpha|)}{\tilde{X}_{\alpha,z,n}} \frac{\tilde{X}_{\alpha,z,n}}{n^{\frac{-\alpha}{1-\alpha}}},$$

the claim follows from a direct application of Slutsky's theorem. This completes the proof. □

3. Discussion

The NB-CPSM is a compound Poisson sampling model generalising the popular LS-CMSM. In this paper, we introduced a compound Poisson perspective of the EP-SM in terms of the NB-CPSM, thus extending the well-known compound Poisson perspective of the E-SM in terms of the LS-CPSM. We conjecture that an analogous perspective holds true for the class of α-stable Poisson–Kingman sampling models (Pitman [23] and Pitman [5]), of which the EP-SM is a noteworthy special case. That is, for $\alpha \in (0,1)$, we conjecture that an α-stable Poisson–Kingman sampling model admits a representation as a randomised NB-CPSM with $\alpha \in (0,1)$ and $z > 0$, where the randomisation acts on z with respect to a scale mixture between a Gamma and a suitable transformation of the Mittag–Leffler distribution. We believe that such a compound Poisson representation would be critical in order to introduce Berry–Esseen type refinements of the large n asymptotic behaviour of K_n under α-stable Poisson–Kingman sampling models. This can be seen in Section 6.1 of Pitman [23] and the references therein. Such a line of research aims to extend the preliminary works of Dolera and Favaro [16,17] on Berry–Esseen type theorems under the EP-SM. Work on this, and on the more general settings induced by normalised random measures (Regazzini et al. [24]) and Poisson–Kingman models (Pitman [23]), is ongoing.

Author Contributions: Formal analysis, E.D. and S.F.; writing—original draft preparation, E.D. and S.F.; writing—review and editing, E.D. and S.F. All authors have read and agreed to the published version of the manuscript.

Funding: This research received funding from the European Research Council (ERC) under the European Union's Horizon 2020 research and innovation programme under grant agreement No 817257.

Institutional Review Board Statement: Not applicable.

Informed Consent Statement: Not applicable.

Data Availability Statement: Not applicable.

Acknowledgments: The authors thank the editor and two anonymous referees for all their comments and suggestions which remarkably improved the original version of the present paper. Emanuele Dolera and Stefano Favaro wish to express their enormous gratitude to Eugenio Regazzini, whose fundamental contributions to the theory of Bayesian statistics have always been a great source of inspiration, transmitting enthusiasm and method for the development of their own research. The authors gratefully acknowledge the financial support from the Italian Ministry of Education, University and Research (MIUR), "Dipartimenti di Eccellenza" grant 2018–2022.

Conflicts of Interest: The authors declare no conflict of interest.

References

1. Perman, M.; Pitman, J.; Yor, M. Size-biased sampling of Poisson point processes and excursions. *Probab. Theory Relat. Fields* **1992**, *92*, 21–39. [CrossRef]
2. Pitman, J. Exchangeable and partially exchangeable random partitions. *Probab. Theory Relat. Fields* **1995**, *102*, 145–158. [CrossRef]
3. Pitman, J.; Yor, M. The two parameter Poisson-Dirichlet distribution derived from a stable subordinator. *Ann. Probab.* **1997**, *25*, 855–900. [CrossRef]
4. Ferguson, T.S. A Bayesian analysis of some nonparametric problems. *Ann. Stat.* **1973**, *1*, 209–230. [CrossRef]
5. Pitman, J. *Combinatorial Stochastic Processes*; Lecture Notes in Mathematics; Springer: Berlin/Heidelberg, Germany, 2006.
6. Ewens, W. The sampling theory or selectively neutral alleles. *Theor. Popul. Biol.* **1972**, *3*, 87–112. [CrossRef]
7. Crane, H. The ubiquitous Ewens sampling formula. *Stat. Sci.* **2016**, *31*, 1–19. [CrossRef]
8. Charalambides, C.A. Distributions of random partitions and their applications. *Methodol. Comput. Appl. Probab.* **2007**, *9*, 163–193. [CrossRef]
9. Korwar, R.M.; Hollander, M. Contributions to the theory of Dirichlet processes. *Ann. Stat.* **1973**, *1*, 705–711. [CrossRef]
10. Wright, E.M. The asymptotic expansion of the generalized Bessel function. *Proc. Lond. Math. Soc.* **1935**, *38*, 257–270. [CrossRef]
11. Charalambides, C.A. *Combinatorial Methods in Discrete Distributions*; Wiley: Hoboken, NJ, USA, 2005.
12. Berg, L. Asymptotische darstellungen für integrale und reihen mit anwendungen. *Math. Nachrichten* **1958**, *17*, 101–135. [CrossRef]
13. Favaro, S.; James, L.F. A note on nonparametric inference for species variety with Gibbs-type priors. *Electron. J. Stat.* **2015**, *9*, 2884–2902. [CrossRef]
14. Gnedin, A.; Pitman, J. Exchangeable Gibbs partitions and Stirling triangles. *J. Math. Sci.* **2006**, *138*, 5674–5685. [CrossRef]
15. Gnedin, A. A species sampling model with finitely many types. *Electron. Commun. Probab.* **2010**, *8*, 79–88. [CrossRef]
16. Dolera, E.; Favaro, S. A Berry—Esseen theorem for Pitman's α—Diversity. *Ann. Appl. Probab.* **2020**, *30*, 847–869. [CrossRef]
17. Dolera, E.; Favaro, S. Rates of convergence in de Finetti's representation theorem, and Hausdorff moment problem. *Bernoulli* **2020**, *26*, 1294–1322. [CrossRef]
18. Favaro, S.; Lijoi, A.; Prünster, I. Asymptotics for a Bayesian nonparametric estimator of species richness. *Bernoulli* **2012**, *18*, 1267–1283 [CrossRef]
19. Favaro, S.; Lijoi, A.; Mena, R.H.; Prünster, I. Bayesian nonparametric inference for species variety with a two parameter Poisson-Dirichlet process prior. *J. R. Stat. Soc. Ser. B* **2009**, *71*, 992–1008. [CrossRef]
20. Mainardi, F.; Mura, A.; Pagnini, G. The M-Wright function in time-fractional diffusion processes: A tutorial survey. *Int. J. Differ. Equat.* **2010**, 104505. [CrossRef]
21. Adell, J.A.; Jodrá, P. Exact Kolmogorov and total variation distances between some familiar discrete distributions. *J. Inequalities Appl.* **2006**, 64307. [CrossRef]
22. Grimmett, G.; Stirzaker, D. *Probability and Random Processes*; Oxford University Press: Oxford, UK, 2001.
23. Pitman, J. Poisson-Kingman partitions. In *Science and Statistics: A Festschrift for Terry Speed*; Goldstein, D.R., Ed.; Institute of Mathematical Statistics: Tachikawa, Japan, 2003.
24. Regazzini, E.; Lijoi, A.; Prünster, I. Distributional results for means of normalized random measures with independent increments. *Ann. Stat.* **2003**, *31*, 560–585. [CrossRef]

Article

Predictive Constructions Based on Measure-Valued Pólya Urn Processes

Sandra Fortini [1,*], Sonia Petrone [1] and Hristo Sariev [2]

[1] Department of Decision Sciences, Bocconi University, 20136 Milano, Italy; sonia.petrone@unibocconi.it
[2] Institute of Mathematics and Informatics, Bulgarian Academy of Sciences, 1113 Sofia, Bulgaria; h.sariev@math.bas.bg
* Correspondence: sandra.fortini@unibocconi.it

Abstract: Measure-valued Pólya urn processes (MVPP) are Markov chains with an additive structure that serve as an extension of the generalized k-color Pólya urn model towards a continuum of possible colors. We prove that, for any MVPP $(\mu_n)_{n \geq 0}$ on a Polish space $\mathbb{X}$, the normalized sequence $(\mu_n/\mu_n(\mathbb{X}))_{n \geq 0}$ agrees with the marginal predictive distributions of some random process $(X_n)_{n \geq 1}$. Moreover, $\mu_n = \mu_{n-1} + R_{X_n}$, $n \geq 1$, where $x \mapsto R_x$ is a random transition kernel on $\mathbb{X}$; thus, if μ_{n-1} represents the contents of an urn, then X_n denotes the color of the ball drawn with distribution $\mu_{n-1}/\mu_{n-1}(\mathbb{X})$ and R_{X_n}—the subsequent reinforcement. In the case $R_{X_n} = W_n \delta_{X_n}$, for some non-negative random weights $W_1, W_2, \ldots$, the process $(X_n)_{n \geq 1}$ is better understood as a randomly reinforced extension of Blackwell and MacQueen's Pólya sequence. We study the asymptotic properties of the predictive distributions and the empirical frequencies of $(X_n)_{n \geq 1}$ under different assumptions on the weights. We also investigate a generalization of the above models via a randomization of the law of the reinforcement.

Keywords: predictive distributions; random probability measures; reinforced processes; Pólya sequences; urn schemes; Bayesian inference; conditional identity in distribution; total variation distance

MSC: 60G57; 60B10; 60G25; 60F05; 60G09

Citation: Fortini, S.; Petrone, S.; Sariev, H. Predictive Constructions Based on Measure-Valued Pólya Urn Processes. *Mathematics* **2021**, *9*, 2845. https://doi.org/10.3390/math9222845

Academic Editors: Emanuele Dolera and Federico Bassetti

Received: 4 October 2021
Accepted: 8 November 2021
Published: 10 November 2021

Publisher's Note: MDPI stays neutral with regard to jurisdictional claims in published maps and institutional affiliations.

Copyright: © 2021 by the authors. Licensee MDPI, Basel, Switzerland. This article is an open access article distributed under the terms and conditions of the Creative Commons Attribution (CC BY) license (https://creativecommons.org/licenses/by/4.0/).

1. Introduction

Let $(X_n)_{n \geq 1}$ be a sequence of homogeneous random observations, taking values in a Polish space $\mathbb{X}$. The central assumption in the Bayesian approach to inductive reasoning is that $(X_n)_{n \geq 1}$ is exchangeable, that is, its law is invariant under finite permutations. Then, by de Finetti's theorem, there exists a random probability measure $\tilde{P}$ on $\mathbb{X}$ such that, given $\tilde{P}$, the random variables $X_1, X_2, \ldots$ are conditionally independent and identically distributed with marginal distribution $\tilde{P}$ (see [1], Section 3), denoted

$$X_n \mid \tilde{P} \stackrel{i.i.d.}{\sim} \tilde{P}. \qquad (1)$$

Furthermore, $\tilde{P}$ is the almost sure (a.s.) weak limit of the predictive distributions and the empirical frequencies,

$$\mathbb{P}(X_{n+1} \in \cdot \mid X_1, \ldots, X_n) \xrightarrow{w} \tilde{P}(\cdot) \quad \text{a.s.} \quad \text{and} \quad \frac{1}{n}\sum_{i=1}^{n} \delta_{X_i}(\cdot) \xrightarrow{w} \tilde{P}(\cdot) \quad \text{a.s.} \qquad (2)$$

The model (1) is completed by choosing a prior distribution for $\tilde{P}$. Inference consists in computing the conditional (posterior) distribution of $\tilde{P}$ given an observed sample $(X_1, \ldots, X_n)$, with most inferential conclusions depending on some average with respect to the posterior distribution; for example, under squared loss, for any measurable set

$B \subseteq \mathbb{X}$, the best estimate of $\tilde{P}(B)$ is the posterior mean, $\mathbb{E}[\tilde{P}(B)|X_1,\ldots,X_n]$. In addition, the posterior mean can be utilized for predictive inference since

$$\mathbb{P}(X_{n+1} \in B | X_1,\ldots,X_n) = \mathbb{E}[\tilde{P}(B)|X_1,\ldots,X_n]. \tag{3}$$

A different modeling strategy uses the Ionescu–Tulcea theorem to define the law of the process from the sequence of predictive distributions, $(\mathbb{P}(X_{n+1} \in \cdot | X_1,\ldots,X_n))_{n\geq 1}$. In that case, one can refer to Theorem 3.1 in [2] for necessary and sufficient conditions on $(\mathbb{P}(X_{n+1} \in \cdot | X_1,\ldots,X_n))_{n\geq 1}$ to be consistent with exchangeability. The predictive approach to model building is deeply rooted in Bayesian statistics, where the parameter $\tilde{P}$ is assigned an auxiliary role and the focus is on observable "facts", see [2–6]. Moreover, using the predictive distributions as primary objects allows one to make predictions instantly or helps ease computations. See [7] for a review on some well-known predictive constructions of priors for Bayesian inference.

In this work, we consider a class of predictive constructions based on measure-valued Pólya urn processes (MVPP). MVPPs have been introduced in the probabilistic literature [8,9] as an extension of k-color urn models, but their implications for (Bayesian) statistics have yet to be explored. A first aim of the paper is thus to show the potential use of MVPPs as predictive constructions in Bayesian inference. In fact, some popular models in Bayesian nonparametric inference can be framed in such a way, see Equation (8). A second aim of the paper is to suggest novel extensions of MVPPs that we believe can offer more flexibility in statistical applications.

MVPPs are essentially measure-valued Markov processes that have an additive structure, with the formal definition being postponed to Section 2.1 (Definition 1). Given an MVPP $(\mu_n)_{n\geq 0}$, we consider a sequence of random observations that are characterized by $\mathbb{P}(X_1 \in \cdot) = \mu_0(\cdot)/\mu_0(\mathbb{X})$ and, for $n \geq 1$,

$$\mathbb{P}(X_{n+1} \in \cdot \mid X_1, \mu_1, \ldots, X_n, \mu_n) = \frac{\mu_n(\cdot)}{\mu_n(\mathbb{X})}. \tag{4}$$

The random measure μ_n is not necessarily measurable with respect to $(X_1,\ldots,X_n)$, so the predictive construction (4) is more flexible than models based solely on the predictive distributions of $(X_n)_{n\geq 1}$; for example, $(\mu_n)_{n\geq 0}$ allows for the presence of latent variables or other sources of observable data (see also [10] for a covariate-based predictive construction). However, (4) can lead to an imbalanced design, which may break the symmetry imposed by exchangeability. Nevertheless, it is still possible that the sequence $(X_n)_{n\geq 1}$ satisfies (2) for some $\tilde{P}$, in which case Lemma 8.2 in [1] implies that $(X_n)_{n\geq 1}$ is asymptotically exchangeable with directing random measure $\tilde{P}$.

In Theorem 1, we show that, taking $(\mu_n)_{n\geq 0}$ as primary, the sequence $(X_n)_{n\geq 1}$ in (4) can be chosen such that

$$\mu_n = \mu_{n-1} + R_{X_n}, \tag{5}$$

where $x \mapsto R_x$ is a measurable map from $\mathbb{X}$ to the space of finite measures on $\mathbb{X}$. Models of the kind (4)–(5) are computationally efficient. Indeed, as new observations become available, predictions can be updated at a constant computational cost and with limited storage of information. If, in addition, $(X_n)_{n\geq 1}$ is asymptotically exchangeable, then (4)–(5) can provide a computationally simple approximation of an exchangeable scheme for Bayesian inference, along the lines in [11].

The recursive formula (5) allows us to interpret the dynamics of MVPPs in terms of an urn sampling scheme, as the name suggests. Let μ_0 be a non-random finite measure on $\mathbb{X}$. Suppose we have an urn whose contents are described by μ_0 in the sense that $\mu_0(B)$ denotes the total mass of balls with colors in $B \subseteq \mathbb{X}$. At time $n = 1$, a ball is extracted at random from the urn, and we denote its color by X_1. The urn is then reinforced according to a replacement rule $(R_x)_{x \in \mathbb{X}}$, so that the updated composition becomes $\mu_1 \equiv \mu_0 + R_{X_1}$. At any time $n > 1$, a ball of color X_n is picked with probability distribution $\mu_{n-1}/\mu_{n-1}(\mathbb{X})$, and the contents of the urn are subsequently reinforced by R_{X_n}. In the case the space of colors is finite, $|\mathbb{X}| = k$, the above procedure is better known as a generalized k-color Pólya urn [12].

We focus our analysis on MVPPs for which R_x is concentrated on x; thus, after each draw, we reinforce only the color of the observed ball. More formally, we consider MVPPs that have a reinforcement measure of the kind $R_{X_n} = W_n \delta_{X_n}$, $n \geq 1$, where W_n is some non-negative random variable. In that case, Equations (4) and (5) become

$$\mathbb{P}(X_{n+1} \in \cdot \mid X_1, W_1, \ldots, X_n, W_n) = \sum_{i=1}^{n} \frac{W_i}{\mu_0(\mathbb{X}) + \sum_{j=1}^{n} W_j} \delta_{X_i}(\cdot) + \frac{\mu_0(\mathbb{X})}{\mu_0(\mathbb{X}) + \sum_{j=1}^{n} W_i} \mu_0'(\cdot), \qquad (6)$$

and

$$\mu_n = \mu_{n-1} + W_n \delta_{X_n}. \qquad (7)$$

A notable example is Blackwell and MacQueen's em Pólya sequence [13], which is a random process $(X_n)_{n \geq 1}$ characterized by $\mathbb{P}(X_1 \in \cdot) = \nu(\cdot)$ and, for $n \geq 1$,

$$\mathbb{P}(X_{n+1} \in \cdot \mid X_1, \ldots, X_n) = \sum_{i=1}^{n} \frac{1}{\theta + n} \delta_{X_i}(\cdot) + \frac{\theta}{\theta + n} \nu(\cdot), \qquad (8)$$

for some probability measure ν on $\mathbb{X}$ and a constant $\theta > 0$. By [13], $(X_n)_{n \geq 1}$ is exchangeable and corresponds to the model (1) with Dirichlet process prior with parameters (θ, ν). It is easily seen that (8) is related to the MVPP $(\mu_n)_{n \geq 0}$ given by $\mu_0 = \theta \nu$ and, for $n \geq 1$,

$$\mu_n = \mu_{n-1} + \delta_{X_n}.$$

Therefore, we will call any MVPP a randomly reinforced Pólya process (RRPP) if it admits representation (6)–(7).

Existing studies on MVPPs look at models that have mostly a balanced design, i.e., $R_x(\mathbb{X}) = r$, $x \in \mathbb{X}$, and assume irreducibility-like conditions for $(R_x)_{x \in \mathbb{X}}$, see [8,9,14,15] and Remark 4 in [16]. In contrast, RRPPs require that $R_x(\{x\}^c) = 0$, and so are excluded from the analysis in those papers. In fact, this difference in reinforcement mechanisms mirrors the dichotomy within k-color urn models, where the replacement R is best described in terms of a matrix with random elements. There, the class of randomly reinforced urns [17] assumes an R with zero off-diagonal elements (i.e., we reinforce only the color of the observed ball), whereas the generalized Pólya urn models require the mean replacement matrix to be irreducible. Similarly to the k-color case, RRPPs need the use of different techniques, which yield completely different results than those in [8,9,14–16]. As an example, Theorem 1 in [16] and our Theorem 2 prove convergence of the kind (2), yet the limit probability measure in [16] is non-random.

The RRPP has been implicitly studied by [17–23], among others, with the focus being on the process $(X_n)_{n \geq 1}$. Those papers deal primarily with the k-color case (with the exception of [18,19,23]) and can be categorized on the basis of their assumptions on $(W_n)_{n \geq 1}$. For example, [18,19,21,22] assume that W_n and $(X_1, W_1, \ldots, X_{n-1}, W_{n-1}, X_n)$ are independent, in which case the process $(X_n)_{n \geq 1}$ is conditionally identically distributed (c.i.d.) [21], that is, conditionally on current information, all future observations are identically distributed. It follows from [21] that c.i.d. processes preserve many of the properties of exchangeable sequences and, in particular, satisfy (2)–(3). In contrast, [17,20,23] assume that the reinforcement W_n depends on the particular color X_n, and prove a version of (2) where $\tilde{P}$ is concentrated on the set of dominant colors for which the expected reinforcement is maximum. In this work, we reconsider the above models in the framework of RRPPs. For the c.i.d. case, we prove results whose analogues have already been established by [23] for the model with dominant colors. In particular, we extend the convergence in (2) to be in total variation and give a unified central limit theorem. We also examine the number of distinct values that are generated by the sequence $(X_n)_{n \geq 1}$.

In some applications, the definition of an MVPP can be too restrictive as it assumes that the probability law of the reinforcement R is known. However, we can envisage situations where the law is itself random, so we extend the definition of an MVPP by introducing a random parameter V. The resulting generalized measure-valued Pólya urn process (GMVPP) turns out to be a mixture of Markov processes and admits representation (4)–(5), conditional on the parameter V. When the reinforcement measure R_x is concentrated on x, we call $(\mu_n)_{n\geq 0}$ a generalized randomly reinforced Pólya process (GRRPP). We give a characterization of GRRPPs with exchangeable weights $(W_n)_{n\geq 1}$ and show that the process $((X_n, W_n))_{n\geq 1}$ is partially conditionally identically distributed (partially c.i.d) [24], that is, conditionally on the past observations and the concurrent observation from the other sequence, the future observations are marginally identically distributed. We also extend some of the results for RRPPs to the generalized setting.

The paper is structured as follows. In Section 2.1, we recall the definition of a measure-valued Pólya urn process and prove representation (4)–(5) for a suitably selected sequence $(X_n)_{n\geq 1}$. Section 2.2 defines a particular subclass of MVPPs, called randomly reinforced Pólya processes (RRPP), which share with exchangeable Pólya sequences the property of reinforcing only the observed color. Section 3 is devoted to the study of the asymptotic properties of RRPPs. In Section 4, we give the definition of GMVPPs and GRRPPs, and obtain basic results.

2. Definitions and a Representation Result

Let $(\mathbb{X}, d)$ be a complete separable metric space, endowed with its Borel σ-field $\mathcal{X}$. Denote by

$$\mathbb{M}_F(\mathbb{X}), \qquad \mathbb{M}_F^*(\mathbb{X}), \qquad \mathbb{M}_P(\mathbb{X}),$$

the collections of measures μ on $\mathbb{X}$ that are finite, finite and non-null, and probability measures, respectively. We regard $\mathbb{M}_F(\mathbb{X})$, $\mathbb{M}_F^*(\mathbb{X})$ and $\mathbb{M}_P(\mathbb{X})$ as measurable spaces equipped with the σ-fields generated by $\mu \mapsto \mu(B)$, $B \in \mathcal{X}$. We further let

$$\mathbb{K}_F(\mathbb{X}, \mathbb{Y}), \qquad \mathbb{K}_P(\mathbb{X}, \mathbb{Y}),$$

be the collections of transition kernels K from $\mathbb{X}$ to $\mathbb{Y}$ that are finite and probability kernels, respectively. Any non-null measure $\mu \in \mathbb{M}_F^*(\mathbb{X})$ has a normalized version $\mu' = \mu/\mu(\mathbb{X})$. If $f: \mathbb{X} \to \mathbb{Y}$ is measurable, then $f^\sharp: \mathbb{M}_F(\mathbb{X}) \to \mathbb{M}_F(\mathbb{Y})$ denotes the induced mapping of measures, $f^\sharp(\mu)(\cdot) = \mu(f^{-1}(\cdot))$, $\mu \in \mathbb{M}_F(\mathbb{X})$.

All random quantities are defined on a common probability space $(\Omega, \mathcal{H}, \mathbb{P})$, which is assumed to be rich enough to support any required randomization. The symbol '$\perp$' will be used to denote independence between random objects, and "$\stackrel{d}{=}$" equality in distribution.

2.1. Measure-Valued Pólya urn Processes

Let $\mu \in \mathbb{M}_F^*(\mathbb{X})$ describe the contents of an urn, as in Section 1. Once a ball is picked at random from μ, the urn is reinforced according to a replacement rule, which is formally a kernel $R \in \mathbb{K}_F(\mathbb{X}, \mathbb{X})$ that maps colors $x \mapsto R_x(\cdot)$ to finite measures; thus,

$$\mu + R_x, \tag{9}$$

represents the updated urn composition if a ball of color x has been observed. In general, R is random and there exists a probability kernel $\mathcal{R} \in \mathbb{K}_P(\mathbb{X}, \mathbb{M}_F(\mathbb{X}))$ such that $R_x \sim \mathcal{R}_x$, $x \in \mathbb{X}$. Then, the distribution of (9) prior to the sampling of the urn is given by

$$\hat{R}_\mu(\cdot) = \int_\mathbb{X} \psi_\mu^\sharp(\mathcal{R}_x)(\cdot)\mu'(dx), \tag{10}$$

where ψ_μ is the measurable map $\nu \mapsto \nu + \mu$ from $\mathbb{M}_F(\mathbb{X})$ to $\mathbb{M}_F^*(\mathbb{X})$. By Lemma 3.3 in [9], $\mu \mapsto \hat{R}_\mu$ is a measurable map from $\mathbb{M}_F^*(\mathbb{X})$ to $\mathbb{M}_P(\mathbb{M}_F^*(\mathbb{X}))$.

Definition 1 (Measure-Valued Pólya Urn Process [9]). *A sequence $(\mu_n)_{n\geq 0}$ of random finite measures on $\mathbb{X}$ is called a measure-valued Pólya urn process (MVPP) with parameters $\mu_0 \in \mathbb{M}_F^*(\mathbb{X})$*

and $\mathcal{R} \in \mathbb{K}_P(\mathbb{X}, \mathbb{M}_F(\mathbb{X}))$ if it is a Markov process with transition kernel $\hat{R}$ given by (10). If, in particular, $\mathcal{R}_x = \delta_{R_x}$ for some $R \in \mathbb{K}_F(\mathbb{X}, \mathbb{X})$, then $(\mu_n)_{n \geq 0}$ is said to be a deterministic MVPP.

The representation theorem below formalizes the idea of MVPP as an urn scheme.

Theorem 1. *A sequence $(\mu_n)_{n \geq 0}$ of random finite measures is an MVPP with parameters $(\mu_0, \mathcal{R})$ if and only if, for every $n \geq 1$,*

$$\mu_n = \mu_{n-1} + R_{X_n} \quad a.s., \tag{11}$$

where $(X_n)_{n \geq 1}$ is a sequence of $\mathbb{X}$-valued random variables such that $X_1 \sim \mu_0'$ and, for $n \geq 2$,

$$\mathbb{P}(X_n \in \cdot \mid X_1, \mu_1, \ldots, X_{n-1}, \mu_{n-1}) = \mu_{n-1}'(\cdot), \tag{12}$$

and R is a random finite transition kernel on $\mathbb{X}$ such that

$$\mathbb{P}(R_{X_n} \in \cdot \mid X_1, \mu_1, \ldots, X_{n-1}, \mu_{n-1}, X_n) = \mathcal{R}_{X_n}(\cdot). \tag{13}$$

Proof. If $(\mu_n)_{n \geq 0}$ satisfies (11)–(13) for every $n \geq 1$, then it holds a.s. that

$$\mathbb{P}(\mu_n \in \cdot \mid \mu_1, \ldots, \mu_{n-1}) = \mathbb{E}[\psi_{\mu_{n-1}}^\sharp(\mathcal{R}_{X_n})(\cdot) \mid \mu_1, \ldots, \mu_{n-1}] = \hat{R}_{\mu_{n-1}}(\cdot).$$

Conversely, suppose $(\mu_n)_{n \geq 0}$ is a MVPP with parameters $(\mu_0, \mathcal{R})$. As $\mathcal{R}$ is a probability kernel from $\mathbb{X}$ to $\mathbb{M}_F(\mathbb{X})$ and $\mathbb{M}_F(\mathbb{X})$ is Polish, then there exists by Lemma 4.22 in [25] a measurable function $f(x, u)$ such that, for every $x \in \mathbb{X}$,

$$f(x, U) \sim \mathcal{R}_x,$$

whenever U is a uniform random variable on $[0, 1]$, denoted $U \sim \text{Unif}[0, 1]$.

Let us prove by induction that there exists a sequence $((X_n, U_n))_{n \geq 1}$ such that $X_1 \sim \mu_0'$, $U_1 \perp X_1$, $U_1 \sim \text{Unif}[0, 1]$, $\mu_1 = \mu_0 + f(X_1, U_1)$ a.s., $(\mu_2, \mu_3, \ldots) \perp (X_1, U_1) \mid \mu_1$, and, for every $n \geq 2$,

(i) $\mathbb{P}(X_n \in \cdot \mid X_1, U_1, \mu_1, \ldots, X_{n-1}, U_{n-1}, \mu_{n-1}) = \mu_{n-1}'(\cdot)$;
(ii) $U_n \sim \text{Unif}[0, 1]$ and $U_n \perp (X_1, U_1, \mu_1, \ldots, X_{n-1}, U_{n-1}, \mu_{n-1}, X_n)$;
(iii) $\mu_n = \mu_{n-1} + f(X_n, U_n)$ a.s.;
(iv) $(\mu_{n+1}, \mu_{n+2}, \ldots) \perp (X_n, U_n) \mid (X_1, U_1, \mu_1, \ldots, X_{n-1}, U_{n-1}, \mu_{n-1}, \mu_n)$;
(v) $\mu_{n+1} \perp (X_1, U_1, \ldots, X_n, U_n) \mid (\mu_1, \ldots, \mu_n)$.

Then, Equations (11)–(13) follow from (i)–(iii) with $R_{X_n} = f(X_n, U_n)$.

Regarding the base case, let $\tilde{X}_1$ and $\tilde{U}_1$ be independent random variables such that $\tilde{U}_1 \sim \text{Unif}[0, 1]$ and $\tilde{X}_1 \sim \mu_0'$. It follows that, for any measurable set $B \subseteq \mathbb{M}_F(\mathbb{X})$,

$$\mathbb{P}(\mu_1 \in B) = \hat{R}_{\mu_0}(B) = \mathbb{E}[\psi_{\mu_0}^\sharp(\mathcal{R}_{\tilde{X}_1})(B)] = \mathbb{P}((\mu_0 + f(\tilde{X}_1, \tilde{U}_1)) \in B);$$

thus, $\mu_1 \stackrel{d}{=} \mu_0 + f(\tilde{X}_1, \tilde{U}_1)$. By Theorem 8.17 in [25], there exist random variables X_1 and U_1 such that

$$(\mu_1, X_1, U_1) \stackrel{d}{=} (\mu_0 + f(\tilde{X}_1, \tilde{U}_1), \tilde{X}_1, \tilde{U}_1),$$

and $(\mu_2, \mu_3, \ldots) \perp (X_1, U_1) \mid \mu_1$. Then, in particular, $(X_1, U_1) \stackrel{d}{=} (\tilde{X}_1, \tilde{U}_1)$ and $(\mu_1, \mu_0 + f(X_1, U_1)) \stackrel{d}{=} (\mu_0 + f(\tilde{X}_1, \tilde{U}_1), \mu_0 + f(\tilde{X}_1, \tilde{U}_1))$, so

$$\mu_1 = \mu_0 + f(X_1, U_1) \quad a.s.$$

Regarding the induction step, assume that (i)–(v) hold true until some $n > 1$. Let $\tilde{X}_{n+1}$ and $\tilde{U}_{n+1}$ be such that $\tilde{U}_{n+1} \sim \text{Unif}[0, 1]$, $\tilde{U}_{n+1} \perp (X_1, U_1, \mu_1, \ldots, X_n, U_n, \mu_n, \tilde{X}_{n+1})$, and

$$\mathbb{P}(\tilde{X}_{n+1} \in \cdot \mid X_1, U_1, \mu_1, \ldots, X_n, U_n, \mu_n) = \mu_n'(\cdot).$$

It follows from (v) that, for any measurable set $B \subseteq \mathbb{M}_F(\mathbb{X})$,

$$\mathbb{P}(\mu_{n+1} \in B | X_1, U_1, \mu_1, \ldots, X_n, U_n, \mu_n) = \mathbb{E}[\psi_{\mu_n}^\sharp(\mathcal{R}_{\tilde{X}_{n+1}})(B)|X_1, U_1, \mu_1, \ldots, X_n, U_n, \mu_n]$$
$$= \mathbb{P}((\mu_n + f(\tilde{X}_{n+1}, \tilde{U}_{n+1})) \in B | X_1, U_1, \mu_1, \ldots, X_n, U_n, \mu_n);$$

thus, $\mu_{n+1} \stackrel{d}{=} \mu_n + f(\tilde{X}_{n+1}, \tilde{U}_{n+1}) \mid X_1, U_1, \mu_1, \ldots, X_n, U_n, \mu_n$. By Theorem 8.17 in [25], there exist random variables X_{n+1} and U_{n+1} such that

$$(\mu_{n+1}, X_1, U_1, \mu_1, \ldots, X_n, U_n, \mu_n, X_{n+1}, U_{n+1})$$
$$\stackrel{d}{=} (\mu_n + f(\tilde{X}_{n+1}, \tilde{U}_{n+1}), X_1, U_1, \mu_1, \ldots, X_n, U_n, \mu_n, \tilde{X}_{n+1}, \tilde{U}_{n+1}),$$

and $(\mu_{n+2}, \mu_{n+3}, \ldots) \perp (X_{n+1}, U_{n+1}) \mid (X_1, U_1, \mu_1, \ldots, X_n, U_n, \mu_n, \mu_{n+1})$. Then, in particular, $U_{n+1} \sim \text{Unif}[0,1]$, $U_{n+1} \perp (X_1, U_1, \mu_1, \ldots, X_n, U_n, \mu_n, X_{n+1})$, and

$$\mathbb{P}(X_{n+1} \in \cdot \mid X_1, U_1, \mu_1, \ldots, X_n, U_n, \mu_n) = \mu_n'(\cdot).$$

Moreover,

$$(\mu_{n+1}, \mu_n + f(X_{n+1}, U_{n+1})) \stackrel{d}{=} (\mu_n + f(\tilde{X}_{n+1}, \tilde{U}_{n+1}), \mu_n + f(\tilde{X}_{n+1}, \tilde{U}_{n+1}));$$

therefore,

$$\mathbb{P}(\mu_{n+1} = \mu_n + f(X_{n+1}, U_{n+1})) = \mathbb{P}(\mu_n + f(\tilde{X}_{n+1}, \tilde{U}_{n+1}) = \mu_n + f(\tilde{X}_{n+1}, \tilde{U}_{n+1})) = 1.$$

By Theorem 8.12 in [25], statement (v) with $n+1$ is equivalent to $\mu_{n+2} \perp (X_1, U_1) \mid (\mu_1, \ldots, \mu_{n+1})$ and $\mu_{n+2} \perp (X_{k+1}, U_{k+1}) \mid (X_1, U_1, \ldots, X_k, U_k, \mu_1, \ldots, \mu_{n+1})$, $k = 1, \ldots, n$. The latter follows from the induction hypothesis since, by (iv), we have $(\mu_{k+2}, \ldots, \mu_{n+2}) \perp (X_{k+1}, U_{k+1}) \mid (X_1, U_1, \ldots, X_k, U_k, \mu_1, \ldots, \mu_{k+1})$ for every $k = 1, \ldots, n$. □

The process $(X_n)_{n \geq 1}$ in Theorem 1 corresponds to the sequence of observed colors from the implied urn sampling scheme. Furthermore, the replacement rule takes the form $R_{X_n} = f(X_n, U_n)$, where f is some measurable function, $U_n \sim \text{Unif}[0,1]$, and $U_n \perp (X_1, U_1, \ldots, X_{n-1}, U_{n-1}, X_n)$, from which it follows that

$$\mu_n = \mu_{n-1} + f(X_n, U_n), \tag{14}$$

and

$$\mathbb{P}(X_{n+1} \in \cdot \mid X_1, \ldots, X_n, (U_m)_{m \geq 1}) = \frac{\mu_0(\cdot) + \sum_{i=1}^n f(X_i, U_i)(\cdot)}{\mu_0(\mathbb{X}) + \sum_{i=1}^n f(X_i, U_i)(\mathbb{X})}. \tag{15}$$

Thus, the sequence $(U_n)_{n \geq 1}$ models the additional randomness in the reinforcement measure R. Janson [9] obtains a rather similar result; Theorem 1.3 in [9] states that any MVPP $(\mu_n)_{n \geq 0}$ can be coupled with a deterministic MVPP $(\bar{\mu}_n)_{n \geq 0}$ on $\mathbb{X} \times [0,1]$ in the sense that

$$\bar{\mu}_n = \mu_n \times \lambda, \tag{16}$$

where λ is the Lebesgue measure on $[0,1]$, and $\mu_n \times \lambda$ is the product measure on $\mathbb{X} \times [0,1]$. In our case, the MVPP defined by $\bar{\mu}_0 = \mu_0 \times \lambda$ and, for $n \geq 1$,

$$\bar{\mu}_n = \bar{\mu}_{n-1} + f(X_n, U_n) \times \lambda,$$

has a non-random replacement rule $R_{x,u} = f(x, u) \times \lambda$ and satisfies (16) on a set of probability one.

2.2. Randomly Reinforced Pólya Processes

It follows from (8) that any Pólya sequence generates a deterministic MVPP through

$$\mu_n = \mu_{n-1} + \delta_{X_n}.$$

Here, we consider a randomly reinforced extension of Pólya sequences in the form of an MVPP with replacement rule $R_x = W(x) \cdot \delta_x$, $x \in \mathbb{X}$, where $W(x)$ is a non-negative random variable.

Definition 2 (Randomly Reinforced Pólya Process). *We call an MVPP with parameters $(\mu_0, \mathcal{R})$ a randomly reinforced Pólya process (RRPP) if there exists $\eta \in \mathbb{K}_P(\mathbb{X}, \mathbb{R}_+)$ such that $\mathcal{R}_x = \xi_x^\sharp(\eta_x)$, $x \in \mathbb{X}$, where $\xi_x : \mathbb{R}_+ \to \mathbb{M}_F(\mathbb{X})$ is the map $w \mapsto w\delta_x$.*

Observe that, for RRPPs, the reinforcement measure $f(x,u)$ in (14)–(15) concentrates its mass on x; thus, we obtain the following variant of the representation result in Theorem 1.

Proposition 1. *Let $(\mu_n)_{n\geq 0}$ be an RRPP with parameters (μ_0, η). Then, there exist a measurable function $h : \mathbb{X} \times [0,1] \to \mathbb{R}_+$ and a sequence $((X_n, U_n))_{n\geq 1}$ such that, using $W_n = h(X_n, U_n)$, we have for every $n \geq 1$ that*

$$\mu_n = \mu_{n-1} + W_n \delta_{X_n} \quad \text{a.s.,} \tag{17}$$

where $X_1 \sim \mu_0'$ and, for $n \geq 1$, $U_n \sim \text{Unif}[0,1]$, $U_n \perp (X_1, U_1, \ldots, X_{n-1}, U_{n-1}, X_n)$, and

$$\mathbb{P}(X_{n+1} \in \cdot \mid X_1, W_1, \ldots, X_n, W_n) = \sum_{i=1}^n \frac{W_i}{\mu_0(\mathbb{X}) + \sum_{j=1}^n W_j} \delta_{X_i}(\cdot) + \frac{\mu_0(\mathbb{X})}{\mu_0(\mathbb{X}) + \sum_{j=1}^n W_j} \mu_0'(\cdot). \tag{18}$$

Moreover,

$$\mathbb{P}(W_n \in \cdot \mid X_1, W_1, \ldots, X_{n-1}, W_{n-1}, X_n) = \eta_{X_n}(\cdot). \tag{19}$$

It follows from (19) that $W(x) \equiv h(x, U) \sim \eta_x$, $x \in \mathbb{X}$, whenever $U \sim \text{Unif}[0,1]$. Then, the random measure

$$R_x = W(x) \cdot \delta_x \tag{20}$$

is such that $R_x \sim \mathcal{R}_x$, where $\mathcal{R}_x$ appears in Definition 2.

3. Asymptotic Properties of RRPP

In this section, we study the asymptotic properties of RRPPs through the sequence $(X_n)_{n\geq 1}$ in the representation (17). We show that the limit behavior of $(\mu_n)_{n\geq 0}$ depends on the relationship between weights and observations. In particular, when $W(x) \equiv W$ in (20) is constant with respect to the color x, the process $(X_n)_{n\geq 1}$ is conditionally identically distributed (c.i.d.) and, for every $A \in \mathcal{X}$, the normalized sequence $(\mu_n'(A))_{n\geq 0}$ is a bounded martingale. We consider the c.i.d. case in Section 3.3. In contrast, if some colors x have a higher expected reinforcement, then they tend to dominate the observation process and, as n grows to infinity, the probability measure μ_n' concentrates its mass on the subset of dominant colors, see Theorem 2.

3.1. Preliminaries

Our focus is on the convergence of the normalized sequence $(\mu_n')_{n\geq 0}$, which by Theorem 1 is a.s. equal to the predictive distributions (18). We also consider the sequence of empirical frequencies of $(X_n)_{n\geq 1}$, defined for $n \geq 1$ by

$$\hat{\mu}_n' = \frac{1}{n} \sum_{i=1}^n \delta_{X_i}.$$

We obtain conditions under which the convergence in (2) extends to convergence in total variation, where the total variation distance between any two probability measures $\alpha, \beta \in \mathbb{M}_P(\mathbb{X})$ is given by

$$d_{TV}(\alpha, \beta) = \sup_{B \in \mathcal{X}} |\alpha(B) - \beta(B)|.$$

To state some of the results, we recall the definition of support of a probability measure $\gamma \in \mathbb{M}_P(\mathbb{R}_+)$,

$$\text{supp}(\gamma) = \{u \geq 0 : \gamma((u - \epsilon, u + \epsilon)) > 0, \forall \epsilon > 0\}.$$

Of particular interest is the conditional probability of observing a new color, given by

$$\theta_n \equiv \mathbb{P}(X_{n+1} \notin \{X_1, \ldots, X_n\} \mid X_1, W_1, \ldots, X_n, W_n) = \frac{\theta}{\theta + \sum_{j=1}^n W_j} \mu_0'(\{X_1, \ldots, X_n\}^c),$$

for $n \geq 1$, where $\theta = \mu_0(\mathbb{X})$. This would inform us on the number of distinct values in a sample $(X_1, \ldots, X_n)$ of size n,

$$L_n = \max\{k \in \{1, \ldots, n\} : X_k \notin \{X_1, \ldots, X_{k-1}\}\},$$

since $\theta_n = \mathbb{P}(L_{n+1} = L_n + 1 | X_1, W_1, \ldots, X_n, W_n)$.

The following modes of convergence are used when we investigate the rate of convergence of the distance between μ_n' and $\hat{\mu}_n$.

Almost sure (a.s.) conditional convergence. Let $\mathcal{G} = (\mathcal{G}_n)_{n \geq 0}$ be a filtration and $\tilde{Q} \in \mathbb{K}_P(\Omega, \mathbb{X})$. A sequence $(Y_n)_{n \geq 1}$ is said to converge to $\tilde{Q}$ in the sense of a.s. conditional convergence w.r.t. $\mathcal{G}$ if the conditional distribution of Y_n, given $\mathcal{G}_n$, converges weakly on a set of probability one to $\tilde{Q}$, that is, as $n \to \infty$,

$$\mathbb{P}(Y_n \in \cdot \mid \mathcal{G}_n) \xrightarrow{w} \tilde{Q}(\cdot) \quad \text{a.s.}$$

We refer to [22] for more details.

Stable convergence. Stable convergence is a strong form of convergence in distribution, albeit weaker than a.s. conditional convergence. A sequence $(Y_n)_{n \geq 1}$ is said to converge stably to $\tilde{Q}$ if

$$\mathbb{E}[V f(Y_n)] \longrightarrow \mathbb{E}\left[V \int_\mathbb{X} f(x) \tilde{Q}(dx)\right],$$

for all continuous bounded functions f and any integrable random variable V. The main application of stable convergence is in central limit theorems that allow for mixing variables in the limit. See [26] for a complete reference on stable convergence.

In the sequel, the stable and a.s. conditional limits will be some Gaussian law, which we denote by $\mathcal{N}(\mu, \sigma^2)$ for parameters (μ, σ^2), where $\mathcal{N}(\mu, 0) = \delta_\mu$.

3.2. RRPP with Dominant Colors

Using (20), let us define, for $x \in \mathbb{X}$,

$$w(x) = \mathbb{E}[W(x)] \quad \text{and} \quad \bar{w} = \sup_{x \in \mathbb{X}} w(x).$$

We further let

$$\mathcal{D} = \{x \in \mathbb{X} : w(x) = \bar{w}\},$$

be the set of dominant colors. The model (18) with $\mathcal{D} \subset \mathbb{X}$ has been studied by [23] under the assumption that $\bar{w}$ is strictly greater than the next largest value of $w(\cdot)$ in the support of $w^\sharp(\mu_0')$. Then, the probability of observing a non-dominant color, $x \in \mathcal{D}^c$, vanishes, and the predictive and the empirical distributions converge in total variation to a common random probability measure, which is concentrated on $\mathcal{D}$. For completeness reasons, we report here the main results from [23].

Theorem 2 ([23], Theorem 3.3). *For any RRPP $(\mu_n)_{n\geq 0}$ that satisfies*

$$W(x) \leq \beta < \infty;$$
$$\tilde{w} \in \text{supp}(w^\sharp(\mu_0')); \qquad (21)$$
$$\tilde{w} > \tilde{w}^c \equiv \sup\{u \geq 0 : u \in \text{supp}(w^\sharp(\mu_0'(\cdot|\mathcal{D}^c)))\},$$

there exists a random probability measure $\tilde{P}$ on $\mathbb{X}$ with $\tilde{P}(\mathcal{D}) = 1$ a.s. such that

$$d_{TV}(\mu_n', \tilde{P}) \xrightarrow{a.s.} 0 \quad \text{and} \quad d_{TV}(\hat{\mu}_n', \tilde{P}) \xrightarrow{a.s.} 0.$$

Under conditions (21), Theorem 3.3 in [23] implies $\sum_{i=1}^n W_i/n \xrightarrow{a.s.} \tilde{w}$. If μ_0 is further diffuse, then $\sum_{n=1}^\infty \theta_n = \infty$ a.s., and so $L_n \xrightarrow{a.s.} \infty$ by Theorem 1 in [27]; thus, by Theorem 1 in [27], Proposition 3.4 in [23] shows that the actual growth rate is that of a Pólya sequence,

$$\frac{L_n}{\log n} \xrightarrow{a.s.} \frac{\theta}{\tilde{w}}. \qquad (22)$$

In addition to the uniform convergence in Theorem 2, the authors in [23] obtain set-wise rates of convergence. To state their result, we introduce, for any $A \in \mathcal{X}$,

$$q_A = \lim_{n\to\infty} \mathbb{E}[W_{n+1}^2 \delta_{X_{n+1}}(A)|X_1, W_1, \ldots, X_n, W_n],$$

which exists a.s. under the assumptions of Theorem 2.

Theorem 3 ([23], Theorem 4.2). *Let $(\mu_n)_{n\geq 0}$ be an RRPP satisfying (21). Suppose $\tilde{w} > 2\tilde{w}^c$. Define*

$$V(A) = \frac{1}{\tilde{w}^2}\{(\tilde{P}(A^c))^2 q_A + (\tilde{P}(A))^2 q_{A^c}\} \quad \text{and} \quad U(A) = V(A) - \tilde{P}(A)\tilde{P}(A^c).$$

Then,

$$\sqrt{n}(\mu_n'(A) - \hat{\mu}_n'(A)) \xrightarrow{stably} \mathcal{N}(0, U(A)),$$

and

$$\sqrt{n}(\mu_n'(A) - \tilde{P}(A)) \xrightarrow{a.s.cond.} \mathcal{N}(0, V(A)) \qquad w.r.t. \ (\mathcal{F}_n^{X,W})_{n\geq 1},$$

where $\mathcal{F}_n^{X,W} = \sigma(X_1, W_1, \ldots, X_n, W_n)$, $n \geq 1$ is the filtration generated by $((X_n, W_n))_{n\geq 1}$.

3.3. RRPP with Independent Weights

Let $(\mu_n)_{n\geq 0}$ be an RRPP with reinforcement distribution $\eta_x \equiv \eta$ that does not depend on x. Using the notation of Section 3.2, we have

$$w(x) \equiv \tilde{w}, \qquad (23)$$

and, thus, $\mathcal{D} = \mathbb{X}$. An equivalent formulation can be given in terms of the sequence of weights $(W_n)_{n\geq 1}$ in Proposition 1, whereby

$$W_n = h(U_n), \qquad (24)$$

for some measurable function h, with $U_n \perp (X_1, U_1, \ldots, X_{n-1}, U_{n-1}, X_n)$ and $U_n \sim \text{Unif}[0,1]$. Then, $W_n \stackrel{i.i.d.}{\sim} \eta$ and $W_n \perp (X_1, \ldots, X_n)$, which implies that $\mathbb{E}[W_1] = \tilde{w}$.

The model (18) with weights (24) has been studied by [18,19,22], among others, where the authors obtain central limit theorems and study the growth rate of L_n when $\tilde{w} < \infty$.

Their results rely on the fact that $(X_n)_{n\geq 1}$ is conditionally identically distributed (c.i.d.) with respect to the filtration generated by $((X_n, W_n))_{n\geq 1}$. By [21], an $\mathbb{X}$-valued random sequence $(Y_n)_{n\geq 1}$ that is adapted to a filtration $(\mathcal{F}_n)_{n\geq 1}$ is said to be c.i.d. with respect to $(\mathcal{F}_n)_{n\geq 1}$ if and only if $(Y_n)_{n\geq 1}$ is identically distributed and, for every $n, k \geq 1$,

$$\mathbb{P}(Y_{n+k} \in \cdot \mid \mathcal{F}_n) = \mathbb{P}(Y_{n+1} \in \cdot \mid \mathcal{F}_n). \tag{25}$$

Proposition 2 ([19], Lemma 6). *For any RRPP $(\mu_n)_{n\geq 0}$ with $\eta_x \equiv \eta$, the observation process $(X_n)_{n\geq 1}$ is c.i.d. with respect to the filtration generated by $((X_n, W_n))_{n\geq 1}$.*

C.i.d. processes preserve many of the properties of exchangeable sequences, see [21]. For example, if $(Y_n)_{n\geq 1}$ is c.i.d., then there exists a random probability measure such that (2)–(3) hold true with respect to the filtration used in the definition (25). It follows for the model in Proposition 2 that there exists $\tilde{P} \in \mathbb{K}_P(\Omega, \mathbb{X})$ such that, for every $A \in \mathcal{X}$,

$$\mu'_n(A) \xrightarrow{a.s.} \tilde{P}(A).$$

In fact, by (25), the sequence $(\mu'_n(A))_{n\geq 0}$ is a bounded martingale. On the other hand, (23) implies that $\mathcal{D} = \mathbb{X}$; therefore, any RRPP with $\eta_x \equiv \eta$ whose weights are bounded, $W_1 \leq \beta < \infty$, satisfies the assumptions of Theorem 2. In that case,

$$d_{TV}(\mu'_n, \tilde{P}) \xrightarrow{a.s.} \tilde{P}.$$

It follows from Theorem 4.2 in [23] that the boundedness condition in (21) is needed to show that (i) $\sum_{i=1}^{n} W_i/n \xrightarrow{a.s.} \bar{w}$; and (ii) μ'_n converge set-wise to $\tilde{P}$, which is non-trivial in that setting. Here, (i) is granted as $(W_n)_{n\geq 1}$ is i.i.d., and (ii) has already been established; thus, we obtain the following result for RRPPs with independent weights.

Theorem 4. *For any RRPP $(\mu_n)_{n\geq 0}$ with $\eta_x \equiv \eta$, there exists a random probability measure $\tilde{P}$ on $\mathbb{X}$ such that*

$$d_{TV}(\mu'_n, \tilde{P}) \xrightarrow{a.s.} 0 \quad \text{and} \quad d_{TV}(\hat{\mu}'_n, \tilde{P}) \xrightarrow{a.s.} 0.$$

Proof. Let $((X_n, W_n))_{n\geq 1}$ be the joint observation process associated to $(\mu_n)_{n\geq 0}$ by Proposition 1. As $\eta_x \equiv \eta$, Equation (19) implies that $W_n \overset{i.i.d.}{\sim} \eta$; thus, by the strong law of large numbers,

$$\frac{1}{n}\sum_{i=1}^{n} W_i \xrightarrow{a.s.} \bar{w} \leq \infty. \tag{26}$$

Let us define, for $n \geq 1$,

$$P_n(\cdot) = \mathbb{P}(X_{n+1} \in \cdot \mid \mathcal{F}_n^{X,W}), \quad \text{where } \mathcal{F}_n^{X,W} = \sigma(X_1, W_1, \ldots, X_n, W_n).$$

By Proposition 2, $(X_n)_{n\geq 1}$ is c.i.d. with respect to $(\mathcal{F}_n^{X,W})_{n\geq 1}$, so there exists by Lemmas 2.1 and 2.4 in [21] a random probability measure $\tilde{P}$ on $\mathbb{X}$ such that, for every $A \in \mathcal{X}$,

$$P_n(A) \xrightarrow{a.s.} \tilde{P}(A). \tag{27}$$

Moreover, $\int_{\mathbb{X}} f(x) P_n(dx) = \mathbb{E}[\int_{\mathbb{X}} f(x) \tilde{P}(dx) | \mathcal{F}_n^{X,W}]$ a.s. for every bounded measurable $f: \mathbb{X} \to \mathbb{R}$. Fix $m \geq 1$. By a monotone class argument, we can show that, for every bounded measurable $f: \mathbb{X}^2 \to \mathbb{R}$,

$$\int_{\mathbb{X}} f(X_m, x) P_n(dx) = \mathbb{E}[\int_{\mathbb{X}} f(X_m, x) \tilde{P}(dx) \mid \mathcal{F}_n^{X,W}] \quad \text{a.s., for all } n > m;$$

thus, $P_n(\{X_m\}) = \mathbb{E}[\tilde{P}(\{X_m\})|\mathcal{F}_n^{X,W}]$ a.s., and so $(P_n(\{X_m\}))_{n>m}$ is a uniformly integrable martingale. It follows from martingale convergence that, as $n \to \infty$,

$$P_n(\{X_m\}) \xrightarrow{a.s.} \tilde{P}(\{X_m\}). \tag{28}$$

Using (26)–(28), we can repeat the argument in the proof of Proposition 3.1 in [23] to show that (i) $d_{TV}(P_n, \tilde{P}) \xrightarrow{a.s.} 0$, and so $d_{TV}(\mu'_n, \tilde{P}) \xrightarrow{a.s.} 0$ by Proposition 1; and (ii) $d_{TV}(\hat{\mu}'_n, \tilde{P}) \xrightarrow{a.s.} 0$. □

Equation (26) implies that $\theta_n \xrightarrow{a.s.} 0$. If, in addition, $\bar{w} < \infty$, then $\sum_{n=1}^{\infty} \theta_n = \infty$ a.s. and $L_n \xrightarrow{a.s.} \infty$. In fact, as long as $\bar{w} < \infty$, the sequence $(L_n)_{n \geq 1}$ grows at the same rate as (22).

Proposition 3 ([18], Lemma 6). *Let $\eta \in \mathbb{M}_P(\mathbb{X})$ and μ_0 be diffuse. If $\bar{w} < \infty$, then*

$$\frac{L_n}{\log n} \xrightarrow{a.s.} \frac{\theta}{\bar{w}}.$$

If $\bar{w} = \infty$, then θ_n may approach zero fast enough that we stop seeing new observations as $n \to \infty$. For example, let us consider random reinforcement with a totally skewed stable distribution $S_\alpha(1, \sigma, 0)$ for $\alpha \in (0, 2]$ and $\sigma > 0$. If $\alpha < 1$, then $\bar{w} = \infty$, and we show that $n^{1/\alpha} \theta_n$ is stochastically bounded, which implies that L_n converges to a finite limit.

Proposition 4. *Let η be a $S_\alpha(1, \sigma, 0)$ distribution with stability parameter $\alpha < 1$, and μ_0 be diffuse. Then, $\theta_n = O_p(n^{-1/\alpha})$ and*

$$\lim_{n \to \infty} L_n < \infty \quad a.s.$$

Proof. From the properties of stable distributions, we obtain $n^{-1/\alpha} \sum_{i=1}^{n} W_i \stackrel{d}{=} W_1$ for every $n \geq 1$ and, as a consequence,

$$\theta_n = n^{-1/\alpha} \frac{\theta}{n^{-1/\alpha}\theta + n^{-1/\alpha} \sum_{i=1}^{n} W_i} \stackrel{d}{=} n^{-1/\alpha} \frac{\theta}{n^{-1/\alpha}\theta + W_1} \leq n^{-1/\alpha} \frac{\theta}{W_1}.$$

By Theorem 5.4.1 in [28], $\mathbb{E}[1/W_1] < \infty$, and so $1/W_1 < \infty$ a.s. It follows for every $M > 0$ that $\mathbb{P}(n^{1/\alpha}\theta_n > M) \leq \mathbb{P}(\theta/W_1 > M)$, which can be made arbitrarily small by taking M large enough. Regarding the second assertion, as $1/\alpha > 1$, we have

$$\mathbb{E}[\lim_{n \to \infty} L_n] = \lim_{n \to \infty} \sum_{i=1}^{n} \mathbb{E}[\mathbf{1}_{\{L_i = L_{i-1}+1\}}] = \sum_{n=1}^{\infty} \mathbb{E}[\theta_n] \leq \sum_{n=1}^{\infty} \frac{\theta}{n^{1/\alpha}} \mathbb{E}[1/W_1] < \infty.$$

□

Proposition 4 can be extended for any fat tailed reinforcement distribution η by means of a generalized central limit theorem (see, e.g., [28] (p. 62)).

The rate of convergence of (18) and $\hat{\mu}'_n$ has already been studied for the model with independent weights under different assumptions, see, e.g., [19] (p. 1363), Examples 4.2 and 4.5 in the technical report to [18], Corollary 4.1 in [22] for $\mathbb{X} = \{0,1\}$. In the next theorem, we combine ideas from [18,20] to give a fairly general result.

Theorem 5. *Let $\eta \in \mathbb{M}_P(\mathbb{R}_+)$. If $\mathbb{E}[W_1^2] < \infty$, then*

$$\sqrt{n}(\mu'_n(A) - \hat{\mu}_n(A)) \xrightarrow{stably} \mathcal{N}(0, U(A)), \quad \text{where } U(A) = \frac{\text{Var}(W_1)}{\mathbb{E}[W_1^2]} \tilde{P}(A)\tilde{P}(A^c). \tag{29}$$

If, in addition, $\mathbb{E}[W_1^4] < \infty$, then, with respect to the filtration generated by $((X_n, W_n))_{n \geq 1}$,

$$\sqrt{n}(\mu'_n(A) - \tilde{P}(A)) \xrightarrow{a.s.cond.} \mathcal{N}(0, V(A)), \quad \text{where } V(A) = \frac{\mathbb{E}[W_1^2]}{\bar{w}^2} \tilde{P}(A)\tilde{P}(A^c). \quad (30)$$

Proof. Let us define, for $n \geq 1$,

$$P_n(\cdot) = \mathbb{P}(X_{n+1} \in \cdot \mid X_1, W_1, \ldots, X_n, W_n).$$

The assertions in Theorem 5 have already been established by [18] when $W_1 \geq \gamma > 0$. In that case, Examples 4.2 and 4.5 in the technical report to [18] show that (29) is a consequence of the fact that

$$\mathbb{E}\left[\max_{1 \leq k \leq n} |Y_{n,k}|\right] \longrightarrow 0 \quad \text{and} \quad \sum_{k=1}^n Y_{n,k}^2 \xrightarrow{p} U(A), \quad (31)$$

where $Y_{n,k} = \frac{1}{\sqrt{n}}\{\delta_{X_k}(A) - kP_k(A) + (k-1)P_{k-1}(A)\}$, and (30) follows from

$$\mathbb{E}\left[\sup_{n \geq 1} \sqrt{n}|P_{n-1}(A) - P_n(A)|\right] < \infty \quad \text{and} \quad n\sum_{k \geq n}(P_{k-1}(A) - P_k(A))^2 \xrightarrow{a.s.} V(A).$$

Replicating the approach of Proposition 9 in [20], we avoid using the assumption $W_1 \geq \gamma > 0$ by conditioning on the sets $H_n = \{2\sum_{i=1}^n W_i \geq n\bar{w}\}$, $n \geq 1$. By (26), $\mathbb{1}_{H_n} \xrightarrow{a.s.} 1$, so (29) follows from (31) with

$$Y_{n,k} = \frac{1}{\sqrt{n}}\mathbb{1}_{H_{k-1}}\{\delta_{X_k}(A) - kP_k(A) + (k-1)P_{k-1}(A)\},$$

whereas (30) is, ultimately, a result of

$$\mathbb{E}\left[\sup_{n \geq 1} \sqrt{n} \cdot \mathbb{1}_{H_n}|P_{n-1}(A) - P_n(A)|\right] < \infty \quad \text{and} \quad n\sum_{k \geq n}(P_{k-1}(A) - P_k(A))^2 \xrightarrow{a.s.} V(A).$$

□

4. Generalized Measure-Valued Pólya Urn Processes

The definition of an MVPP assumes that the law of the reinforcement $\mathcal{R}$ is fixed, yet, in some situations, $\mathcal{R}$ can itself be random (e.g., RRPP with exchangeable weights, see Section 4.1). To avoid measurability issues, we assume a parametric model for $\mathcal{R}$, with the parameter taking values in a Polish space $\mathbb{V}$.

Definition 3 (Generalized Measure-Valued Pólya Urn Process). *Let V be a $\mathbb{V}$-valued random variable. A sequence $(\mu_n)_{n \geq 0}$ of random finite measure on $\mathbb{X}$ is called a generalized measure-valued Pólya urn process (GMVPP) with uncertainty parameter V, initial state $\mu_0 \in \mathbb{M}_F^*(\mathbb{X})$ and replacement rule $\mathcal{R} \in \mathbb{K}_P(\mathbb{V} \times \mathbb{X}, \mathbb{M}_F(\mathbb{X}))$ if $\mu_1 \mid V \sim \hat{R}_{\mu_0}^V$, and, for every $n \geq 2$,*

$$\mathbb{P}(\mu_n \in \cdot \mid V, \mu_1, \ldots, \mu_{n-1}) = \hat{R}_{\mu_{n-1}}^V(\cdot),$$

where $\hat{R}$ is the transition probability kernel from $\mathbb{V} \times \mathbb{M}_F^*(\mathbb{X})$ to $\mathbb{M}_F^*(\mathbb{X})$ given by

$$(v, \mu) \mapsto \hat{R}_\mu^v(\cdot) = \int_{\mathbb{X}} \psi_\mu^\sharp(\mathcal{R}(v, x))(\cdot)\mu'(dx),$$

and ψ_μ is the map $\nu \mapsto \nu + \mu$.

It follows from Definition 3 that any GMVPP is a mixture of Markov chains with initial state μ_0 and transition kernel $\hat{\mathcal{R}}^V$. A separate modeling approach, which we do not examine here, defines a measure-valued Markov chain with transition kernel

$$\mu \mapsto \int_{\mathbb{X}} \psi_\mu^\sharp(\mathcal{R}(\mu,x))(\cdot)\mu'(dx).$$

In fact, some of the predictive constructions in [11,29] can be framed in such a way.

Theorem 1 extends to GMVPPs, provided that we condition all quantities on the parameter V. As a consequence, there exists a measurable function f from $\mathbb{V} \times \mathbb{X} \times [0,1]$ to $\mathbb{M}_F(\mathbb{X})$ and a random sequence $((X_n, U_n))_{n \geq 1}$ such that

$$\mu_n = \mu_{n-1} + f(V, X_n, U_n) \quad \text{a.s.,} \tag{32}$$

where $U_n \sim \text{Unif}[0,1]$, $U_n \perp (V, X_1, U_1, \ldots, X_{n-1}, U_{n-1}, X_n)$, $X_1 \mid (V, (U_m)_{m \geq 1}) \sim \mu_0'$, and, for $n \geq 1$,

$$\mathbb{P}(X_{n+1} \in \cdot \mid V, X_1, \ldots, X_n, (U_m)_{m \geq 1}) = \frac{\mu_0(\cdot) + \sum_{i=1}^n f(V, X_i, U_i)(\cdot)}{\mu_0(\mathbb{X}) + \sum_{i=1}^n f(V, X_i, U_i)(\mathbb{X})}, \tag{33}$$

and

$$\mathbb{P}(f(V, X_n, U_n) \in \cdot \mid V, X_1, U_1, \ldots, X_{n-1}, U_{n-1}, X_n) = \mathcal{R}(V, X_n)(\cdot). \tag{34}$$

The definition of a randomly reinforced Pólya process is similarly generalized to cover the case of a random reinforcement distribution η.

Definition 4 (Generalized Randomly Reinforced Pólya Process). *We call a GMVPP with parameters $(V, \mu_0, \mathcal{R})$ a generalized randomly reinforced Pólya process (GRRPP) if there exists $\eta \in \mathbb{K}_P(\mathbb{V} \times \mathbb{X}, \mathbb{R}_+)$ such that $\mathcal{R}(v,x) = \xi_x^\sharp(\eta(v,x))$, where $\xi_x : \mathbb{R}_+ \to \mathbb{M}_F(\mathbb{X})$ is the map $w \mapsto w\delta_x$.*

For GRRPPs, the function f in the representation (32)–(34) can be written as

$$f(v,x,u) = h(v,x,u) \cdot \delta_x,$$

where h is a measurable function from $\mathbb{V} \times \mathbb{X} \times [0,1]$ to $\mathbb{R}_+$ such that $h(v,x,U) \sim \eta(v,x)$ for all $v \in \mathbb{V}$ and $x \in \mathbb{X}$, whenever $U \sim \text{Unif}[0,1]$. Letting $W_n = h(V, X_n, U_n)$, we obtain

$$\mu_n = \mu_{n-1} + W_n \delta_{X_n} \quad \text{a.s.,} \tag{35}$$

where

$$\mathbb{P}(X_{n+1} \in \cdot \mid V, X_1, \ldots, X_n, (U_m)_{m \geq 1}) = \frac{\mu_0(\cdot) + \sum_{i=1}^n W_i \delta_{X_i}(\cdot)}{\mu_0(\mathbb{X}) + \sum_{i=1}^n W_i}, \tag{36}$$

and

$$\mathbb{P}(W_n \in \cdot \mid V, X_1, U_1, \ldots, X_{n-1}, U_{n-1}, X_n) = \eta(V, X_n)(\cdot). \tag{37}$$

The weights W_n in (36) allow us to incorporate additional information about the observations $(X_n)_{n \geq 1}$. As an example, consider the problem of computer-based classification, where the output usually includes confidence scores, which reflect the software's confidence that the classifications are correct. In analyzing the number and dimension of the types already discovered, or the probability of detecting a new type, a typical procedure would take into account only those classifications whose confidence scores are above a certain threshold. Alternatively, we could adopt a Bayesian perspective and weigh each classification according to its confidence score. Denoting by $((X_n, W_n))_{n \geq 1}$ the sequence of classifications and confidence scores, we would model the distribution of the next classification by (36).

4.1. GRRPP with Exchangeable Weights

Let $(\mu_n)_{n\geq 0}$ be a GRRPP with reinforcement distribution $\eta(v)$ that does not depend on x. Then,
$$W_n = h(V, U_n),$$
for some measurable function $h(v, u)$. The next result shows that the sequence $(W_n)_{n\geq 1}$ is exchangeable with directing random measure $\tilde{\eta} \equiv \eta(V)$. Moreover, $(\mu_n)_{n\geq 0}$ is completely parameterized by $(\mu_0, \tilde{\eta})$.

Theorem 6. *A sequence $(\mu_n)_{n\geq 0}$ of random finite measures is a GRRPP with parameters $(\mu_0, \tilde{\eta})$ for $\tilde{\eta} \in \mathbb{K}_P(\Omega, \mathbb{R}_+)$ if and only if $\mu_0 = \theta v$ and, for every $n \geq 1$,*
$$\mu_n = \mu_{n-1} + W_n \delta_{X_n} \quad a.s.,$$
where $\theta \in (0, \infty)$, $v \in \mathbb{M}_P(\mathbb{X})$, $(W_n)_{n\geq 1}$ is an exchangeable process with directing random measure $\tilde{\eta}$, and $(X_n)_{n\geq 1}$ is a sequence of $\mathbb{X}$-valued random variables such that $X_1 \mid (W_k)_{k\geq 1} \sim v$ and, for $n \geq 1$,
$$\mathbb{P}(X_{n+1} \in \cdot \mid X_1, \ldots, X_n, (W_k)_{k\geq 1}) = \sum_{i=1}^n \frac{W_i}{\theta + \sum_{j=1}^n W_j} \delta_{X_i}(\cdot) + \frac{\theta}{\theta + \sum_{j=1}^n W_j} v(\cdot). \tag{38}$$

Proof. Let $(\mu_n)_{n\geq 0}$ be a GRRPP with parameters $(\mu_0, \tilde{\eta})$, and consider the representation (35)–(37). Put $\theta = \mu_0(\mathbb{X})$ and $v = \mu_0'$. It follows from (37) that
$$W_n \mid \tilde{\eta} \overset{i.i.d.}{\sim} \tilde{\eta};$$
thus, $(W_n)_{n\geq 1}$ is exchangeable. Moreover, $W_n = h(V, U_n)$, $n \geq 1$, so (38) follows from (36).

Conversely, suppose $\mu_n = \mu_{n-1} + W_n \delta_{X_n}$, where the process $((X_n, W_n))_{n\geq 1}$ is as described. It follows from (38) and Theorem 8.12 in [25] that
$$(W_k)_{k\geq 1} \perp X_1 \quad \text{and} \quad (W_{n+k})_{k\geq 1} \perp (X_1, \ldots, X_{n+1}) \mid (W_1, \ldots, W_n), \ n \geq 1. \tag{39}$$

Since $(W_n)_{n\geq 1}$ is exchangeable with directing random measure $\tilde{\eta}$, we have
$$W_n \mid \tilde{\eta} \overset{i.i.d.}{\sim} \tilde{\eta}. \tag{40}$$

Furthermore, $\tilde{\eta}$ is measurable with respect to the tail σ-field of $(W_n)_{n\geq 1}$, so, by (39),
$$\tilde{\eta} \perp X_1 \quad \text{and} \quad \tilde{\eta} \perp (X_1, \ldots, X_{n+1}) \mid (W_1, \ldots, W_n), \ n \geq 1. \tag{41}$$

Using (39)–(41), we can show that
$$W_1 \perp X_1 \mid \tilde{\eta} \quad \text{and} \quad W_{n+1} \perp (X_1, W_1, \ldots, X_n, W_n, X_{n+1}) \mid \tilde{\eta}, \ n \geq 1.$$

Then, $\mathbb{P}(\mu_1 \in \cdot \mid \tilde{\eta}) = \mathbb{P}(\mu_0 + W_1 \delta_{X_1} \in \cdot \mid \tilde{\eta}) = \int_{\mathbb{X}} \psi^\sharp_{\mu_0}(\xi^\sharp_x(\tilde{\eta}))(\cdot) \mu_0'(dx)$ and, for $n \geq 2$,
$$\mathbb{P}(\mu_n \in \cdot \mid \tilde{\eta}, \mu_1, \ldots, \mu_{n-1})$$
$$= \mathbb{E}\big[\mathbb{P}(\mu_{n-1} + W_n \delta_{X_n} \in \cdot \mid \tilde{\eta}, X_1, \ldots, W_{n-1}, X_n) \big| \tilde{\eta}, \mu_1, \ldots, \mu_{n-1}\big]$$
$$= \mathbb{E}\big[\mathbb{E}[\psi^\sharp_{\mu_{n-1}}(\xi^\sharp_{X_n}(\tilde{\eta}))(\cdot) \mid X_1, \ldots, X_{n-1}, (W_m)_{m\geq 1}] \big| \tilde{\eta}, \mu_1, \ldots, \mu_{n-1}\big]$$
$$= \int_{\mathbb{X}} \psi^\sharp_{\mu_{n-1}}(\xi^\sharp_x(\tilde{\eta}))(\cdot) \mu_{n-1}'(dx).$$

□

It follows from the proof of Theorem 6 that $(X_1, W_1) \sim \mu'_0 \times \mathbb{E}[\tilde{\eta}]$ and, for $n \geq 1$,

$$\mathbb{P}((X_{n+1}, W_{n+1}) \in \cdot \mid X_1, W_1, \ldots, X_n, W_n) = (\mu'_n \times \mathbb{E}[\tilde{\eta}|W_1, \ldots, W_n])(\cdot). \tag{42}$$

As μ'_n and $\mathbb{E}[\tilde{\eta}|W_1, \ldots, W_n]$ are both symmetric with respect to $((X_1, W_1), \ldots, (X_n, W_n))$, then (42) is a symmetric function of $((X_1, W_1), \ldots, (X_n, W_n))$. This is a necessary but not sufficient condition for $((X_n, W_n))_{n \geq 1}$ to be exchangeable, see Proposition 3.2 and Example 3.1 in [2]. In Proposition 5, we show that $((X_n, W_n))_{n \geq 1}$ is exchangeable if and only if either μ'_0 is degenerate or the weights are a.s. identical. On the other hand, for every $n, k \geq 1$, the sequence $((X_n, W_n))_{n \geq 1}$ satisfies

$$\mathbb{P}(W_k \in \cdot \mid X_1) = \mathbb{P}(W_1 \in \cdot \mid X_1), \qquad \mathbb{P}(X_k \in \cdot \mid W_1) = \mathbb{P}(X_1 \in \cdot \mid W_1), \tag{43}$$

and

$$\begin{aligned} \mathbb{P}(W_{n+k} \in \cdot | X_1, W_1, \ldots, X_n, W_n, X_{n+1}) &= \mathbb{P}(W_{n+1} \in \cdot | X_1, W_1, \ldots, X_n, W_n, X_{n+1}), \\ \mathbb{P}(X_{n+k} \in \cdot | X_1, W_1, \ldots, X_n, W_n, W_{n+1}) &= \mathbb{P}(X_{n+1} \in \cdot | X_1, W_1, \ldots, X_n, W_n, W_{n+1}). \end{aligned} \tag{44}$$

By [24], Equations (43) and (44) are defining a process that is partially conditionally identity distributed (partially c.i.d.). Analogously to the c.i.d. case, partially c.i.d. processes preserve many of the properties of partially exchangeable sequences, see [24].

Proposition 5. *Under the conditions of Theorem 6, $((X_n, W_n))_{n \geq 1}$ is partially c.i.d. Moreover, $((X_n, W_n))_{n \geq 1}$ is exchangeable if and only if either μ'_0 is degenerate or $W_n = W_1$ a.s., $n \geq 1$. In that case, $((X_n, W_n))_{n \geq 1}$ is partially exchangeable.*

Proof. It follows that $((X_n, W_n))_{n \geq 1}$ is partially c.i.d. if and only if $X_2 \stackrel{d}{=} X_1 \mid W_1$, $W_2 \stackrel{d}{=} W_1 \mid X_1$, and (44) is true for every $n \geq 1$ with $k = 2$. By hypothesis, $(W_n)_{n \geq 1}$ is exchangeable and $(W_n)_{n \geq 1} \perp X_1$, so $W_2 \stackrel{d}{=} W_1 \mid X_1$. Moreover, applying (39) repeatedly, we obtain

$$\begin{aligned} \mathbb{P}(W_{n+2} \in \cdot \mid X_1, W_1, \ldots, X_n, W_n, X_{n+1}) \\ &= \mathbb{E}\big[\mathbb{P}(W_{n+2} \in \cdot \mid X_1, \ldots, W_{n+1}, X_{n+2}) | X_1, W_1, \ldots, X_n, W_n, X_{n+1}\big] \\ &= \mathbb{E}\big[\mathbb{P}(W_{n+2} \in \cdot \mid W_1, \ldots, W_{n+1}) | W_1, \ldots, W_n\big] \\ &= \mathbb{P}(W_{n+1} \in \cdot \mid W_1, \ldots, W_n) = \mathbb{P}(W_{n+1} \in \cdot \mid X_1, W_1, \ldots, X_n, W_n, X_{n+1}). \end{aligned}$$

On the other hand, by (38),

$$\begin{aligned} \mathbb{P}(X_{n+2} \in \cdot \mid X_1, W_1, \ldots, X_n, W_n, W_{n+1}) &= \mathbb{E}\big[\mu'_{n+1}(\cdot) \mid X_1, W_1, \ldots, X_n, W_n, W_{n+1}\big] \\ &= \frac{\mu_n(\cdot) + W_{n+1} \cdot \mu'_n(\cdot)}{\mu_{n+1}(\mathbb{X})} = \mu'_n(\cdot) \\ &= \mathbb{P}(X_{n+1} \in \cdot \mid X_1, W_1, \ldots, X_n, W_n, W_{n+1}). \end{aligned}$$

Analogously, $\mathbb{P}(X_2 \in \cdot \mid W_1) = \mu_1(\cdot) = \mathbb{P}(X_1 \in \cdot \mid W_1)$, which completes the proof of the first part.

If μ'_0 is degenerate, then $((X_n, W_n))_{n \geq 1}$ is trivially exchangeable. If $W_n = W_1$ a.s. instead, then one can show that $((X_n, W_n))_{n \geq 1}$ satisfies condition (b) of Proposition 3.2 in [2], which, together with the symmetry of (42), implies by Theorem 3.1 in [2] that $((X_n, W_n))_{n \geq 1}$ is exchangeable.

Conversely, suppose that $((X_n, W_n))_{n \geq 1}$ is exchangeable. As $((X_n, W_n))_{n \geq 1}$ is partially c.i.d., the predictive distributions (42) converge to a product random measure [24]. It follows from de Finetti's theorem that $((X_n, W_n))_{n \geq 1}$ is partially exchangeable, so, in particular,

$$(X_1, W_1, X_2, W_2) \stackrel{d}{=} (X_1, W_2, X_2, W_1).$$

However, $W_2 \perp X_2 \mid (X_1, W_1)$ from (36), so $W_1 \perp X_2 \mid (X_1, W_2)$. Thus, for every bounded measurable function $\tilde{f}$, there exists a measurable function $g_{\tilde{f}}$ such that

$$\mathbb{E}[\tilde{f}(X_2) \mid X_1, W_1, W_2] = g_{\tilde{f}}(X_1, W_2) \quad \text{a.s.}$$

Integrating $\tilde{f}(X_2)$ with respect to (38) and rearranging the terms, we obtain

$$W_1\big(\tilde{f}(X_1) - g_{\tilde{f}}(X_1, W_2)\big) = \theta\big(g_{\tilde{f}}(X_1, W_2) - \mathbb{E}[\tilde{f}(X_1)]\big) \quad \text{a.s.}$$

Assume that μ_0' is non-degenerate. Then, there is an $\tilde{f}$ such that $\mathbb{P}\big(\tilde{f}(X_1) = \mathbb{E}[\tilde{f}(X_1)]\big) = 0$; e.g., take $\tilde{f} = \mathbb{1}_B$ for some $B \in \mathcal{X}$ such that $0 < \mathbb{P}(X_1 \in B) < 1$. It follows that

$$\mathbb{P}\big(\tilde{f}(X_1) = g_{\tilde{f}}(X_1, W_2) = 0\big) = \mathbb{P}\big(\tilde{f}(X_1) = \mathbb{E}[\tilde{f}(X_2) \mid X_1, W_1, W_2]\big)$$
$$= \mathbb{P}\big(\tilde{f}(X_1) = \mathbb{E}[\tilde{f}(X_1)]\big) = 0;$$

therefore,

$$W_1 = \frac{\theta\big(g_{\tilde{f}}(X_1, W_2) - \mathbb{E}[\tilde{f}(X_1)]\big)}{\tilde{f}(X_1) - g_{\tilde{f}}(X_1, W_2)} \quad \text{a.s.}$$

In other words, there exists a measurable function $\tilde{h}$ such that $W_1 = \tilde{h}(X_1, W_2)$ a.s., and so $W_2 = \tilde{h}(X_1, W_1)$ a.s. by partial exchangeability. It follows from $X_1 \perp (W_1, W_2)$ that, for every $A \in \mathcal{B}(\mathbb{R}_+)$,

$$\mathbb{P}(W_2 \in A \mid W_1) = \mathbb{P}(W_2 \in A \mid X_1, W_1) = \mathbb{1}_A(W_2) \quad \text{a.s.};$$

thus, $W_2 = W_1$ a.s. and, from exchangeability, $W_n = W_1$ a.s., $n \geq 1$. □

4.2. Asymptotic Properties of GRRPP with Exchangeable Weights

It follows from (38) that the GRRPP with exchangeable weights is a mixture of RRPPs with independent weights, with the mixing distribution affecting only the sequence $(W_n)_{n \geq 1}$. Thus, we expect that the results in Section 3.3 carry over to this more general setting. In this section, we concentrate on the behavior of θ_n and the sequence $(L_n)_{n \geq 1}$.

Assume that $\mathbb{P}(W_1 > 0 \mid \tilde{\eta}) > 0$. If $\mathbb{E}[W_1] < \infty$, then $0 < \mathbb{E}[W_1 \mid \tilde{\eta}] < \infty$ a.s., and, by the law of large numbers for exchangeable random variables (see [1], Section 2),

$$\frac{1}{n} \sum_{i=1}^n W_i \xrightarrow{a.s.} \mathbb{E}[W_1 \mid \tilde{\eta}] \in (0, +\infty).$$

Then, if μ_0 is diffuse, $n \cdot \theta_n \xrightarrow{a.s.} \theta / \mathbb{E}[W_1 \mid \tilde{\eta}]$ and $\sum_{i=1}^n \theta_n = \infty$ a.s., so Theorem 1 in [27] implies

$$\frac{L_n}{\log n} = \frac{L_n}{\sum_{k=1}^n \theta_k} \Big(\frac{1}{\log n} \sum_{k=1}^n \frac{1}{k}(k \cdot \theta_k)\Big) \xrightarrow{a.s.} \frac{\theta}{\mathbb{E}[W_1 \mid \tilde{\eta}]}.$$

If $\mathbb{E}[W_1] = \infty$, then L_n may converge to a finite limit, as $n \to \infty$. For example, let us consider a strictly stable reinforcement distribution as in Proposition 4.

Proposition 6. *Let $(\mu_n)_{n \geq 0}$ be a GRRPP with parameters (V, μ_0, η) such that V is a strictly positive random variable with $\mathbb{E}[V^{-1}] < \infty$, μ_0 is diffuse, and $\eta(v)$, $v > 0$ is a $S_\alpha(1, v, 0)$ distribution with stability parameter $\alpha < 1$. Then, $\theta_n = O_P(n^{-1/\alpha})$ and*

$$\lim_{n \to \infty} L_n < \infty \quad \text{a.s.}$$

Proof. It follows from how the weights in the representation (35) are chosen that we can take
$$W_n = VF^{-1}(U_n),$$
where $U_n \sim \text{Unif}[0,1]$, $U_n \perp (V, X_1, U_1, \ldots, X_{n-1}, U_{n-1}, X_n)$, and F^{-1} is the inverse of the $S_\alpha(1,1,0)$ distribution function. Then,
$$\theta_n = \frac{\theta}{\theta + \sum_{i=1}^n W_i} \leq n^{-1/\alpha} \frac{\theta}{V n^{-1/\alpha} \sum_{i=1}^n F^{-1}(U_i)} \stackrel{d}{=} n^{-1/\alpha} \frac{\theta}{VY},$$
for some $Y \sim S_\alpha(1,1,0)$ such that $Y \perp V$. It follows for every $M > 0$ that $\mathbb{P}(n^{1/\alpha} \theta_n > M) \leq \mathbb{P}(\theta/VY > M)$, which can be made arbitrarily small by taking M large enough. Regarding the second assertion, as $1/\alpha > 1$ and $\mathbb{E}[\theta/(VY)] < \infty$ by Theorem 5.4.1 in [28], we have
$$\mathbb{E}[\lim_{n\to\infty} L_n] = \lim_{n\to\infty} \sum_{i=1}^n \mathbb{E}[\mathbb{1}_{\{L_i = L_{i-1}+1\}}] = \sum_{n=1}^\infty \mathbb{E}[\theta_n] \leq \sum_{n=1}^\infty \frac{\theta}{n^{1/\alpha}} \mathbb{E}[1/VY] < \infty.$$

□

Extensions of Proposition 6 can be obtained by exploiting the central limit theorems for exchangeable random variables, which are found in [30,31].

5. Discussion

In this paper, we study the extension of randomly reinforced urns [17] to an unbounded set of possible colors. The resulting measure-valued urn process provides a predictive characterization of the law of an asymptotically exchangeable sequence of random variables, which corresponds to the observation process of an implied urn sampling scheme. In fact, the model (6)–(7) fits into a line of recent research, which explores efficient predictive constructions for fast online prediction or approximately-Bayesian solutions, see [11,29,32] and references therein. To that end, one direction for future work is to generalize the functional relationship in (7) and/or, as one referee suggested, to consider finitely-additive measures, along the lines discussed in [33].

We investigate the asymptotic properties of the sequences of predictive distributions and empirical frequencies of the observation process, and prove their convergence in total variation distance to a common random limit. The rate of convergence of their difference is given set-wise; so, another possible direction for future research is to consider a stronger distance. As far as we know, the topic of merging of the predictive and empirical distributions is largely unexplored. Within the relevant literature, we mention the works of [4,34], where the authors study the rate of convergence of the Wasserstein or Prokhorov distances under exchangeability, and the papers by Berti et al. [21,35], who consider the c.i.d. case and regard the difference between the predictive and empirical measures as a map in the space of real bounded functions.

Author Contributions: Formal analysis, S.F., S.P., H.S.; writing—original draft preparation, S.F., S.P., H.S.; writing—review and editing, S.F., S.P., H.S. All authors have read and agreed to the published version of the manuscript.

Funding: This project has received funding from the European Research Council (ERC) under the European Union's Horizon 2020 research and innovation programme (grant agreement No. 817257). H.S. was partially supported by the Bulgarian Ministry of Education and Science under the National Research Programme "Young scientists and postdoctoral students" approved by DCM No. 577/17.08.2018.

Institutional Review Board Statement: Not applicable.

Informed Consent Statement: Not applicable.

Data Availability Statement: Not applicable.

Acknowledgments: We wish to express our sincere gratitude to Regazzini for his deeply inspiring ideas and for instilling in us his passion for research. We thank the four anonymous referees for the valuable comments.

Conflicts of Interest: The authors declare no conflict of interest.

References

1. Aldous, D.J. Exchangeability and related topics. *École D'Été De Probab. De St.-Flour XIII 1983* **1985**, *1117*, 1–198.
2. Fortini, S.; Ladelli, L.; Regazzini, E. Exchangeability, predictive distributions and parametric models. *Sankhya Ser. A* **2000**, *62*, 86–109.
3. Cifarelli, D.M.; Regazzini, E. De Finetti's contribution to probability and statistics. *Statist. Sci.* **1996**, *11*, 253–282. [CrossRef]
4. Cifarelli, D.M.; Dolera, E.; Regazzini, E. Frequentistic approximations to Bayesian prevision of exchangeable random elements. *Int. J. Approx. Reason.* **2016**, *78*, 138–152. [CrossRef]
5. Fortini, S.; Petrone, S. Predictive distribution (de Finetti's view). In *Wiley StatsRef: Statistics Reference Online*; Wiley Online Library: 2014; pp. 1–9. Available online: https://onlinelibrary.wiley.com/doi/full/10.1002/9781118445112.stat07831 (accessed on 4 October 2021).
6. Regazzini, E. Old and recent results on the relationship between predictive inference and statistical modeling either in nonparametric or parametric form. In *Bayesian Statistics 6*; Oxford University Press: Oxford, UK, 1999; pp. 571–588.
7. Fortini, S.; Petrone, S. Predictive construction of priors in Bayesian nonparametrics. *Braz. J. Probab. Stat.* **2012**, *26*, 423–449. [CrossRef]
8. Mailler, C.; Marckert, J.F. Measure-valued Pólya urn processes. *Electron. Commun. Probab.* **2017**, *22*, 33. [CrossRef]
9. Janson, S. Random replacements in Pólya urns with infinitely many colours. *Electron. Commun. Probab.* **2019**, *24*, 11. [CrossRef]
10. Aletti, G.; Ghiglietti, A.; Rosenberger, W.F. Nonparametric covariate-adjusted reponse-adaptive design based on a functional urn model. *Ann. Stat.* **2018**, *46*, 3838–3866. [CrossRef]
11. Fortini, S.; Petrone, S. Quasi-Bayes properties of a procedure for sequential learning in mixture models. *J. R. Stat. Soc. Ser. B* **2020**, *82*, 1087–1114. [CrossRef]
12. Zhang, L.X.; Hu, F.; Cheung, S.H.; Chan, W.S. Immigrated urn models—Theoretical properties and applications. *Ann. Stat.* **2011**, *39*, 643–671. [CrossRef]
13. Blackwell, D.; MacQueen, J.B. Ferguson distributions via Pólya urn schemes. *Ann. Stat.* **1973**, *1*, 353–355. [CrossRef]
14. Bandyopadhyay, A.; Thacker, D. Pólya urn schemes with infinitely many colors. *Bernoulli* **2017**, *23*, 3243–3267. [CrossRef]
15. Janson, S. A.s. convergence for infinite colour Pólya urns associated with random walks. *Ark. Mat.* **2021**, *59*, 87–123. [CrossRef]
16. Mailler, C.; Villemonais, D. Stochastic approximation on non-compact measure spaces and application to measure-valued Pólya processes. *Ann. Appl. Probab.* **2020**, *30*, 2393–2438. [CrossRef]
17. Muliere, P.; Paganoni, A.M.; Secchi, P. A randomly reinforced urn. *J. Stat. Plan. Inference* **2006**, *136*, 1853–1874. [CrossRef]
18. Bassetti, F.; Crimaldi, I.; Leisen, F. Conditionally identically distributed species sampling sequences. *Adv. Appl. Probab.* **2010**, *42*, 433–459. [CrossRef]
19. Berti, P.; Crimaldi, I.; Pratelli, L.; Rigo, P. Rate of convergence of predictive distributions for dependent data. *Bernoulli* **2009**, *15*, 1351–1367. [CrossRef]
20. Berti, P.; Crimaldi, I.; Pratelli, L.; Rigo, P. Central limit theorems for multicolor urns with dominated colors. *Stoch. Process. Appl.* **2010**, *120*, 1473–1491. [CrossRef]
21. Berti, P.; Pratelli, L.; Rigo, P. Limit theorems for a class of identically distributed random variables. *Ann. Probab.* **2004**, *32*, 2029–2052. [CrossRef]
22. Crimaldi, I. An almost sure conditional convergence result and an application to a generalized Pólya urn. *Int. Math. Forum* **2009**, *4*, 1139–1156.
23. Sariev, H.; Fortini, S.; Petrone, S. Infinite-Color Randomly Reinforced Urns with Dominant Colors. 2021. Preprint. Available online: https://arxiv.org/abs/2106.04307 (accessed on 4 October 2021).
24. Fortini, S.; Petrone, S.; Sporysheva, P. On a notion of partially conditionally identically distributed sequences. *Stoch. Process. Appl.* **2018**, *128*, 819–846. [CrossRef]
25. Kallenberg, O. *Foundations of Modern Probability*, 3rd ed.; Springer: New York, NY, USA, 2021.

26. Häusler, E.; Luschgy, H. *Stable Convergence and Stable Limit Theorems*; Springer: Cham, Switzerland, 2015.
27. Dubins, L.; Freedman, D. A sharper form of the Borel-Cantelli lemma and the strong law. *Ann. Math. Stat.* **1965**, *36*, 800–807. [CrossRef]
28. Uchaikin, V.V.; Zolotarev, V.M. *Chance and Stability: Stable Distributions and Their Applications*; Walter de Gruyter: Berlin, Germany, 2011.
29. Fong, E.; Holmes, C.; Walker, S. Martingale Posterior Distributions 2021. Preprint. Available online: https://arxiv.org/abs/2103.15671 (accessed on 4 October 2021).
30. Fortini, S.; Ladelli, L.; Regazzini, E. A central limit problem for partially exchangeable random variables. *Theory Probab. Appl.* **1997**, *41*, 224–246. [CrossRef]
31. Fortini, S.; Ladelli, L.; Regazzini, E. Central limit theorem with exchangeable summands and mixtures of stable laws as limits. *Boll. Unione Mat. Ital.* **2012**, *5*, 515–542.
32. Berti, P.; Dreassi, E.; Pratelli, L.; Rigo, P. A class of models for Bayesian predictive inference. *Bernoulli* **2021**, *27*, 702–726. [CrossRef]
33. de Cooman, G.; Bock, J.D.; Diniz, M.A. Coherent predictive inference under exchangeability with imprecise probabilities. *J. Artif. Intell. Res.* **2015**, *52*, 1–95. [CrossRef]
34. Dolera, E.; Regazzini, E. Uniform rates of the Glivenko-Cantelli convergence and their use in approximating Bayesian inferences. *Bernoulli* **2019**, *25*, 2982–3015. [CrossRef]
35. Berti, P.; Pratelli, L.; Rigo, P. Limit theorems for empirical processes based on dependent data. *Electron. J. Probab.* **2012**, *17*, 1–18. [CrossRef]

Article

On Johnson's "Sufficientness" Postulates for Feature-Sampling Models

Federico Camerlenghi [1,2,3,*] and Stefano Favaro [2,4,5]

1. Department of Economics, Management and Statistics, University of Milano-Bicocca, Piazza dell'Ateneo Nuovo 1, 20126 Milano, Italy
2. Collegio Carlo Alberto, Piazza V. Arbarello 8, 10122 Torino, Italy; stefano.favaro@unito.it
3. BIDSA, Bocconi University, via Röntgen 1, 20136 Milano, Italy
4. Department of Economics, Social Studies, Applied Mathematics and Statistics, University of Torino, Corso Unione Sovietica 218/bis, 10134 Torino, Italy
5. IMATI-CNR "Enrico Magenes", 20133 Milano, Italy
* Correspondence: federico.camerlenghi@unimib.it

Abstract: In the 1920s, the English philosopher W.E. Johnson introduced a characterization of the symmetric Dirichlet prior distribution in terms of its predictive distribution. This is typically referred to as Johnson's "sufficientness" postulate, and it has been the subject of many contributions in Bayesian statistics, leading to predictive characterization for infinite-dimensional generalizations of the Dirichlet distribution, i.e., species-sampling models. In this paper, we review "sufficientness" postulates for species-sampling models, and then investigate analogous predictive characterizations for the more general feature-sampling models. In particular, we present a "sufficientness" postulate for a class of feature-sampling models referred to as Scaled Processes (SPs), and then discuss analogous characterizations in the general setup of feature-sampling models.

Keywords: Bayesian nonparametrics; exchangeability; feature-sampling model; de Finetti theorem; Johnson's "sufficientness" postulate; predictive distribution; scaled process prior; species-sampling model

1. Introduction

Exchangeability (de Finetti [1]) provides a natural modeling assumption in a large variety of statistical problems, and it amounts to the assumption that the order in which observations are recorded is not relevant. Consider a sequence of random variables $(Z_j)_{j\geq 1}$ defined on a common probability space $(\Omega, \mathscr{A}, \mathbb{P})$ and taking values in an arbitrary space, which is assumed to be Polish. The sequence $(Z_j)_{j\geq 1}$ is exchangeable if and only if

$$(Z_1, \ldots, Z_n) \stackrel{\mathrm{d}}{=} (Z_{\sigma(1)}, \ldots, Z_{\sigma(n)})$$

for any permutation σ of the set $\{1, \ldots, n\}$ and any $n \geq 1$. By virtue of the celebrated de Finetti representation theorem, exchangeability of $(Z_j)_{j\geq 1}$ is tantamount to asserting the existence of a random element $\tilde{\mu}$, defined on a (parameter) space Θ, such that, conditionally on $\tilde{\mu}$, the Z_js are independent and identically distributed with common distribution $p_{\tilde{\mu}}$, i.e.,

$$\begin{aligned} Z_j \mid \tilde{\mu} &\stackrel{\mathrm{iid}}{\sim} p_{\tilde{\mu}} \quad j \geq 1 \\ \tilde{\mu} &\sim \mathscr{M}, \end{aligned} \quad (1)$$

where $\mathscr{M}$ is the distribution of $\tilde{\mu}$. In a Bayesian setting, $\mathscr{M}$ takes on the interpretation of a prior distribution for the parameter object of interest. In this sense, the de Finetti representation theorem is a natural framework for Bayesian statistics. For mathematical convenience,

Θ is assumed to be a Polish space, equipped with the Borel σ-algebra $\mathscr{B}(\Theta)$. Hereafter, with the term *parameter*, we refer to both a finite- and an infinite-dimensional object.

Within the framework of exchangeability (1), a critical role is played by the predictive distributions, namely, the conditional distributions of the $(n+1)$th observation Z_{n+1} given $\mathbf{Z}_n := (Z_1, \ldots, Z_n)$. The problem of characterizing prior distributions $\mathscr{M}$ in terms of their predictive distributions has a long history in Bayesian statistics, starting from the seminal work of the English philosopher Johnson [2] who provided a predictive characterization of the symmetric Dirichlet prior distribution. Such a characterization is typically referred to as Johnson's "sufficientness" postulate. Species-sampling models (Pitman [3]) provide arguably the most popular infinite-dimensional generalization of the Dirichlet distribution. They form a broad class of nonparametric prior models that correspond to the assumption that $p_{\tilde{\mu}}$ in (1) is an almost surely discrete random probability measure

$$\tilde{p} = \sum_{i \geq 1} \tilde{p}_i \delta_{\tilde{z}_i}, \qquad (2)$$

where: (i) $(\tilde{p}_i)_{i \geq 1}$ are non-negative random weights almost surely summing up to 1; (ii) $(\tilde{z}_i)_{i \geq 1}$ are random species' labels, independent of $(\tilde{p}_i)_{i \geq 1}$, and i.i.d. with common (non-atomic) distribution P. The term *species* refers to the fact that the law of $\tilde{p}$ is a prior distribution for the unknown species composition $(\tilde{p}_i)_{i \geq 1}$ of a population of individuals Z_js, with Z_j belonging to a species $\tilde{z}_i$ with probability $\tilde{p}_i$ for $j, i \geq 1$. In the context of species-sampling models, Regazzini [4] and Lo [5] provided a "sufficientness" postulate for the Dirichlet process (Ferguson [6]). Such a characterization was then extended by Zabell [7] to the Pitman–Yor process (Perman et al. [8], Pitman and Yor [9]) and by Bacallado et al. [10] to the more general Gibbs-type prior models (Gnedin and Pitman [11]).

In this paper, we introduce and discuss Johnson's "sufficientness" postulates in the feature-sampling setting, which generalizes the species-sampling setting by allowing each individual of the population to belong to multiple species, now called features. We point out that feature-sampling models are extremely important in different areas of application; see, e.g., Griffiths and Ghahramani [12], Ayed et al. [13] and the references therein. Under the framework of exchangeability (1), the feature-sampling setting assumes that

$$Z_j | \tilde{\mu} = \sum_{i \geq 1} A_{j,i} \delta_{\tilde{w}_i} \sim p_{\tilde{\mu}}, \qquad (3)$$

and

$$\tilde{\mu} = \sum_{i \geq 1} \tilde{p}_i \delta_{\tilde{w}_i}$$

where: (i) conditionally on $\tilde{\mu}$, $(A_{j,i})_{i \geq 1}$ are independent Bernoulli random variables with parameters $(\tilde{p}_i)_{i \geq 1}$; (ii) $(\tilde{p}_i)_{i \geq 1}$ are $(0,1)$-valued random weights; (iii) $(\tilde{w}_i)_{i \geq 1}$ are random features' labels, independent of $(\tilde{p}_i)_{i \geq 1}$, and i.i.d. with common (non-atomic) distribution P. That is, individual Z_j displays feature $\tilde{w}_i$ if and only if $A_{j,i} = 1$, which happens with probability $\tilde{p}_i$. For example, if, conditionally on $\tilde{\mu}$, Z_j displays only two features, say $\tilde{w}_1$ and $\tilde{w}_5$, it equals the random measure $\delta_{\tilde{w}_1} + \delta_{\tilde{w}_5}$. The distribution $p_{\tilde{\mu}}$ is the law of a Bernoulli process with parameter $\tilde{\mu}$, which is denoted by $\mathrm{BeP}(\tilde{\mu})$, whereas the law of $\tilde{\mu}$ is a nonparametric prior distribution for the unknown feature probabilities $(\tilde{p}_i)_{i \geq 1}$, i.e., a feature-sampling model. Here, we investigate the problem of characterizing prior distributions for $\tilde{\mu}$ in terms of their predictive distributions, with the goal of providing "sufficientness" postulates for feature-sampling models. We discuss such a problem and present partial results for a class of feature-sampling models referred to as Scaled Process (SP) priors for $\tilde{\mu}$ (James et al. [14], Camerlenghi et al. [15]). With these results, we aim at stimulating future research in this field to obtain "sufficientness" postulates for general feature-sampling models.

The paper is structured as follows. In Section 2, we present a brief review on Johnson's "sufficientness" postulates for species-sampling models. Section 3 focuses on nonparametric prior models for the Bernoulli process, i.e., feature-sampling models; we review their definitions, properties, and sampling structures. In Section 4, we present a "sufficientness" postulate for SPs. Section 5 concludes the paper by discussing our results and conjecturing analogous results for more general classes of feature-sampling models.

2. Species-Sampling Models

To introduce species-sampling models, we assume that the observations are $\mathbb{Z}$-valued random elements, and $\mathbb{Z}$ is supposed to be a Polish space whose Borel σ-algebra is denoted by $\mathscr{Z}$. Thus, $\mathbb{Z}$ contains all the possible species' labels of the populations. When we deal with species-sampling models, the hierarchical formulation (1) specializes as

$$Z_j | \tilde{p} \stackrel{iid}{\sim} \tilde{p} \quad j \geq 1$$
$$\tilde{p} \sim \mathscr{M} \tag{4}$$

where $\tilde{p} = \sum_{i \geq 1} \tilde{p}_i \delta_{\tilde{z}_i}$ is an almost surely discrete random probability measure on $\mathbb{Z}$, and $\mathscr{M}$ denotes its law. We also remind the reader that: (i) $(\tilde{p}_i)_{i \geq 1}$ are non-negative random weights almost surely summing up to 1; (ii) $(\tilde{z}_i)_{i \geq 1}$ are random species' labels, independent of $(\tilde{p}_i)_{i \geq 1}$, and i.i.d. as a common (non-atomic) distribution P. Using the terminology of Pitman [3], the discrete random probability measure $\tilde{p}$ is a *species-sampling model*. In Bayesian nonparametrics, popular examples of species-sampling models are: the Dirichlet process (Ferguson [6]), the Pitman–Yor process (Perman et al. [8], Pitman and Yor [9]), and the normalized generalized Gamma process (Brix [16], Lijoi et al. [17]). These are examples belonging to a peculiar subclass of species-sampling models, which are referred to as Gibbs-type prior models (Gnedin and Pitman [11], De Blasi et al. [18]). More general subclasses of species-sampling models are, e.g., the homogeneous normalized random measures (Regazzini et al. [19]) and the Poisson–Kingman models (Pitman [20,21]). We refer to Lijoi and Prünster [22] and Ghosal and van der Vaart [23] for a detailed and stimulating account on species-sampling models and their use in Bayesian nonparametrics.

Because of the almost sure discreteness of $\tilde{p}$ in (4), a random sample $\mathbf{Z}_n := (Z_1, \ldots, Z_n)$ from $\tilde{p}$ features ties, that is, $\mathbb{P}(Z_{j_1} = Z_{j_2}) > 0$ if $j_1 \neq j_2$. Thus, $\mathbf{Z}_n$ induces a random partition of the set $\{1, \ldots, n\}$ into $K_n = k \leq n$ blocks, labeled by $Z_1^*, \ldots, Z_{K_n}^*$, with corresponding frequencies $(N_{n,1}, \ldots, N_{n,K_n}) = (n_1, \ldots, n_k)$, such that $N_{i,n} \geq 1$ and $\sum_{1 \leq i \leq K_n} N_{i,n} = n$. From Pitman [3], the predictive distribution of $\tilde{p}$ is of the form

$$\mathbb{P}(Z_{n+1} \in A | \mathbf{Z}_n) = g(n,k,\mathbf{n})P(A) + \sum_{i=1}^{k} f_i(n,k,\mathbf{n})\delta_{Z_i^*}(A), \quad A \in \mathscr{Z}, \tag{5}$$

for any $n \geq 1$, having set $\mathbf{n} = (n_1, \ldots, n_k)$, with g and f_i being arbitrary non-negative functions that satisfy the constraint $g(n,k,\mathbf{n}) + \sum_{i=1}^{k} f_i(n,k,\mathbf{n}) = 1$. The predictive distribution (5) admits the following interpretation: (i) $g(n,k,\mathbf{n})$ corresponds to the probability that Z_{n+1} is a new species, that is, a species not observed in $\mathbf{Z}_n$; (ii) $f_i(n,k,\mathbf{n})$ corresponds to the probability that Z_{n+1} is a species Z_i^* in $\mathbf{Z}_n$. The functions g and f_i completely determine the distribution of the exchangeable sequence $(Z_j)_{j \geq 1}$ and, in turn, the distribution of the random partition of $\mathbb{N}$ induced by $(Z_j)_{j \geq 1}$. Predictive distributions of popular species-sampling models, e.g., the Dirichlet process, the Pitman–Yor process, and the normalized generalized Gamma process, are of the form (5) for suitable specification of the functions g and f_i. We refer to Pitman [21] for a detailed account of random partitions induced by species-sampling models and generalizations thereof.

Here, we recall the predictive distribution of Gibbs-type prior models (Gnedin and Pitman [11], De Blasi et al. [18]). Let us first introduce the definition of these processes.

Definition 1. Let $\sigma \in (-\infty, 1)$ and let P be a (non-atomic) distribution on $(\mathbb{Z}, \mathscr{Z})$. A Gibbs-type prior model is a species-sampling model with a predictive distribution of the form

$$\mathbb{P}(Z_{n+1} \in A | \mathbf{Z}_n) = \frac{V_{n+1,k+1}}{V_{n,k}} P(A) + \frac{V_{n+1,k}}{V_{n,k}} \sum_{i=1}^{k} (n_i - \sigma) \delta_{Z_i^*}(A), \quad A \in \mathscr{Z}, \qquad (6)$$

for any $n \geq 1$, where $\{V_{n,k} : n \geq 1, 1 \leq k \leq n\}$ is a collection of non-negative weights that satisfy the recurrence relation $V_{n,k} = (n - \sigma k) V_{n+1,k} + V_{n+1,k+1}$ for all $k = 1, \ldots, n$, $n \geq 1$, with the proviso $V_{1,1} = 1$.

Note that the Dirichlet process is a Gibbs-type prior model that corresponds to

$$V_{n,k} = \frac{\theta^k}{(\theta)_n}$$

for $\theta > 0$, where we have denoted by $(a)_b = \Gamma(a+b)/\Gamma(a)$ the Pochhammer symbol for the rising factorials. Moreover, the Pitman–Yor process is a Gibbs-type prior model corresponding to

$$V_{n,k} = \frac{\prod_{i=0}^{k-1}(\theta + i\sigma)}{(\theta)_n}$$

for $\sigma \in (0,1)$ and $\theta > -\alpha$. We refer to Pitman [20] for other examples of Gibbs-type prior models and for a detailed account of the $V_{n,k}$s; see also Pitman [21] and the references therein.

Because of de Finetti's representation theorem, there exists a one-to-one correspondence between the functions g and f_i in the predictive distribution (5) and the law $\mathscr{M}$ of $\tilde{p}$, i.e., the de Finetti measure. This is at the basis of Johnson's "sufficientness" postulates, characterizing species-sampling models through their predictive distributions. Regazzini [4] and, later, Lo [5] provided the first "sufficientness" postulate for species-sampling models, showing that the Dirichlet process is the unique species-sampling model for which the function g depends on $\mathbf{Z}_n$ only through n, and the function f_i depends on $\mathbf{Z}_n$ only through n and n_i for $i \geq 1$. Such a result was extended in Zabell [24], providing the following "sufficientness" postulate for the Pitman–Yor process: The Pitman–Yor process is the unique species-sampling model for which the function g depends on $\mathbf{Z}_n$ only through n and k, and the function f_i depends on $\mathbf{Z}_n$ only through n and n_i for $i \geq 1$. Bacallado et al. [10] discussed the "sufficientness" postulate in the more general setting of Gibbs-type prior models, showing that Gibbs-type prior models are the sole species-sampling models for which the function g depends on $\mathbf{Z}_n$ only through n and k, and the function f_i depends on $\mathbf{Z}_n$ only through n, k, and n_i. This result shows a critical difference—at the sampling level—between the Pitman–Yor process and Gibbs-type prior models, which lies in the inclusion of the sampling information on the observed number of distinct species in the probability of observing, at the $(n+1)$-th draw, a species already observed in the sample.

3. Feature-Sampling Models

Feature-sampling models generalize species-sampling models by allowing each individual to belong to more than one species, which are now called features. To introduce feature-sampling models, we consider a space of features $\mathbb{W}$, which is assumed to be a Polish space, and we denote by $\mathscr{W}$ its Borel σ-field. Thus, $\mathbb{W}$ contains all the possible features' labels of the population. Observations are represented through the counting measure (3), whose parameter $\tilde{\mu}$ is an almost surely discrete measure with masses in $(0,1)$. When we deal with feature-sampling models, the hierarchical formulation (1) specializes as

$$Z_j | \tilde{\mu} \stackrel{\text{iid}}{\sim} \text{BeP}(\tilde{\mu}) \qquad (7)$$
$$\tilde{\mu} \sim \mathscr{M}$$

where $\tilde{\mu} = \sum_{i\geq 1} \tilde{p}_i \delta_{\tilde{w}_i}$ is an almost surely discrete random measure on $\mathbb{W}$, and $\mathcal{M}$ denotes its law. We also remind the reader that: (i) conditionally on $\tilde{\mu}$, $(A_{j,i})_{i\geq 1}$ are independent Bernoulli random variables with parameters $(\tilde{p}_i)_{i\geq 1}$; (ii) $(\tilde{p}_i)_{i\geq 1}$ are $(0,1)$-valued random weights; (iii) $(\tilde{w}_i)_{i\geq 1}$ are random features' labels, independent of $(\tilde{p}_i)_{i\geq 1}$, and i.i.d. with common (non-atomic) distribution P. Completely random measures (CRMs) (Daley and Vere-Jones [25], Kingman [26]) provide a popular class of nonparametric priors $\mathcal{M}$, the most common examples of which are the Beta process prior and the stable Beta process prior (Teh and Gorur [27], James [28]); see also Broderick et al. [29] and the references therein for other examples of CRM priors and generalizations thereof. Recently, Camerlenghi et al. [15] investigated an alternative class of nonparametric priors $\mathcal{M}$, generalizing CRM priors and referring to these as Scaled Processes (SPs). SP priors first appeared in the work of James [28].

We assume a random sample $\mathbf{Z}_n := (Z_1, \ldots, Z_n)$ to be modeled as in (7), and we introduce the predictive distribution of $\tilde{\mu}$, that is, the conditional probability of Z_{n+1} given $\mathbf{Z}_n$. Note that, because of the pure discreteness of $\tilde{\mu}$, the observations $\mathbf{Z}_n$ may share a random number of distinct features, say $K_n = k$, denoted here as $W_1^*, \ldots, W_{K_n}^*$, and each feature W_i^* is displayed exactly by $M_{n,i} = m_i$ of the n individuals as $i = 1, \ldots, k$. Since the features' labels are immaterial and i.i.d. form the base measure P, the conditional distribution of Z_{n+1}, given $\mathbf{Z}_n$, may be equivalently characterized through the vector $(Y_{n+1}, A_{n+1,1}^*, \ldots, A_{n+1,K_n}^*)$, where: (i) Y_{n+1} is the number of new features displayed by the $(n+1)$th individual, namely, hitherto unobserved out of the sample $\mathbf{Z}_n$; (ii) $A_{n+1,i}^*$ is a $\{0,1\}$-valued random variable for any $i = 1, \ldots, K_n$, and $A_{n+1,i}^* = 1$ if the $(n+1)$th individual displays feature W_i^*; it equals 0 otherwise. Hence, the predictive distribution of $\tilde{\mu}$ is

$$\mathbb{P}((Y_{n+1}, A_{n+1,1}^*, \ldots, A_{n+1,K_n}^*) = (y, a_1, \ldots, a_{K_n})|\mathbf{Z}_n) = f(y, a_1, \ldots, a_k; n, k, \mathbf{m}) \quad (8)$$

where we denote by f a probability distribution evaluated at $(y, a_1, \ldots, a_k)$, and where n, k and $\mathbf{m} := (m_1, \ldots, m_k)$ is the sampling information. In the rest of this section, we specify the function f under the assumption of a CRM prior and an SP prior, showing its dependence on n, K_n, and $(M_{n,1}, \ldots, M_{n,K_n})$. In particular, we show how SP priors allow one to enrich the predictive distribution of CRM priors by including additional sampling information in terms of the number of distinct features and their corresponding frequencies.

3.1. Priors Based on CRMs

Let $\mathsf{M}_{\mathbb{W}}$ denote the space of all bounded and finite measures on $(\mathbb{W}, \mathscr{W})$, that is to say, $\mu \in \mathsf{M}_{\mathbb{W}}$ iff $\mu(A) < +\infty$ for any bounded set $A \in \mathscr{W}$. Here, we recall the definition of a Completely Random Measure (CRM) (see, e.g., Daley and Vere-Jones [25]).

Definition 2. *A Completely Random Measure (CRM) $\tilde{\mu}$ on $(\mathbb{W}, \mathscr{W})$ is a random element taking values in the space $\mathsf{M}_{\mathbb{W}}$ such that the random variables $\tilde{\mu}(A_1), \ldots, \tilde{\mu}(A_n)$ are independent for any choice of bounded and disjoint sets $A_1, \ldots, A_n \in \mathscr{W}$ and for any $n \geq 1$.*

We remind the reader that Kingman [26] proved that a CRM may be decomposed as the sum of a deterministic drift and a purely atomic component. In Bayesian nonparametrics, it is common to consider purely atomic CRMs without fixed points of discontinuity, that is to say, $\tilde{\mu}$ may be represented as $\tilde{\mu} := \sum_{i\geq 1} \tilde{\eta}_i \delta_{\tilde{w}_i}$, where $(\tilde{\eta}_i)_{i\geq 1}$ is a sequence of random atoms and $(\tilde{w}_i)_{i\geq 1}$ are the random locations. An appealing property of purely atomic CRMs is the availability of their Laplace functional; indeed, for any measurable function $f : \mathbb{W} \to \mathbb{R}^+$, one has

$$\mathbb{E}\left[e^{-\int_{\mathbb{W}} f(w)\tilde{\mu}(\mathrm{d}w)}\right] = \exp\left\{-\int_{\mathbb{W}\times\mathbb{R}^+}(1-e^{-sf(w)})\nu(\mathrm{d}w, \mathrm{d}s)\right\} \quad (9)$$

where ν is a measure on $\mathbb{W} \times \mathbb{R}^+$ called the Lévy intensity of the CRM $\tilde{\mu}$, and it is such that

$$\nu(\{w\} \times \mathbb{R}^+) = 0 \quad \forall w \in \mathbb{W}, \quad \text{and} \quad \int_{A \times \mathbb{R}^+} \min\{s,1\} \nu(\mathrm{d}w, \mathrm{d}s) < \infty \qquad (10)$$

for any bounded Borel set A. Here, we focus on homogeneous CRMs by assuming that the atoms $\tilde{\eta}_i$s and the locations $\tilde{w}_i$s are independent; in this case, the Lévy measure may be written as

$$\nu(\mathrm{d}w, \mathrm{d}s) = \lambda(s)\mathrm{d}s P(\mathrm{d}w)$$

for some measurable function $\lambda : \mathbb{R}^+ \to \mathbb{R}^+$ and a probability measure P on $(\mathbb{W}, \mathscr{W})$, called the *base measure*, which is assumed to be diffuse. In this case, the distribution of $\tilde{\mu}$ will be denoted as $\mathrm{CRM}(\lambda; P)$, and the second integrability condition in (10) reduces to the following:

$$\int_{\mathbb{R}^+} \min\{s,1\} \lambda(s) \mathrm{d}s < +\infty. \qquad (11)$$

In the feature-sampling framework, $\tilde{\mu}$ may be used as a prior distribution if the sequence of atoms $(\tilde{\eta}_i)_{i \geq 1}$ is in between $[0,1]$, which happens if the Lévy intensity has support on $\mathbb{W} \times [0,1]$. A noteworthy example, widely used in this setting, is the stable Beta process prior (Teh and Gorur [27]). It is defined as a CRM with Lévy intensity

$$\lambda(s) = \alpha \cdot \frac{\Gamma(1+c)}{\Gamma(1-\sigma)\Gamma(c+\sigma)} s^{-1-\sigma}(1-s)^{c+\sigma-1} \mathbb{1}_{(0,1)}(s) \qquad (12)$$

where $c > 0$, $\sigma \in (0,1)$, and $\alpha > 0$ (James [28], Masoero et al. [30]). Now, we describe the predictive distribution for an arbitrary CRM $\tilde{\mu}$. For the sake of clarity, we fix the following notation:

$$\mathrm{Poiss}(y; C) := \frac{C^y e^{-C}}{y!}, y \in \mathbb{N} \quad \text{and} \quad \mathrm{Bern}(a; p) := p^a (1-p)^{1-a}, a \in \{0,1\}$$

to denote the probability mass functions of a Poisson with parameter $C > 0$ and a Bernoulli random variable with parameter $p \in [0,1]$, respectively. We refer to James [28] for a detailed posterior analysis of CRM priors; see also Broderick et al. [29] and the references therein.

Theorem 1 (James [28]). *Let $Z_1, Z_2, \ldots$ be exchangeable random variables modeled as in (7), where $\mathscr{M}$ equals $\mathrm{CRM}(\lambda; P)$. If $\mathbf{Z}_n$ is a random sample that displays $K_n = k$ distinct features $\{W_1^*, \ldots, W_{K_n}^*\}$, and feature W_i^* appears exactly $M_{n,i} = m_i$ times in the samples, such as $i = 1, \ldots, K_n$, then*

$$\mathbb{P}((Y_{n+1}, A_{n+1,1}^*, \ldots, A_{n+1,K_n}^*) = (y, a_1, \ldots, a_{K_n}) | \mathbf{Z}_n)$$
$$= \mathrm{Poiss}\left(y; \int_0^1 s(1-s)^n \lambda(s) \mathrm{d}s\right) \prod_{i=1}^k \mathrm{Bern}(a_i; p_i^*) \qquad (13)$$

being

$$p_i^* := \frac{\int_0^1 s^{m_i+1}(1-s)^{n-m_i} \lambda(s) \mathrm{d}s}{\int_0^1 s^{m_i}(1-s)^{n-m_i} \lambda(s) \mathrm{d}s}.$$

Proof. We consider James [28] (Proposition 3.2) for Bernoulli product models (see also Camerlenghi et al. [15] (Proposition 1)); thus, the distribution of Z_{n+1}, given $\mathbf{Z}_n$, equals the distribution of

$$Z'_{n+1} + \sum_{i=1}^{K_n} A_{n+1,i}^* \delta_{W_i^*}, \qquad (14)$$

where $Z'_{n+1}|\tilde{\mu}' = \sum_{i\geq 1} A'_{n+1,i}\delta_{\tilde{w}'_i} \sim \text{BeP}(\tilde{\mu}')$ such that $\tilde{\mu}' \sim \text{CRM}((1-s)^n\lambda; P)$, and $A^*_{n+1,1}, \ldots, A^*_{n+1,K_n}$ are Bernoulli random variables with parameters $J_1, \ldots, J_{K_n}$, respectively, such that each J_i is a random variable whose distribution is with a density function of the form

$$f_{J_i}(s) \propto (1-s)^{n-m_i} s^{m_i} \lambda(s).$$

By exploiting the previous predictive characterization, we can derive the posterior distribution of Y_{n+1} given Z_n by means of a direct application of the Laplace functional. Indeed, the distribution of $Y_{n+1}|Z_n$ equals $\sum_{i\geq 1} A'_{n+1,i}$. Thus, for any $t \in \mathbb{R}$, we have the following:

$$\mathbb{E}[e^{-tY_{n+1}}|Z_n] = \mathbb{E}[e^{-t\sum_{i\geq 1} A'_{n+1,i}}] = \mathbb{E}\left[\prod_{i\geq 1} e^{-tA'_{n+1,i}}\right] = \mathbb{E}\left[\mathbb{E}\left[\prod_{i\geq 1} e^{-tA'_{n+1,i}} \mid \tilde{\mu}'\right]\right]$$

$$= \mathbb{E}\left[\prod_{i\geq 1}\left(e^{-t}\tilde{\eta}'_i + (1-\tilde{\eta}'_i)\right)\right],$$

where we used the representation $\tilde{\mu}' = \sum_{i\geq 1} \tilde{\eta}'_i \delta_{\tilde{w}'_i}$ and the fact that the $A_{n+1,i}$s are independent Bernoulli random variables conditionally on $\tilde{\mu}'$. We now use the Laplace functional for $\tilde{\mu}'$ to get

$$\mathbb{E}[e^{-tY_{n+1}}|Z_n] = \mathbb{E}\left[\exp\left\{\sum_{i\geq 1}\log(1+\tilde{\eta}'_i(e^{-t}-1))\right\}\right]$$

$$= \exp\left\{-(1-e^{-t})\int_0^1 (1-s)^n s\lambda(s)\mathrm{d}s\right\}.$$

As a direct consequence, the posterior distribution of Y_{n+1} given Z_n is a Poisson distribution with mean $\int_0^1 (1-s)^n s\lambda(s)\mathrm{d}s$. Again, by exploiting the predictive representation (14), the posterior distribution of $A^*_{n+1,i}$, as $i = 1, \ldots, K_n$, is a Bernoulli with the following mean:

$$\mathbb{E}[J_i] = \int_0^1 s f_{J_i}(s) \mathrm{d}s = \frac{\int_0^1 (1-s)^{n-m_i} s^{m_i+1} \lambda(s) \mathrm{d}s}{\int_0^1 (1-s)^{n-m_i} s^{m_i} \lambda(s) \mathrm{d}s}.$$

□

Corollary 1. *Let $Z_1, Z_2, \ldots$ be exchangeable random variables modeled as in (7), where $\mathcal{M}$ is the law of the stable Beta process. If Z_n is a random sample that displays $K_n = k$ distinct features $\{W_1^*, \ldots, W_{K_n}^*\}$, and feature W_i^* appears exactly $M_{n,i} = m_i$ times in the samples, such as $i = 1, \ldots, K_n$, then*

$$\mathbb{P}((Y_{n+1}, A^*_{n+1,1}, \ldots, A^*_{n+1,K_n}) = (y, a_1, \ldots, a_{K_n})|Z_n)$$
$$= \text{Poiss}\left(y; \alpha \frac{(c+\sigma)_n}{(c+1)_n}\right) \prod_{i=1}^k \text{Bern}\left(a_i; \frac{m_i - \sigma}{n+c}\right), \quad (15)$$

where $(x)_y = \Gamma(x+y)/\Gamma(x)$ denotes the Pochhammer symbol for $x, y > 0$.

Proof. It is sufficient to specialize Theorem 1 for the stable Beta process. In particular, from Theorem 1, the posterior distribution of Y_{n+1} given Z_n is a Poisson distribution with mean

$$\int_0^1 s(1-s)^n \lambda(s)\mathrm{d}s \stackrel{(12)}{=} \frac{\alpha \Gamma(1+c)}{\Gamma(1-\sigma)\Gamma(c+\sigma)} \int_0^1 s^{-\sigma}(1-s)^{n+c+\sigma}\mathrm{d}s = \alpha \frac{(c+\sigma)_n}{(c+1)_n}.$$

Moreover, the parameters of the Bernoulli random variables $A^*_{n+1,1}, \ldots, A^*_{n+1,K_n}$ are equal to

$$p_i^* = \frac{\int_0^1 s^{m_i+1}(1-s)^{n-m_i}\lambda(s)ds}{\int_0^1 s^{m_i}(1-s)^{n-m_i}\lambda(s)ds} \stackrel{(12)}{=} \frac{B(m_i+1-\sigma, c+\sigma+n-m_i)}{B(m_i-\sigma, c+\sigma+n-m_i)} = \frac{m_i-\sigma}{n+c}$$

as $i = 1, \ldots, K_n$. □

3.2. SP Priors

From Theorem 1, under CRM priors, the distribution of the number of new features Y_{n+1} is a Poisson distribution that depends on the sampling information only through the sample size n. Moreover, the probability of observing a feature already observed in the sample, say W_i^*, depends only on the sample size n and the frequency m_i of feature W_i^* out of the initial sample. Camerlenghi et al. [15] showed that SP priors allow one to enrich the predictive structure of CRM priors, including additional sampling information in the probability of discovering new features. To introduce SP priors, consider a CRM $\tilde{\mu} = \sum_{i\geq 1} \tilde{\tau}_i \delta_{\tilde{w}_i}$ on $\mathbb{W}$, where $(\tilde{\tau}_i)_{i\geq 1}$ are positive random atoms and $(\tilde{w}_i)_{i\geq 1}$ are i.i.d. random atoms, with Lévy intensity $\nu(dw, ds) = \lambda(s)ds P(dw)$ satisfying

$$\int_0^\infty \min\{s, 1\}\lambda(s)ds < +\infty. \tag{16}$$

Consider the ordered jumps $\Delta_1 > \Delta_2 > \cdots$ of the CRM $\tilde{\mu}$ and define the random measure

$$\tilde{\mu}_{\Delta_1} = \sum_{i\geq 1} \frac{\Delta_{i+1}}{\Delta_1} \delta_{\tilde{w}_i}$$

normalizing $\tilde{\mu}$ by the largest jump. The definition of SPs follows with a suitable change in the measure of Δ_1 (James et al. [14], Camerlenghi et al. [15]). Let us denote by $\mathscr{L}(\cdot, a)$ a regular version of the conditional probability distribution of $(\Delta_{i+1}/\Delta_1)_{i\geq 1}$ given $\Delta_1 = a$. Now denote by Ψ_1 a positive random variable with density function f_{Ψ_1} on $\mathbb{R}^+$ and define

$$\mathscr{L}(\cdot) := \int_{\mathbb{R}^+} \mathscr{L}(\cdot, a) f_{\Psi_1}(a) da$$

The distribution of $(\Delta_{i+1}/\Delta_1)_{i\geq 1}$ is obtained by mixing $\mathscr{L}(\cdot, a)$ with respect to the density function f_{Ψ_1}. Thus, we are ready to define an SP.

Definition 3. *A Scaled Process (SP) prior on* $(\mathbb{W}, \mathscr{W})$ *is defined as the almost surely discrete random measure*

$$\tilde{\mu}_{\Psi_1} := \sum_{i\geq 1} \tilde{\eta}_i \delta_{\tilde{w}_i}, \tag{17}$$

where $(\tilde{\eta}_i)_{i\geq 1}$ *has distribution* $\mathscr{L}$ *and* $(\tilde{w}_i)_{i\geq 1}$ *is a sequence of independent random variables with common distribution* P, *also independent of* $(\tilde{\eta}_i)_{i\geq 1}$. *We will write* $\tilde{\mu}_{\Psi_1} \sim \mathrm{SP}(\nu, f_{\Psi_1})$.

A thoughtful account with a complete posterior analysis for SPs is given in Camerlenghi et al. [15]. Here, we characterize the predictive distribution (8) of SPs.

Theorem 2 (Camerlenghi et al. [15], James [28]). *Let* $Z_1, Z_2, \ldots$ *be exchangeable random variables modeled as in (7), where* $\mathscr{M}$ *equals* $\mathrm{SP}(\nu, f_{\Psi_1})$. *If* $\mathbf{Z}_n$ *is a random sample that displays* $K_n = k$ *distinct features* $\{W_1^*, \ldots, W_{K_n}^*\}$, *and feature* W_i^* *appears exactly* $M_{n,i} = m_i$ *times in the samples, such as* $i = 1, \ldots, K_n$, *then the conditional distribution of* Ψ_1, *given* $\mathbf{Z}_n$, *has posterior density:*

$$f_{\Psi_1|\mathbf{Z}_n}(a) \propto e^{-\sum_{i=1}^n \int_0^1 s(1-s)^{n-1}a\lambda(as)ds} \prod_{i=1}^k \int_0^1 s^{m_i}(1-s)^{n-m_i}a\lambda(as)ds f_{\Psi_1}(a). \tag{18}$$

Moreover, conditionally on Z_n and Ψ_1,

$$\mathbb{P}((Y_{n+1}, A^*_{n+1,1}, \ldots, A^*_{n+1,K_n}) = (y, a_1, \ldots, a_{K_n}) | Z_n, \Psi_1)$$
$$= \text{Poiss}\left(y; \int_0^1 s\Psi_1(1-s)^n \lambda(s\Psi_1) ds\right) \prod_{i=1}^k \text{Bern}(a_i; p_i^*(\Psi_1)) \quad (19)$$

being

$$p_i^*(\Psi_1) := \frac{\int_0^1 s^{m_i+1}(1-s)^{n-m_i} \lambda(s\Psi_1) ds}{\int_0^1 s^{m_i}(1-s)^{n-m_i} \lambda(s\Psi_1) ds}.$$

Proof. The representation of the predictive distribution (19) follows from Camerlenghi et al. [15] (Proposition 2). Indeed, the posterior distribution of the largest jump directly follows from [15] (Equation (4)). In addition, the authors of [15] (Proposition 2) showed that the conditional distribution of Z_{n+1}, given Z_n and Ψ_1, equals the distribution of the following counting measure:

$$Z'_{n+1} + \sum_{i=1}^{K_n} A^*_{n+1,i} \delta_{W_i^*}, \quad (20)$$

where $Z'_{n+1} | \tilde{\mu}' = \sum_{i \geq 1} A'_{n+1,i} \delta_{\tilde{w}'_i} \sim \text{BeP}(\tilde{\mu}'_{\Psi_1})$ and $\tilde{\mu}'_{\Psi_1}$ is a CRM with Lévy intensity of the form

$$\nu'_{\Psi_1}(dw, ds) = (1-s)^n \Psi_1 \lambda(\Psi_1 s) \mathbb{1}_{(0,1)}(s) ds P(dw).$$

Moreover, $A^*_{n+1,1}, \ldots, A^*_{n+1,K_n}$ are Bernoulli random variables with parameters $J_1, \ldots, J_{K_n}$, respectively, such that conditionally on Ψ_1, each J_i has a distribution with a density function of the form

$$f_{J_i|\Psi_1}(s) \propto (1-s)^{n-m_i} s^{m_i} \Psi_1 \lambda(\Psi_1 s) \quad \text{on } (0,1).$$

As in the proof of Theorem 1, we show that the distribution of $Y_{n+1}|(\Psi_1, Z_n)$ equals $\sum_{i \geq 1} A'_{n+1,i}$. Thus, by the evaluation of the Laplace functional, one may easily realize that the last random sum has a Poisson distribution with mean $\int_0^1 (1-s)^n s \Psi_1 \lambda(\Psi_1 s) ds$. Moreover, by exploiting the posterior representation (20), the variables $A^*_{n+1,i}$, such as $i = 1, \ldots, K_n$, conditionally on Z_n and Ψ, are independent and Bernoulli distributed with mean

$$\mathbb{E}[J_i | \Psi_1] = \int_0^1 s f_{J_i|\Psi_1}(s) ds = \frac{\int_0^1 (1-s)^{n-m_i} s^{m_i+1} \Psi_1 \lambda(s\Psi_1) ds}{\int_0^1 (1-s)^{n-m_i} s^{m_i} \Psi_1 \lambda(s\Psi_1) ds}.$$

□

Remark 1. *According to (18), the conditional distribution of Ψ_1 given Z_n may include the whole sampling information, depending on the specification of ν and f_{Ψ_1}, and hence, the conditional distribution of Y_{n+1} given Z_n may also include such sampling information. As a corollary of Theorem 2, the conditional distribution of Y_{n+1} given Z_n is a mixture of Poisson distributions that may include the whole sampling information; in particular, the amount of sampling information in the posterior distribution is uniquely determined by the mixing distribution, namely by the conditional distribution of Ψ_1, given Z_n.*

Hereafter, we specialize Theorem 2 for the stable SP, that is, a peculiar SP defined through a CRM with a Lévy intensity ν such that $\lambda(s) = \sigma s^{-1-\sigma}$ for a parameter $\sigma \in (0,1)$. We refer to Camerlenghi et al. [15] for a detailed posterior analysis of the stable SP prior.

Corollary 2. *Let $Z_1, Z_2, \ldots$ be exchangeable random variables modeled as in (7), where $\mathscr{M}$ equals $SP(\nu, f_{\Psi_1})$, with $\lambda(s) = \sigma s^{-1-\sigma}$ for some $\sigma \in (0,1)$. If Z_n is a random sample that displays $K_n = k$ distinct features $\{W_1^*, \ldots, W_{K_n}^*\}$, and feature W_i^* appears exactly $M_{n,i} = m_i$ times* in the samples, such as $i = 1, \ldots, K_n$, then the conditional distribution of Ψ_1, given Z_n, has posterior density:

$$f_{\Psi_1|Z_n}(a) \propto a^{-k\sigma} e^{-\sigma a^{-\sigma} \sum_{i=1}^{n} B(1-\sigma,i)} f_{\Psi_1}(a) \tag{21}$$

having denoted by $B(\,\cdot\,,\,\cdot\,)$ the classical Euler Beta function. Moreover, conditionally on Z_n and Ψ_1,

$$\mathbb{P}((Y_{n+1}, A_{n+1,1}^*, \ldots, A_{n+1,K_n}^*) = (y, a_1, \ldots, a_{K_n})|Z_n, \Psi_1)$$
$$= \mathrm{Poiss}\left(y; \sigma \Psi_1^{-\sigma} B(1-\sigma, n+1)\right) \prod_{i=1}^{k} \mathrm{Bern}\left(a_i; \frac{m_i - \sigma}{n - \sigma + 1}\right). \tag{22}$$

Proof. The proof is a plain application of Theorem 2 under the choice $\lambda(s) = \sigma s^{-1-\sigma}$. □

4. Predictive Characterizations for SPs

In this section, we introduce and discuss Johnson's "sufficientness" postulates in the context of feature-sampling models under the class of SP priors. According to Theorem 1, if the feature-sampling model is a CRM prior, then the conditional distribution of Y_{n+1}, given Z_n, is a Poisson distribution that depends on the sampling information Z_n only through the sample size n. Moreover, the conditional probability of generating an old feature W_i^* given Z_n depends on the sampling information Z_n only through n and m_i. As shown in Theorem 2, SP priors enrich the predictive structure of CRM priors through the conditional distribution of the latent variable Ψ_1 given the observable sample Z_n. In the next theorem, we characterize the class of SP priors for which the conditional distribution of Y_{n+1} given Z_n depends on the sampling information only through n.

Theorem 3. Let $Z_1, Z_2, \ldots$ be exchangeable random variables modeled as in (7), where $\mathcal{M}$ equals $\mathrm{SP}(\nu, f_{\Psi_1})$ and $\nu(dw, ds) = \lambda(ds)ds P(dw)$. Moreover, suppose that Z_n is a random sample that displays $K_n = k$ distinct features $\{W_1^*, \ldots, W_{K_n}^*\}$, and feature W_i^* appears exactly $M_{n,i} = m_i$ times in the samples, such as $i = 1, \ldots, K_n$. If $f_{\Psi_1}: (0,r) \to \mathbb{R}^+$ is a continuous function on the compact support $(0,r)$ with $r > 0$, and the function $\lambda: \mathbb{R}^+ \to \mathbb{R}^+$ is continuous on its domain, then the conditional distribution of the latent variable Ψ_1 given Z_n depends on the sampling information Z_n only through n if and only if $\lambda(s) = Cs^{-1}$ on $(0,r)$ for some constant $C > 0$.

Proof. First of all, if f_{Ψ_1} is defined on the compact support $(0,r)$ and if $\lambda(s) = Cs^{-1}$ on $(0,r)$ for some constant $C > 0$, then it is easy to see that the posterior distribution of Ψ_1 in (18) depends only on n and not on the other sample statistics. We now show the reverse implication. The posterior density of Ψ_1, conditionally on Z_n, satisfies (18), and it is proportional to

$$f_{\Psi_1|Z_n}(a) \propto \prod_{i=1}^{n} e^{-\phi_i(a)} \prod_{i=1}^{K_n} \int_0^1 s^{m_i}(1-s)^{n-m_i} a\lambda(as)ds\, f_{\Psi_1}(a),$$

where $\phi_i(a) = \int_0^1 s(1-s)^{i-1} a\lambda(as)ds$. Then, there exists $c(m_1, \ldots, m_k, k, n)$ such that it holds that

$$f_{\Psi_1|Z_n}(a) = \frac{\prod_{i=1}^{n} e^{-\phi_i(a)} \prod_{i=1}^{K_n} \int_0^1 s^{m_i}(1-s)^{n-m_i} a\lambda(as)ds\, f_{\Psi_1}(a)}{c(m_1, \ldots, m_k, k, n)}. \tag{23}$$

Because of the assumptions imposed, the distribution of $\Psi_1|Z_n$ does not depend on K_n, nor on the corresponding sample frequencies $M_{n,1}, \ldots, M_{n,K_n}$. Accordingly, the function

$$f_1(a, n) := f_{\Psi_1|Z_n}^{-1}(a) \prod_{i=1}^{n} e^{-\phi_i(a)}, \quad a \in (0, r), \tag{24}$$

depends only on a and n, but not on k and $(m_1, \ldots, m_k)$. Then, putting together (23) and (24), it holds that

$$f_1(a,n) \cdot \prod_{i=1}^{k} \int_0^1 s^{m_i}(1-s)^{n-m_i} a\lambda(as) ds = c(m_1, \ldots, m_k, n, k) \quad \forall a \in (0,r), \quad (25)$$

where c is the normalizing factor, and it does not depend on the variable a. By choosing $m_1 = \ldots = m_k = n \in \mathbb{N}$, thanks to Equation (25), we can state that the following function:

$$f_1(a,n) \left(\int_0^1 s^n a\lambda(as) ds \right)^k, \quad (26)$$

which is defined for any $a \in (0,r)$ and does not depend on a, but only on k and n. Since the previous assertion is true for any $k \geq 1$, one may select $k = 1$, thus obtaining the following identity:

$$f_1(a,n) = c^* \left(\int_0^1 s^n a\lambda(as) ds \right)^{-1} \quad (27)$$

for some constant c^*, independent of a, but that may depend on n. Substituting (27) into (26), we obtain that

$$c^* \left(\int_0^1 s^n a\lambda(as) ds \right)^{k-1} \quad (28)$$

is a function that does not depend on a, but only on n and k. As a consequence, we have that

$$\int_0^1 s^n a\lambda(as) ds = \int_0^a \frac{s^n}{a^n} \lambda(s) ds = C^{**}$$

for a suitable constant C^{**}, which does not depend on $a \in (0,r)$. To conclude, we take a derivative of the previous expression with respect to a, and this allows us to show that

$$a^n \lambda(a) = n a^{n-1} C^{**},$$

namely, $\lambda(a) = C/a$ for $a \in (0,r)$, where C is a positive constant. This is a Lévy intensity; indeed, it satisfies the condition (11). Outside the interval $(0,r)$, λ may be defined arbitrarily; indeed, the values of λ on $[r, +\infty)$ do not affect the posterior distribution of Ψ_1 (18). □

Remark 2. *Note that in Theorem 3, we have supposed that f_{Ψ_1} has a compact support on $(0,r)$; thus, we are interested in defining λ on $(0,r)$; outside the interval, λ can be defined arbitrarily because it does not affect the posterior distribution (18) of Ψ_1. From the proof of Theorem 3, it becomes apparent that if the support of f_{Ψ_1} is the entire positive real line $\mathbb{R}^+$, the posterior distribution of the largest jump depends only on n if and only if $\lambda(s) = Cs^{-1}$ on $\mathbb{R}^+$ for some constant $C > 0$. However, in this case, λ does not meet the integrability condition (11); hence, this can only considered a limiting case. It is interesting to observe that such a limiting situation, with the additional assumption $f_{\Psi_1} = f_{\Delta_1}$, corresponds to the Beta process case with $\sigma = 0$ and $c = 1$ (Griffiths and Ghahramani [12]).*

Now, we characterize SPs for which the posterior distribution of Ψ_1 depends only on n and K_n, but not on the sample frequencies of the different features m. Here, we assume that f_{Ψ_1} has full support a priori. The following characterization has been provided in Camerlenghi et al. [15] (Theorem 3), but for completeness, we report the proof.

Theorem 4 (Camerlenghi et al. [15]). *Let $Z_1, Z_2, \ldots$ be exchangeable random variables modeled as in (7), where $\mathcal{M}$ equals $SP(\nu, f_{\Psi_1})$ and $\nu(dw, ds) = \lambda(ds) ds P(dw)$. Suppose that Z_n is a random sample that displays $K_n = k$ distinct features $\{W_1^*, \ldots, W_{K_n}^*\}$, and feature W_i^* appears exactly $M_{n,i} = m_i$ times in the sample, such as $i = 1, \ldots, K_n$. If $f_{\Psi_1} : \mathbb{R}^+ \to \mathbb{R}^+$ is a strictly* positive function on $\mathbb{R}^+$ and continuously differentiable, and λ is continuously differentiable, then the conditional distribution of the latent variable Ψ_1, given $\mathbf{Z}_n$, depends on $\mathbf{Z}_n$ only through n and K_n if and only if $\lambda(s) = Cs^{-1-\sigma}$ on $\mathbb{R}^+$ for some constant $C > 0$ and $\sigma \in (0,1)$.

Proof. By arguing as in the proof of Theorem 3, the posterior density of Ψ_1 given $\mathbf{Z}_n$ is proportional to

$$\prod_{i=1}^n e^{-\phi_i(a)} \prod_{i=1}^k \int_0^1 s^{m_i}(1-s)^{n-m_i} a\lambda(as) ds \, f_{\Psi_1}(a),$$

where $\phi_i(a) = \int_0^1 s(1-s)^{i-1} a\lambda(as) ds$. Then, there exists $c(m_1, \ldots, m_k, n, k)$ such that it holds that

$$f_{\Psi_1 | \mathbf{Z}_n}(a) = \frac{\prod_{i=1}^n e^{-\phi_i(a)} \prod_{i=1}^k \int_0^1 s^{m_i}(1-s)^{n-m_i} a\lambda(as) ds \, f_{\Psi_1}(a)}{c(m_1, \ldots, m_k, n, k)}.$$

As a consequence,

$$f_{\Psi_1 | \mathbf{Z}_n}^{-1}(a) \prod_{i=1}^n e^{-\phi_i(a)} \prod_{i=1}^k \int_0^1 s^{m_i}(1-s)^{n-m_i} a\lambda(as) ds \, f_{\Psi_1}(a) = c(m_1, \ldots, m_k, n, k). \quad (29)$$

If the density function $f_{\Psi_1|\mathbf{Z}_n}(a)$ does not depend on $m_1, \ldots, m_k$, then the following function

$$f_{\Psi_1 | \mathbf{Z}_n}^{-1}(a) \prod_{i=1}^n e^{-\phi_i(a)} f_{\Psi_1}(a) = f_1(a, k, n)$$

depends only on k, n and a, but not on the frequency counts. Therefore, (29) boils down to

$$f_1(a, k, n) \cdot \prod_{i=1}^k \int_0^1 s^{m_i}(1-s)^{n-m_i} a\lambda(as) ds = c(m_1, \ldots, m_k, n, k). \quad (30)$$

where the function on the right-hand side of (30) is independent of a for any choice of the vector of sampling information $(m_1, \ldots, m_k, n, k)$. Now, since the vector $(m_1, \ldots, m_k, n, k)$ can be chosen arbitrarily, we can make the choice $m_1 = \cdots = m_k = m > 0$, such that the function

$$\left[w(a, k, n) \int_0^1 s^m (1-s)^{n-m} a\lambda(as) ds \right]^k \quad (31)$$

does not depend on $a \in \mathbb{R}^+$, where $w(a, k, n) = \sqrt[k]{f_1(a, k, n)}$. Moreover, suppose that $m = n$; thus,

$$w(a, k, n) \int_0^1 s^n a\lambda(as) ds \quad (32)$$

does not depend on $a \in \mathbb{R}^+$, which implies that

$$w(a, k, n) = c^* \left(\int_0^1 s^n a\lambda(as) ds \right)^{-1} \quad (33)$$

for a constant $c^* > 0$ with respect to a, which can only depend on k and n. By substituting (33) into (31), we obtain

$$\left[\frac{c^*}{\int_0^1 s^n \lambda(as) ds} \cdot \int_0^1 s^m (1-s)^{n-m} \lambda(as) ds \right]^k,$$

which is independent of $a \in \mathbb{R}^+$. Now, it is possible to choose $m = n - 1$ in the previous function. Therefore, there exists a constant c^{**} independent of a such that the following identity holds:

$$\int_0^1 s^{n-1}\lambda(as)ds - \int_0^1 s^n \lambda(as)ds = c^{**}\int_0^1 s^n \lambda(as)ds.$$

By taking the derivative of the previous equation two times with respect to a, one obtains

$$\lambda(a)(1 - nc^{**}) = a\lambda'(a)c^{**},$$

which is an ordinary differential equation in λ that can be solved by separation of variables. In particular, we obtain

$$\lambda(a) = Ca^{(1-nc^{**})/c^{**}}, \quad \text{for } C > 0. \tag{34}$$

To conclude, observe that the exponent of a in (34) should satisfy the integrability condition (11) for homogeneous CRMs. Accordingly, it is easy to see that we must consider

$$\lambda(a) = C\frac{1}{a^{1+\sigma}}$$

where $C > 0$ and $\sigma \in (0, 1)$. The reverse implication of the theorem is trivially satisfied; hence, the proof is completed. □

We recall from Theorem 2 that the conditional distribution of Ψ_1 given Z_n uniquely determines the amount of sampling information included in the conditional distribution of the number of new features Y_{n+1} given Z_n. Such sampling information may range from the whole information, in terms of n, K_n, and $(M_{1,n}, \ldots, M_{K_n,n})$, to the sole information on the sample size n. According to Theorem 4, the stable SP prior of Corollary 2 is the sole SP prior for which the conditional distribution of the number of new features Y_{n+1} given Z_n depends on the sampling information Z_n only on n and K_n. Moreover, according to Theorem 3, the Beta process prior is the sole SP prior for which the conditional distribution of the number of new features Y_{n+1} given Z_n depends on the sampling information Z_n only on n. In particular, Theorems 3 and 4 show that the Beta process prior and the stable SP prior may be considered, to some extent, the feature sampling counterparts of the Dirichlet process prior the Pitman–Yor process prior.

5. Discussion and Conclusions

In this paper, we have introduced and discussed Johnson's "sufficientness" postulates in the context of feature-sampling models. "Sufficientness" postulates have been investigated extensively in the context of species-sampling models, providing an effective classification of species-sampling models on the basis of the form of their corresponding predictive distributions. Here, we made a first step towards the problem of providing an analogous classification for feature-sampling models. In particular, we obtained Johnson's "sufficientness" postulates when the class of feature-sampling models is restricted to the class of scaled process priors. However, the results presented in the paper remain preliminary, and do not at all provide a complete answer to the characterization problem within the general class of feature-sampling models. This problem remains open.

Within the feature-sampling setting, the predictive distribution is of the form (8), though for the purpose of providing "sufficientness" postulates, one may focus on feature-sampling models exhibiting a general predictive distribution of the following type:

$$\mathbb{P}((Y_{n+1}, A^*_{n+1,1}, \ldots, A^*_{n+1,K_n}) = (y, a_1, \ldots, a_{K_n})|Z_n)$$
$$= g(y; n, k, \boldsymbol{m}) \prod_{i=1}^{k} f_i(a_i; n, k, \boldsymbol{m}). \tag{35}$$

Note that (35) is a probability distribution, and it must satisfy a consistency condition, as usual. Among all the feature-sampling models whose predictive distribution can be written in the form (35), we are interested in characterizing nonparametric priors such that: (i) The function g depends on the sampling information only through n, and the function f_i depends only on (n, m_i); (ii) g depends only on (n, k) and f_i depends only on (n, m_i); (iii) g depends only on (n, k) and f_i depends only on (n, k, m_i). In our view, these characterizations may provide a complete picture of sufficientness postulates within the feature setting, and they are also fundamental to guiding the selection of the prior distribution. We conjecture that CRMs are the nonparametric priors satisfying the characterization (i), the SP with a stable Lévy measure is an example of prior satisfying (ii), and no examples satisfying (iii) have been considered in the current literature. Results in this direction are in Battiston et al. [31], where the authors characterize an exchangeable feature allocation probability function (Broderick et al. [32]) in product forms; this could be a stimulating point of departure to study the characterization problem depicted above.

Author Contributions: Writing–original draft, F.C. and S.F.; writing–review and editing, F.C. and S.F. The authors contributed equally to this work. All authors have read and agreed to the published version of the manuscript.

Funding: This research received funding from the European Research Council (ERC) under the European Union's Horizon 2020 research and innovation program under grant agreement No. 817257.

Institutional Review Board Statement: Not applicable.

Informed Consent Statement: Not applicable.

Data Availability Statement: Not applicable.

Acknowledgments: F.C. is extremely grateful to Eugenio Regazzini for the time spent at the Department of Mathematics of University of Pavia during his Ph.D. studies in Mathematical Statistics; F.C. wants to especially thank Eugenio Regazzini for having introduced him to the study of Bayesian Statistics with a stimulating Ph.D. course held together with Antonio Lijoi. S.F. wishes to express his gratitude to Eugenio Regazzini, whose fundamental contributions to Bayesian statistics have always been a great source of inspiration, transmitting enthusiasm and methods for the development of his own research. The authors gratefully acknowledge the financial support from the Italian Ministry of Education, University, and Research (MIUR), "Dipartimenti di Eccellenza" grant 2018-2022. F.C. is a member of the *Gruppo Nazionale per l'Analisi Matematica, la Probabilità e le loro Applicazioni* (GNAMPA) of the *Istituto Nazionale di Alta Matematica* (INdAM).

Conflicts of Interest: The authors declare no conflict of interest.

References

1. de Finetti, B. La prévision: Ses lois logiques, ses sources subjectives. *Ann. Inst. H. Poincaré* **1937**, *7*, 1–68.
2. Johnson, W.E. Probability: The Deductive and Inductive Problems. *Mind* **1932**, *41*, 409–423. [CrossRef]
3. Pitman, J. Some developments of the Blackwell-MacQueen urn scheme. In *Statistics, Probability and Game Theory*; IMS Lecture Notes Monograph Series; Institute of Mathematical Statistics: Hayward, CA, USA, 1996; Volume 30, pp. 245–267. [CrossRef]
4. Regazzini, E. Intorno ad alcune questioni relative alla definizione del premio secondo la teoria della credibilità. *Giornale dell'Istituto Italiano degli Attuari* **1978**, *41*, 77–89.
5. Lo, A.Y. A characterization of the Dirichlet process. *Stat. Probab. Lett.* **1991**, *12*, 185–187. [CrossRef]
6. Ferguson, T.S. A Bayesian analysis of some nonparametric problems. *Ann. Statist.* **1973**, *1*, 209–230. [CrossRef]
7. Zabell, S.L. Symmetry and its discontents. In *Cambridge Studies in Probability, Induction, and Decision Theory*; Essays on the history of inductive probability, with a preface by Brian Skyrms; Cambridge University Press: New York, NY, USA, 2005; p. xii+279.
8. Perman, M.; Pitman, J.; Yor, M. Size-biased sampling of Poisson point processes and excursions. *Probab. Theory Relat. Fields* **1992**, *92*, 21–39. [CrossRef]
9. Pitman, J.; Yor, M. The two-parameter Poisson-Dirichlet distribution derived from a stable subordinator. *Ann. Probab.* **1997**, *25*, 855–900. [CrossRef]
10. Bacallado, S.; Battiston, M.; Favaro, S.; Trippa, L. Sufficientness postulates for Gibbs-type priors and hierarchical generalizations. *Stat. Sci.* **2017**, *32*, 487–500. [CrossRef]
11. Gnedin, A.; Pitman, J. Exchangeable Gibbs partitions and Stirling triangles. *Zap. Nauchn. Sem. S.-Peterburg. Otdel. Mat. Inst. Steklov. (POMI)* **2005**, *325*, 83–102. 244–245. [CrossRef]

12. Griffiths, T.L.; Ghahramani, Z. The Indian buffet process: An introduction and review. *J. Mach. Learn. Res.* **2011**, *12*, 1185–1224.
13. Ayed, F.; Battiston, M.; Camerlenghi, F.; Favaro, S. Consistent estimation of small masses in feature sampling. *J. Mach. Learn. Res.* **2021**, *22*, 1–28.
14. James, L.F.; Orbanz, P.; Teh, Y.W. Scaled subordinators and generalizations of the Indian buffet process. *arXiv* **2015**, arXiv:1510.07309.
15. Camerlenghi, F.; Favaro, S.; Masoero, L.; Broderick, T. Scaled process priors for Bayesian nonparametric estimation of the unseen genetic variation. *arXiv* **2021**, arXiv:2106.15480.
16. Brix, A. Generalized gamma measures and shot-noise Cox processes. *Adv. Appl. Probab.* **1999**, *31*, 929–953. [CrossRef]
17. Lijoi, A.; Mena, R.H.; Prünster, I. Controlling the reinforcement in Bayesian non-parametric mixture models. *J. R. Stat. Soc. Ser. B Stat. Methodol.* **2007**, *69*, 715–740. [CrossRef]
18. De Blasi, P.; Favaro, S.; Lijoi, A.; Mena, R.H.; Prunster, I.; Ruggiero, M. Are Gibbs-type priors the most natural generalization of the Dirichlet process? *IEEE Trans. Pattern Anal. Mach. Intell.* **2015**, *37*, 212–229. [CrossRef]
19. Regazzini, E.; Lijoi, A.; Prünster, I. Distributional results for means of normalized random measures with independent increments. *Ann. Stat.* **2003**, *31*, 560–585. [CrossRef]
20. Pitman, J. *Poisson-Kingman Partitions*; Lecture Notes-Monograph Series; Institute of Mathematical Statistics: Beachwood, OH, USA, 2003; pp. 1–34.
21. Pitman, J. *Combinatorial Stochastic Processes*; Lecture Notes in Mathematics; Lectures from the 32nd Summer School on Probability Theory held in Saint-Flour, 7–24 July 2002, with a foreword by Jean Picard; Springer: Berlin, Germany, 2006; Volume 1875, p. x+256.
22. Lijoi, A.; Prünster, I. Models beyond the Dirichlet process. In *Bayesian Nonparametrics*; Hjort, N.L., Holmes, C., Müller, P., Walker, S., Eds.; Cambridge University Press: Cambridge, UK, 2010; pp. 80–136.
23. Ghosal, S.; van der Vaart, A. *Fundamentals of Nonparametric Bayesian Inference*; Cambridge Series in Statistical and Probabilistic Mathematics; Cambridge University Press: Cambridge, UK, 2017; Volume 44, p. xxiv+646.
24. Zabell, S.L. The continuum of inductive methods revisited. In *The Cosmos of Science: Essays of Exploration*; University of Pittsburgh Press: Pittsburgh, PA, USA, 1997; pp. 351–385.
25. Daley, D.J.; Vere-Jones, D. *An Introduction to the Theory of Point Processes: Volume II: General Theory and Structure (Probability and Its Applications)*, 2nd ed.; Springer: New York, NY, USA, 2008; p. xviii+573.
26. Kingman, J. Completely random measures. *Pac. J. Math.* **1967**, *21*, 59–78. [CrossRef]
27. Teh, Y.; Gorur, D. Indian buffet processes with power-law behavior. *Adv. Neural Inf. Process. Syst.* **2009**, *22*, 1838–1846.
28. James, L.F. Bayesian Poisson calculus for latent feature modeling via generalized Indian buffet process priors. *Ann. Stat.* **2017**, *45*, 2016–2045. [CrossRef]
29. Broderick, T.; Wilson, A.C.; Jordan, M.I. Posteriors, conjugacy, and exponential families for completely random measures. *Bernoulli* **2018**, *24*, 3181–3221. [CrossRef]
30. Masoero, L.; Camerlenghi, F.; Favaro, S.; Broderick, T. More for less: Predicting and maximizing genomic variant discovery via Bayesian nonparametrics. *Biometrika* **2021**, asab012, [CrossRef]
31. Battiston, M.; Favaro, S.; Roy, D.M.; Teh, Y.W. A characterization of product-form exchangeable feature probability functions. *Ann. Appl. Probab.* **2018**, *28*, 1423–1448. [CrossRef]
32. Broderick, T.; Pitman, J.; Jordan, M.I. Feature allocations, probability functions, and paintboxes. *Bayesian Anal.* **2013**, *8*, 801–836. [CrossRef]

Article

The Rescaled Pólya Urn and the Wright—Fisher Process with Mutation

Giacomo Aletti [1,†] **and Irene Crimaldi** [2,*,†]

1. Environmental Science and Policy Department, Università degli Studi di Milano, 20133 Milan, Italy; giacomo.aletti@unimi.it
2. IMT School for Advanced Studies Lucca, 55100 Lucca, Italy
* Correspondence: irene.crimaldi@imtlucca.it
† These authors contributed equally to this work.

Abstract: In recent papers the authors introduce, study and apply a variant of the Eggenberger—Pólya urn, called the "rescaled" Pólya urn, which, for a suitable choice of the model parameters, exhibits a reinforcement mechanism mainly based on the last observations, a random persistent fluctuation of the predictive mean and the almost sure convergence of the empirical mean to a deterministic limit. In this work, motivated by some empirical evidence, we show that the multidimensional Wright—Fisher diffusion with mutation can be obtained as a suitable limit of the predictive means associated to a family of rescaled Pólya urns.

Keywords: Pólya urn; predictive mean; urn model; Wright—Fisher diffusion

1. Introduction

The well-known standard Eggenberger—Pólya urn [1,2] works as follows. An urn initially contains $N_{0,i}$ balls of color i, for $i = 1, \ldots, k$, and at each time-step, a ball is drawn from the urn and then it is returned into the urn together with $\alpha > 0$ additional balls of the same color (here and in the following, the expression "number of balls" is not to be understood literally, but all the quantities are real numbers, not necessarily integers). Hence, denoting by $N_{n,i}$ the number of balls of color i inside the urn at time-step n, we have

$$N_{n,i} = N_{n-1,i} + \alpha \xi_{n,i} \qquad \text{for } n \geq 1,$$

where $\xi_{n,i} = 1$ if the drawn ball at time-step n is of color i, and $\xi_{n,i} = 0$ otherwise. The parameter α tunes the reinforcement mechanism: the greater the α, the greater the dependence of $N_{n,i}$ on $\sum_{h=1}^{n} \xi_{h,i}$.

In [3–5], the rescaled Pólya (RP) urn has been introduced, studied, generalized and applied. This model differs from the original one by the introduction of a parameter β such that

$$N_{n,i} = b_i + B_{n,i} \qquad \text{with}$$
$$B_{n+1,i} = \beta B_{n,i} + \alpha \xi_{n+1,i} \qquad n \geq 0.$$

Therefore, at time-step 0, the urn contains $b_i + B_{0,i} > 0$ balls of color i and the parameters $\alpha > 0$ and $\beta \geq 0$ regulate the reinforcement mechanism. More precisely, the term $\beta B_{n,i}$ connects $N_{n+1,i}$ to the "configuration" at time-step n by means of the "scaling" parameter β, and the term $\alpha \xi_{n+1,i}$ connects $N_{n+1,i}$ to the outcome of the drawing at time-step $n + 1$ by means of the parameter α. The case $\beta = 1$ corresponds to the standard Eggenberger—Pólya urn with an initial number $N_{0,i} = b_i + B_{0,i}$ of balls of color i. When $\beta < 1$, the RP urn model shows the following three characteristics:

(i) A reinforcement mechanism mainly based on the last observations;
(ii) A random persistent fluctuation of the predictive mean $\psi_{n,i} = E[\xi_{n+1,i} = 1 | \xi_{h,j}, 0 \leq h \leq n, 1 \leq j \leq k]$;

Citation: Aletti, G.; Crimaldi, I. The Rescaled Pólya Urn and the Wright—Fisher Process with Mutation. *Mathematics* **2021**, *9*, 2909. https://doi.org/10.3390/math9222909

Academic Editors: Emanuele Dolera and Federico Bassetti

Received: 8 October 2021
Accepted: 13 November 2021
Published: 15 November 2021

Publisher's Note: MDPI stays neutral with regard to jurisdictional claims in published maps and institutional affiliations.

Copyright: © 2021 by the authors. Licensee MDPI, Basel, Switzerland. This article is an open access article distributed under the terms and conditions of the Creative Commons Attribution (CC BY) license (https://creativecommons.org/licenses/by/4.0/).

(iii) The almost sure convergence of the empirical mean $\sum_{n=1}^{N} \xi_{n,i}/N$ to the deterministic limit $p_i = b_i / \sum_{i=1}^{n} b_i$, and a chi-squared goodness of fit result for the long-term probability distribution $\{p_1, \ldots, p_k\}$.

Regarding point (iii), we specifically have that the chi-squared statistics

$$\chi^2 = N \sum_{i=1}^{k} \frac{(O_i/N - p_i)^2}{p_i},$$

where N is the sample size and $O_i = \sum_{n=1}^{N} \xi_{n,i}$ the number of sampled observations equal to i, is asymptotically distributed as $\chi^2(k-1)\lambda$, with $\lambda > 1$. Therefore, the presence of correlation among observations attenuates the effect of N, which multiplies the chi-squared distance between the observed frequencies and the expected probabilities. This is a key feature for statistical applications in the framework of a "big sample", where a small value of the chi-squared distance might be significant, and hence a correction related to the correlation between observations is required. In [3,5], a possible application in the context of clustered data was described, with independence between clusters and correlation due to a reinforcement mechanism inside each cluster.

In [4], the RP urn was applied as a good model for the evolution of the sentiment associated with Twitter posts. Precisely, we analyzed three data sets: (i) the "COVID-19 epidemic" data set covers the period from 21 February to 20 April to 2020 and includes tweets in Italian about the COVID-19 epidemic; (ii) the "Migration debate" data set refers to the period from 23 January to 22 February 2019 and the collected posts are related to the Italian debate on migration; (iii) the "10 days of traffic" data set collects the entire traffic of posts in Italian in the period from 1 September to 10 September 2019. For every post, the relative sentiment, that is, the positive or negative connotation of the text, was computed using the polyglot python module developed in [6], which provides a numerical value $v \in [-1, 1]$ for the sentiment of a post (for a survey on sentiment analysis, also known as opinion mining, we refer to [7] and references therein). We fixed a threshold T so that a tweet with $v > T$ was classified as a tweet with a positive sentiment and one with $v < -T$ was classified as a tweet with a negative sentiment. Tweets with a value $v \in [-T, T]$ were discarded. We took the following different values for T: $T = 0$, $T = 0.35$ and $T = 0.5$. We applied the RP urn model, ordering the tweets according to their creation time and taking each tweet with a positive/negative classification as an extraction in the urn model. More specifically, we applied the RP model with $k = 2$: the time series of the tweets represents the time series of the extractions from the urn, that is, the random variables $\xi_{n,1}$. The event $\{\xi_{n,1} = 1\}$ means that tweet n exhibits a positive sentiment, while $\{\xi_{n,1} = 0\}$ means that tweet n exhibits a negative sentiment. For all the considered data sets, the estimated values of β were strictly smaller than 1, but very near to 1 (details about the parameters estimation can be found in [4]). Note that the RP urn dynamics with such a value for β cannot be approximated by the standard Pólya urn ($\beta = 1$), because one would lose the fluctuations of the predictive means and the possibility of touching the barriers $\{0, 1\}$. In this work, we show that the law of such an RP urn process can be approximated by a Wright—Fisher diffusion with mutation. More precisely, we prove that the multidimensional Wright—Fisher diffusion with mutation can be obtained as a suitable limit of the predictive means associated with a family of RP urns with $\beta \in [0, 1)$, $\beta \to 1$. As an example, in Figure 1, for the data set "COVID-19 epidemic", we show the plot of the process $(\psi_{n,1})_n$, reconstructed from the data (details about the reconstruction process can be found in [4]) and rescaled in time as $t = n(1 - \beta)^2$, the plot of a simulated (by the Euler–Maruyama method) trajectory of the Wright—Fisher process, the plot of the approximation of this trajectory by means of the RP urn and the approximation of the data process by means of the standard Pólya urn.

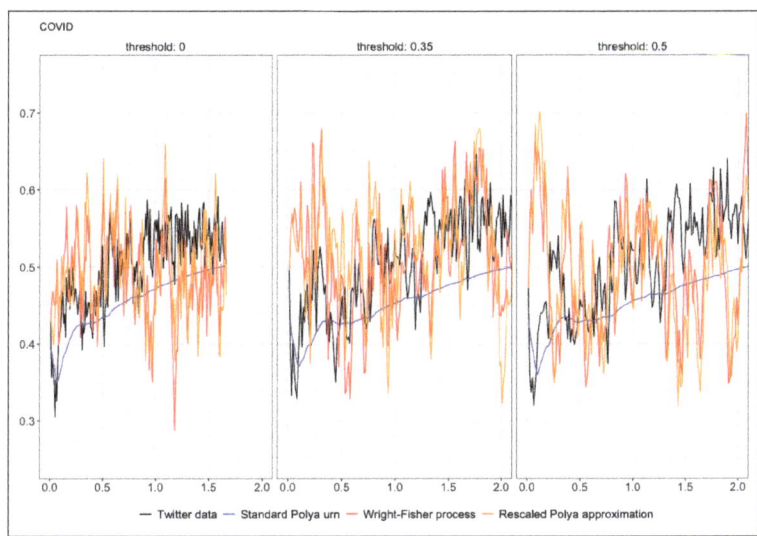

Figure 1. "COVID-19 epidemic" Twitter data set: the black line is the process $(\psi_{n,1})_n$, reconstructed from the data and rescaled in time as $t = n(1 - \beta)^2$; the red line is a simulated trajectory of the Wright—Fisher process; the orange line is the approximation of this trajectory by means of the RP urn and the blue line is the approximation of the data process by means of the standard Pólya urn. The numbers 0, 0.35 and 0.5 refer to the values chosen for the threshold T. The corresponding estimated values for $1 - \beta$ are: 0.000776 (8×10^{-4}), 0.00115 (11×10^{-4}) and 0.00130 (13×10^{-4}).

The Wright–Fisher (WF) class of diffusion processes models the evolution of the relative frequency of a genetic variant, or allele, in a large randomly mating population with a finite number k of genetic variants. When $k = 2$, the WF diffusion obeys the one-dimensional stochastic differential equation

$$dX_t = F(X_t)dt + \sqrt{X_t(1 - X_t)}dW_t, \qquad X_0 = x_0, t \in [0, T]. \qquad (1)$$

The drift coefficient, $F : [0,1] \to R$, can include a variety of evolutionary forces such as mutation and selection. For example, $F(x) = p_1 - (p_1 + p_2)x = p_1(1 - x) - p_2 x$ describes a process with recurrent mutation between the two alleles, governed by the mutation rates $p_1 > 0$ and $p_2 > 0$. The drift vanishes when $x = p_1/(p_1 + p_2)$ which is an attracting point for the dynamics. Equation (1) can be generalized to the case $k > 2$. The WF diffusion processes are widely employed in Bayesian statistics, as models for time-evolving priors [8–11] and as a discrete-time finite-population construction method of the two-parameter Poisson–Dirichlet diffusion [12]. They have been applied in genetics [13–18], in biophysics [19,20], in filtering theory [21,22] and in finance [23,24].

The benefit coming from the proven limit result is twofold. First, the known properties of the WF process can give a description of the RP urn when the parameter β is strictly smaller than one, but very near to one. Second, the given result might furnish the theoretical base for a new simulation method of the WF process. Indeed, the simulation from Equation (1) is highly nontrivial because there is no known closed form expression for the transition function of the diffusion, even in the simple case with null drift [25].

The rest of the paper is organized as follows. In Section 2, we set up our notation and we formally define the RP urn model. Section 3 provides the main result of this work, that is, the convergence result of a suitable family of predictive means associated with RP urns with $\beta \to 1$. In Section 4, employing the boundary classification of the WF diffusion with mutation and connecting it to the parameters of the RP urn model, we introduce an RP urn with a value of β very near to 1 the notion of recessive subsets of colors and the notion of dominant color. These two concepts are related to the possibility of reaching the barriers 0 and 1 by the predictive means of the urn process. Finally, Section 5 summarizes the work and concludes it.

2. The Rescaled Pólya Urn

For a vector $x = (x_1, \ldots, x_k)^\top \in \mathbb{R}^k$, we set $|x| = \sum_{i=1}^k |x_i|$ and $\|x\|^2 = x^\top x = \sum_{i=1}^k |x_i|^2$. Moreover we denote by $\mathbf{1}$ and $\mathbf{0}$ the vectors with all the components equal to 1 and equal to 0, respectively.

Let $\alpha > 0$ and $\beta \geq 0$. At time-step 0, the urn contains $b_i + B_{0,i} > 0$ distinct balls of color i, with $i = 1, \ldots, k$. We set $\boldsymbol{b} = (b_1, \ldots, b_k)^\top$ and $\boldsymbol{B_0} = (B_{0,1}, \ldots, B_{0,k})^\top$. We suppose $b = |\boldsymbol{b}| > 0$ and we set $\boldsymbol{p} = \frac{\boldsymbol{b}}{b}$. At each time-step $(n+1) \geq 1$, a ball is drawn at random from the urn and we define the random vector $\boldsymbol{\xi_{n+1}} = (\xi_{n+1,1}, \ldots, \xi_{n+1,k})^\top$ as

$$\xi_{n+1,i} = \begin{cases} 1 & \text{when the drawn ball at time-step } n+1 \text{ is of color } i \\ 0 & \text{otherwise.} \end{cases}$$

The number of balls inside the urn is updated as follows:

$$\boldsymbol{N_{n+1}} = \boldsymbol{b} + \boldsymbol{B_{n+1}} \quad \text{with} \quad \boldsymbol{B_{n+1}} = \beta \boldsymbol{B_n} + \alpha \boldsymbol{\xi_{n+1}}, \tag{2}$$

which gives

$$\boldsymbol{B_n} = \beta^n \boldsymbol{B_0} + \alpha \beta^n \sum_{h=1}^n \beta^{-h} \boldsymbol{\xi_h}. \tag{3}$$

Similarly, from the equality

$$|\boldsymbol{B_{n+1}}| = \beta |\boldsymbol{B_n}| + \alpha,$$

we get, using $\sum_{h=0}^{n-1} x^h = (1 - x^n)/(1 - x)$,

$$|\boldsymbol{B_n}| = \beta^n |\boldsymbol{B_0}| + \alpha \sum_{h=1}^n \beta^{n-h} = \beta^n \left(|\boldsymbol{B_0}| - \frac{\alpha}{1-\beta} \right) + \frac{\alpha}{1-\beta}. \tag{4}$$

Setting $r_n^* = |\boldsymbol{N_n}| = b + |\boldsymbol{B_n}|$, that is the total number of balls inside the urn at time-step n, we get the relations

$$r_{n+1}^* = r_n^* + (\beta - 1)|\boldsymbol{B_n}| + \alpha \tag{5}$$

and

$$r_n^* = b + \frac{\alpha}{1-\beta} + \beta^n \left(|\boldsymbol{B_0}| - \frac{\alpha}{1-\beta} \right). \tag{6}$$

Denoting by $\mathcal{F}_0$ the trivial σ-field and setting $\mathcal{F}_n = \sigma(\boldsymbol{\xi_1}, \ldots, \boldsymbol{\xi_n})$ for $n \geq 1$, the conditional probabilities $\boldsymbol{\psi_n} = (\psi_{n,1}, \ldots, \psi_{n,k})^\top$ of the extraction process, also called predictive means, are

$$\boldsymbol{\psi_n} = E[\boldsymbol{\xi_{n+1}} | \mathcal{F}_n] = \frac{\boldsymbol{N_n}}{|\boldsymbol{N_n}|} = \frac{\boldsymbol{b} + \boldsymbol{B_n}}{r_n^*} \quad n \geq 0 \tag{7}$$

and, from (3) and (4), we have

$$\boldsymbol{\psi_n} = \frac{\boldsymbol{b} + \beta^n \boldsymbol{B_0} + \alpha \sum_{h=1}^n \beta^{n-h} \boldsymbol{\xi_h}}{b + \frac{\alpha}{1-\beta} + \beta^n \left(|\boldsymbol{B_0}| - \frac{\alpha}{1-\beta} \right)}. \tag{8}$$

The dependence of $\boldsymbol{\psi_n}$ on $\boldsymbol{\xi_h}$ is regulated by the factor $f(h,n) = \alpha \beta^{n-h}$, with $1 \leq h \leq n$, $n \geq 0$. In the case of the standard Eggenberger—Pólya urn (i.e., the case $\beta = 1$), each observation ξ_h has the same "weight" $f(h,n) = \alpha$. Instead, when $\beta < 1$ the factor $f(h,n)$ increases with h, and the main contribution is given by the most recent drawings. The case $\beta = 0$ is an extreme case, for which ψ_n depends only on the last drawing ξ_n.

By means of (7), together with (2) and (5), we get

$$\psi_{n+1} - \psi_n = -\frac{(1-\beta)}{r^*_{n+1}} b(\psi_n - p) + \frac{\alpha}{r^*_{n+1}}(\xi_{n+1} - \psi_n). \tag{9}$$

Setting $\Delta M_{n+1} = \xi_{n+1} - \psi_n$ and letting $\epsilon_n = b(1-\beta)/r^*_{n+1}$ and $\delta_n = \alpha/r^*_{n+1}$, from (9) we obtain

$$\psi_{n+1} - \psi_n = -\epsilon_n(\psi_n - p) + \delta_n \Delta M_{n+1}. \tag{10}$$

3. Main Result

Consider the RP urn with parameters $\alpha > 0$, $\beta \in [0,1)$, $b > 0$ and B_0 such that $|B_0| = r(\beta) = \alpha/(1-\beta)$. Consequently, the total number of balls in the urn along the time-steps is constantly equal to $r^*(\beta) = b + r(\beta)$ and if we denote by $\psi^{(\beta)} = (\psi_n^{(\beta)})_n$ the predictive means corresponding to the fixed value β, we have the dynamics

$$\psi_n^{(\beta)} - \psi_{n-1}^{(\beta)} = -\epsilon(\beta)(\psi_{n-1}^{(\beta)} - p) + \delta(\beta)\Delta M_n^{(\beta)}, \tag{11}$$

where

$$\epsilon(\beta) = \frac{b(1-\beta)^2}{\alpha + b(1-\beta)}, \qquad \delta(\beta) = \frac{\alpha(1-\beta)}{\alpha + b(1-\beta)} \tag{12}$$

and $\Delta M_n^{(\beta)} = \xi_n^{(\beta)} - \psi_{n-1}^{(\beta)}$. Note that we have $\epsilon(\beta) \sim c\delta(\beta)^2$ for $\beta \to 1$, with $c = b/\alpha > 0$. Finally, we define $X^{(\beta)} = (X_t^{(\beta)})_{t \geq 0}$, where

$$X_t^{(\beta)} = \psi_{\lfloor t/(1-\beta)^2 \rfloor}^{(\beta)} \iff X_t^{(\beta)} = \psi_{n-1}^{(\beta)}, \, t \in [(n-1)(1-\beta)^2, n(1-\beta)^2). \tag{13}$$

The following result holds true:

Theorem 1. *Suppose that $X_0^{(\beta)}$ weakly converges towards some process X_0 when $\beta \to 1$. Then, for $\beta \to 1$, the family of stochastic processes $\{X^{(\beta)}, \beta \in [0,1)\}$ weakly converges towards the k-alleles Wright–Fisher diffusion $X = (X_t)_{t \geq 0}$, with type-independent mutation kernel given by p and with dynamics*

$$dX_t = -b\frac{X_t - p}{\alpha}dt + \Sigma(X_t)dW_t, \tag{14}$$

with $\Sigma(X_t)\Sigma(X_t)^\top = \left(\mathrm{diag}(X_t) - X_t X_t^\top\right)$ and $\mathbf{1}^\top \Sigma(X_t) = \mathbf{0}^\top$, that is,

$$\Sigma(X_t)_{ij} = \begin{cases} 0 & \text{if } X_{t,i}X_{t,j} = 0 \text{ or } i < j \\ \sqrt{X_{t,i}\frac{\sum_{l=i+1}^k X_{t,l}}{\sum_{l=i}^k X_{t,l}}} & \text{if } i = j \text{ and } X_{t,i}X_{t,j} \neq 0 \\ -X_{t,i}\sqrt{\frac{X_{t,j}}{\sum_{l=j}^k X_{t,l}\sum_{l=j+1}^k X_{t,l}}} & \text{if } i > j \text{ and } X_{t,i}X_{t,j} \neq 0. \end{cases} \tag{15}$$

Proof. Fix a sequence (β_n), with $\beta_n \in [0,1)$ and $\beta_n \to 1$. The sequence of processes $\{X^{(\beta_n)}, n \in \mathbb{N}\}$ is bounded, hence we have to prove the tightness of the sequence in the space $D^k[0,\infty)$ of right-continuous functions with the usual Skorohod topology, and the characterization of the law of the unique limit process.

For any $f \in C_b^2$, define

$$\gamma_n^{(\beta,f)}(x) = \widehat{A}^{(\beta)} f((n-1)(1-\beta)^2)(x)$$

$$= E\left[\frac{f(X_{n(1-\beta)^2}^{(\beta)}) - f(X_{(n-1)(1-\beta)^2}^{(\beta)})}{(1-\beta)^2}\Big| X_{(n-1)(1-\beta)^2}^{(\beta)} = x\right]$$

$$= E\left[\frac{f(\psi_n^{(\beta)}) - f(\psi_{n-1}^{(\beta)})}{(1-\beta)^2}\Big| \psi_{n-1}^{(\beta)} = x\right]$$

by $\psi_n^{(\beta)} - \psi_{n-1}^{(\beta)} = -\epsilon(\beta)(\psi_{n-1}^{(\beta)} - p) + \delta(\beta)\Delta M_n^{(\beta)}$

$$\frac{1}{(1-\beta)^2}\left(E\left[f(x) + \sum_i \frac{\partial f}{\partial x_i}(x)(-\epsilon(\beta)(x_i - p_i) + \delta(\beta)\Delta M_{n,i}^{(\beta)})\right.\right. \tag{16}$$

$$\left.\left. + \tfrac{1}{2}\delta(\beta)^2 \sum_{ij} \frac{\partial^2 f}{\partial x_i \partial x_j}(x)\Delta M_{n,i}^{(\beta)}\Delta M_{n,j}^{(\beta)} + O((1-\beta)^3)\Big|\mathcal{F}_{n-1}\right] - f(x)\right)$$

$$= -\frac{b}{\alpha + b(1-\beta)} \sum_i \frac{\partial f}{\partial x_i}(x)(x_i - p_i) + \frac{1}{2} \frac{a^2}{(\alpha + b(1-\beta))^2} \sum_{ij} \frac{\partial^2 f}{\partial x_i \partial x_j}(x)(x_i \mathbb{1}_{i=j} - x_i x_j)$$

$$+ O(1-\beta).$$

We note that, for any $f \in C_b^2$, the partial derivatives in (16) are uniformly bounded, as x belongs to the compact simplex $S = \{x_i \geq 0, \sum_i x_i = 1\}$. The family $\{\gamma_n^{(\beta,f)}(x), n \in \mathbb{N}, \beta < 1, x \in S\}$ is then uniformly integrable. Thus, as a consequence of [26] (Theorem 4) (or [27] (ch. 7.4.3, Theorem 4.3, p. 236)), we have that the sequence of processes $\{X^{(\beta_n)}, n \in \mathbb{N}\}$ is tight in the space of right-continuous functions with the usual Skorohod topology. Since, for any n and t, $X_t^{(\beta_n)} \in S$, then $\mathbf{1}^\top \Sigma(X_t) = \mathbf{0}^\top$. Moreover, the generator of the limit process is determined by the limit

$$Af(t)(x) = \lim_{n \to \infty} \gamma_{\lfloor t/(1-\beta)^2 \rfloor}^{(\beta_n, f)}(x)$$

$$= -\frac{b}{\alpha} \sum_i \frac{\partial f}{\partial x_i}(x)(x_i - p_i) + \frac{1}{2} \sum_{ij} \frac{\partial^2 f}{\partial x_i \partial x_j}(x)(x_i \mathbb{1}_{i=j} - x_i x_j).$$

Hence, the weak limit of the sequence of the bounded processes $X^{(\beta_n)}$ is the diffusion process

$$dX_t = -b\frac{X_t - p}{\alpha}dt + \Sigma(X_t)dW_t, \qquad \Sigma(X_t)\Sigma(X_t)^\top = \left(\text{diag}(X_t) - X_t X_t^\top\right).$$

The expression (15) follows from [28] (Corollary 3). □

Remark 1 (Limiting ergodic distribution). *Since the simplex has dimension $k-1$ with respect to the Lebesgue measure, it is convenient to change the notations. Let T^{k-1} be the $k-1$-dimensional simplex defined by*

$$T^{k-1} := \{y \in \mathbb{R}^{k-1} : y_1 \geq 0, \ldots, y_{k-1} \geq 0, 1 - y_1 - y_2 - \cdots - y_{k-1} \geq 0\},$$

where, with the old definition, we have $x_i = y_i, i < k$ and $x_k := 1 - y_1 - y_2 - \cdots - y_{k-1}$. Obviously, there is a one-to-one natural correspondence between T^{k-1} and the simplex $\{x \in \mathbb{R}^k : x_1 \geq 0, \ldots, x_k \geq 0, \sum_i x_i = 1\}$ defined by

$$y = (y_1, \ldots, y_{k-1}) \quad \longleftrightarrow \quad (y_1, \ldots, y_{k-1}, 1 - y_1 - y_2 - \cdots - y_{k-1}) = (x_1, \ldots, x_{k-1}, x_k) = x.$$

The Markov diffusion process X_t in (14) may be redefined as $Y_t = (X_{t,1}, \ldots, X_{t,k-1})$ on $y \in T^{k-1}$ with the corresponding generator

$$Lf(y) = -\frac{b}{\alpha} \sum_{i=1}^{k-1} \frac{\partial f}{\partial y_i}(y)(y_i - p_i) + \frac{1}{2} \sum_{i,j=1}^{k-1} \frac{\partial^2 f}{\partial y_i \partial y_j}(y)(y_i \mathbb{1}_{i=j} - y_i y_j). \tag{17}$$

The Kolmogorov forward equation for the density $p(\mathbf{y}, t)$ of the limiting process $\mathbf{Y}_t$ is

$$\frac{\partial}{\partial t} p(\mathbf{y}, t) = \frac{1}{2} \left(\frac{b}{\alpha} \sum_{i=1}^{k-1} \frac{\partial}{\partial y_i} \left(p(\mathbf{y}, t)(y_i - p_i) \right) \right.$$
$$\left. + \sum_{i=1}^{k-1} \frac{\partial^2}{\partial y_i^2} \left(y_i(1 - y_i) p(\mathbf{y}, t) \right) - 2 \sum_{1 \leq i < j \leq k-1} \frac{\partial^2}{\partial y_i \partial y_j} \left(y_i y_j p(\mathbf{y}, t) \right) \right). \quad (18)$$

Therefore, it is not hard to show that the limit invariant ergodic distribution is

$$p(\mathbf{y}) = \frac{1}{B(2\frac{b}{\alpha} \mathbf{p})} (1 - y_1 - \cdots - y_{k-1})^{\frac{2b(1-p_1-\cdots-p_{k-1})}{\alpha} - 1} \prod_{i=1}^{k-1} y_i^{\frac{2bp_i}{\alpha} - 1}, \quad (19)$$

because it satisfies (18) (see also [29]). The above distribution is the Dirichlet distribution $Dir(2\frac{b}{\alpha} \mathbf{p})$ as a function of $\mathbf{x} = (\mathbf{y}, 1 - y_1 - \cdots - y_{k-1})$.

Remark 2 (Transition density of the limit process). *The transition density $p(\mathbf{y}_0, \mathbf{y}; t)$ is defined by*

$$P(\mathbf{Y}_t \in S | \mathbf{Y}_0 = \mathbf{y}_0) = \int_{S \cap T^{k-1}} p(\mathbf{y}_0, \mathbf{y}; t) d\mathbf{y}$$

and it can be represented in terms of series of orthogonal polynomials [30] as shown in [31]. Moreover, we refer to [9,32,33] for the explicit form of the reproducing kernel orthogonal polynomials.

4. Recessive and Dominant Colors in an RP Urn with β Near to 1

Let $J = \{J_1, \ldots, J_{k_J}\}$ be a partition of $\{1, \ldots, k\}$, in that $J_l \neq \varnothing$, $J_{i_1} \cap J_{i_2} = \varnothing$, and $\cup_{l=1}^{k_J} = \{1, \ldots, k\}$. Here k_J denotes the cardinality of J. Define the k_J-dimensional objects $(\boldsymbol{\psi}_n^{(\beta,J)})_n$, $(\boldsymbol{\zeta}_n^{(\beta,J)})_n$ and $\mathbf{p}^{(J)}$ as

$$\left. \begin{array}{l} \psi_{n,i}^{(\beta,J)} = \sum_{l \in J_i} \psi_{n,l}^{(\beta)} \\ \zeta_{n,i}^{(\beta,J)} = \sum_{l \in J_i} \zeta_{n,l}^{(\varepsilon)} \\ p_i^{(J)} = \sum_{l \in J_i} p_l \end{array} \right\} \text{ for } i = 1, \ldots, k_J,$$

and $X_t^{(\beta,J)} = \boldsymbol{\psi}_{\lfloor t/(1-\beta)^2 \rfloor}^{(\beta,J)}$. With these definitions, from (11), we immediately get that $(\boldsymbol{\psi}_n^{(\beta,J)})_n$ is a k_J-dimensional RP urn following the dynamics

$$\boldsymbol{\psi}_n^{(\beta,J)} - \boldsymbol{\psi}_{n-1}^{(\beta,J)} = -\epsilon(\beta)(\boldsymbol{\psi}_{n-1}^{(\beta,J)} - \mathbf{p}^{(J)}) + \delta(\beta)(\boldsymbol{\zeta}_n^{(\beta,J)} - \boldsymbol{\psi}_{n-1}^{(\beta,J)}) \quad (20)$$

and that Theorem 1 holds for $X_t^{(\beta,J)}$. Consequently, the convergence to the Wright—Fisher diffusion still holds if we group together some components of the process. For instance, when we consider two groups of components, we have the following result:

Corollary 1. *Let $J = \{J, J^c\}$ with $J \neq \varnothing$, $J^c \neq \varnothing$. Under the hypothesis of Theorem 1, each component of the sequence of processes $X_t^{(\beta,J)}$ converges, for $\beta \to 1$, to the one-dimensional diffusion process with values in $[0,1]$ that satisfies the SDE*

$$dX_{t,i}^{(J)} = -b \frac{X_{t,i}^{(J)} - p_i}{\alpha} dt + (-1)^{i+1} \sqrt{X_{t,i}^{(J)}(1 - X_{t,i}^{(J)})} dW_t.$$

In addition, $X_{t,1}^{(J)} = \sum_{l \in J} X_{t,l}$ and $X_{t,2}^{(J)} = \sum_{l \in J^c} X_{t,l}$.

Now, if we further specialize the grouping choice to $J = (\{i\}, \{1, \ldots, i-1, i+1, \ldots, k\})$, we get:

Corollary 2. *Under the conditions of Theorem 1 the i-th component of the sequence of processes $X^{(\beta)}$ converges, for $\beta \to 1$, to the one-dimensional diffusion $(X_{t,i})_{t \geq 0}$ with values in $[0,1]$ satisfying the SDE*

$$dX_{t,i} = -b \frac{X_{t,i} - p_i}{\alpha} dt + \sqrt{X_{t,i}(1 - X_{t,i})} dW_t.$$

For instance, the above two results are useful in order to translate the well-known classification of the boundaries of the WF process with mutation [34] (p. 239, Example 8) (see also [35]) to the RP urn model when the parameter β is strictly smaller than 1, but very near to 1. Indeed, Corollary 1 implies that $Z_t = \sum_{l \in J} X_{t,l}$ satisfies the SDE

$$dZ_t = -b \frac{Z_t - \sum_{l \in J} p_l}{\alpha} dt + \sqrt{Z_t(1 - Z_t)} dW_t$$

$$= \left(-\frac{b}{\alpha}\left(1 - \sum_{l \in J} p_l\right) Z_t + \frac{b}{\alpha} \sum_{l \in J} p_l(1 - Z_t) \right) dt + \sqrt{Z_t(1 - Z_t)} dW_t.$$

Setting $a_0 = \frac{b}{\alpha} \sum_{l \in J} p_l$ and $a_1 = \frac{b}{\alpha} - a_0$ and noting that $\cap_{i \in J}\{X_{t,i} = 0\} = \{Z_t = 0\}$, we obtain:

(1) $a_0 < 1/2$, i.e., $\sum_{l \in J} p_l < \frac{\alpha}{2b}$, if and only if $P(\exists t \colon \cap_{i \in J} \{X_{t,i} = 0\}) = 1$;
(2) $a_0 \geq 1/2$, i.e., $\sum_{l \in J} p_l \geq \frac{\alpha}{2b}$, if and only if $P(\exists t \colon \cap_{i \in J} \{X_{t,i} = 0\}) = 0$.

With the same spirit, Corollary 2 states that $Z_t = 1 - X_{t,i}$ satisfies the SDE

$$dZ_t = -b \frac{Z_t - \sum_{l \neq i} p_l}{\alpha} dt + \sqrt{(1 - Z_t) Z_t} dW_t$$

$$= \left(-\frac{b}{\alpha} p_i Z_t + \frac{b}{\alpha}(1 - p_i)(1 - Z_t) \right) dt + \sqrt{Z_t(1 - Z_t)} dW_t.$$

Setting $a_0 = \frac{b}{\alpha}(1 - p_i)$ and $a_1 = \frac{b}{\alpha} - a_0$, we get:

(3) $a_0 < 1/2$, i.e., $p_i > 1 - \frac{\alpha}{2b}$, if and only if $P(\exists t \colon \{X_{t,i} = 1\}) = 1$;
(4) $a_0 \geq 1/2$, i.e., $p_i \leq 1 - \frac{\alpha}{2b}$, if and only if $P(\exists t \colon \{X_{t,i} = 1\}) = 0$.

Therefore, for an RP urn with $\beta < 1$, but very near to 1, we can give the following definition:

Definition 1. *We call recessive a non-empty subset $J \subsetneq \{1, \ldots, k\}$ of colors such that $\sum_{l \in J} p_l < \frac{\alpha}{2b}$. We call dominant a color $i \in \{1, \ldots, k\}$ such that $\{1, \ldots, k\} \setminus \{i\}$ is recessive.*

Obviously, every subset of a recessive set is recessive. Moreover, when $\frac{\alpha}{b} > 2(1 - \min_i p_i)$, every set $J \subsetneq \{1, \ldots, k\}$ is recessive. The terms "recessive" and "dominant" are justified by the fact that, recalling properties (1)–(4) of the WF process, if a set of colors is recessive, then we can observe that at some times the corresponding predictive means of the urn process are very near to zero. On the contrary, when a color is dominant, we can observe that at some times the corresponding predictive mean of the urn process is very near to one. In Figure 2, we plot the process $(\psi_{n,1})$ related to the simulation of an RP urn with $k = 2$, $\alpha/b = 1$ and $p = 0.75$, where it is possible to observe the excursions near the barrier 1.

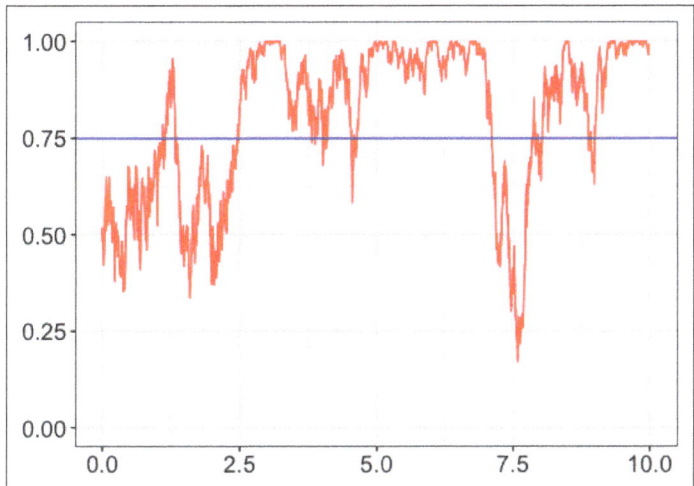

Figure 2. Simulation: plot of the process $(\psi_{n,1})$ related to the simulation of an RP urn with $k = 2$, $\alpha/b = 1$ and $p = 0.75$.

5. Conclusions

We have proven that the multidimensional WF diffusion with mutation can be obtained as the limit of the predictive means associated with a family of RP urns with $\beta < 1$, $\beta \to 1$. As a consequence, the known properties of the WF process can give a description of the RP urn when the parameter β is strictly smaller than 1, but very near to 1. For instance, starting from the known classification of the boundaries for the WF process and connecting it to the model parameters of the RP urn, we have obtained for an RP urn with a value of β very near to one, the notion of recessive subsets of colors and the notion of a dominant color. These two concepts are related to the possibility of reaching the barriers 0 and 1 by the predictive means of the urn process. Other classical problems, together with the corresponding known results for the WF process, can be found in [31]. These results can be used in order to give an approximated answer to the considered problems in the case of an RP urn with a value of β near 1.

Author Contributions: Both authors contributed equally to this work. All authors have read and agreed to the published version of the manuscript.

Funding: Irene Crimaldi is partially supported by the Italian "Programma di Attività Integrata" (PAI), project "TOol for Fighting FakEs" (TOFFE) funded by the IMT School for Advanced Studies Lucca. This research received funding from the European Research Council (ERC) under the European Union's Horizon 2020 research and innovation programme under grant agreement No 817257.

Institutional Review Board Statement: Not applicable.

Informed Consent Statement: Not applicable.

Acknowledgments: Both authors sincerely thank the organizers of the present special issue for their invitation to contribute and Fabio Saracco for having collected and shared with them the analyzed Twitter data sets. Giacomo Aletti is a member of the Italian group "Gruppo Nazionale per il Calcolo Scientifico" of the Italian Institute "Istituto Nazionale di Alta Matematica". Irene Crimaldi is a member of the Italian group "Gruppo Nazionale per l'Analisi Matematica, la Probabilità e le loro Applicazioni" of the Italian Institute "Istituto Nazionale di Alta Matematica".

Conflicts of Interest: The authors declare no conflict of interest.

References

1. Eggenberger, F.; Pólya, G. Über die Statistik verketteter Vorgänge. *ZAMM-J. Appl. Math. Mech./Z. Angew. Math. Mech.* **1923**, *3*, 279–289. [CrossRef]
2. Mahmoud, H.M. *Pólya Urn Models*; Texts in Statistical Science Series; CRC Press: Boca Raton, FL, USA, 2009.
3. Aletti, G.; Crimaldi, I. The Rescaled Pólya Urn: Local reinforcement and chi-squared goodness of fit test. *Adv. Appl. Probab.* Available online: https://iris.imtlucca.it/handle/20.500.11771/19197#.YZNznboRVPZ (accessed on 1 November 2021).
4. Aletti, G.; Crimaldi, I.; Saracco, F. A model for the Twitter sentiment curve. *PLoS ONE* **2021**, *16*, e0249634. [CrossRef]
5. Aletti, G.; Crimaldi, I. Generalized Rescaled Pólya urn and its statistical applications. *arXiv* **2021**, arXiv:2010.06373.
6. Chen, Y.; Skiena, S. Building sentiment lexicons for all major languages. In Proceedings of the 52nd Annual Meeting of the Association for Computational Linguistics (Short Papers), Baltimore, MD, USA, 22–27 June 2014; pp. 383–389.
7. Chakraborty, K.; Bhattacharyya, S.; Bag, R. A Survey of Sentiment Analysis from Social Media Data. *IEEE Trans. Comput. Soc. Syst.* **2020**, *7*, 450–464. [CrossRef]
8. Favaro, S.; Ruggiero, M.; Walker, S.G. On a Gibbs sampler based random process in Bayesian nonparametrics. *Electron. J. Stat.* **2009**, *3*, 1556–1566. [CrossRef]
9. Griffiths, R.C.; Spanò, D. Diffusion processes and coalescent trees. In *Probability and Mathematical Genetics, Papers in Honour of Sir John Kingman*; Bingham, N.H., Goldie, C.M., Eds.; LMS Lecture Note Series; Cambridge University Press: Cambridge, UK, 2010; Volume 378, pp. 358–375.
10. Mena, R.; Ruggiero, M. Dynamic density estimation with diffusive Dirichlet mixtures. *Bernoulli* **2016**, *22*, 901–926. [CrossRef]
11. Walker, S.G.; Hatjispyros, S.J.; Nicoleris, T. A Fleming-Viot process and Bayesian nonparametrics. *Ann. Appl. Probab.* **2007**, *17*, 67–80. [CrossRef]
12. Costantini, C.; De Blasi, P.; Ethier, S.; Ruggiero, M.; Spanò, D. Wright-Fisher construction of the two-parameter Poisson-Dirichlet diffusion. *Ann. Appl. Probab.* **2017**, *27*, 1923–1950. [CrossRef]
13. Bollback, J.P.; York, T.L.; Nielsen, R. Estimation of $2N_e s$ from temporal allele frequency data. *Genetics* **2008**, *179*, 497–502. [CrossRef]
14. Gutenkunst, R.N.; Hernandez, R.D.; Williamson, S.H.; Bustamante, C.D. Inferring the Joint Demographic History of Multiple Populations from Multidimensional SNP Frequency Data. *PLoS Genet.* **2009**, *5*, e1000695. [CrossRef] [PubMed]
15. Malaspinas, A.S.; Malaspinas, O.; Evans, S.N.; Slatkin, M. Estimating allele age and selection coefficient from time-serial data. *Genetics* **2012**, *192*, 599–607. [CrossRef]
16. Schraiber, J.; Griffiths, R.C.; Evans, S.N. Analysis and rejection sampling of Wright-Fisher diffusion bridges. *Theor. Popul. Biol.* **2013**, *89*, 64–74. [CrossRef] [PubMed]
17. Williamson, S.H.; Hernandez, R.; Fledel-Alon, A.; Zhu, L.; Bustamante, C.D. Simultaneous inference of selection and population growth from patterns of variation in the human genome. *Proc. Natl. Acad. Sci. USA* **2005**, *102*, 7882–7887. [CrossRef]
18. Zhao, L.; Lascoux, M.; Overall, A.D.J.; Waxman, D. The characteristic trajectory of a fixing allele: a consequence of fictitious selection that arises from conditioning. *Genetics* **2013**, *195*, 993–1006. [CrossRef]
19. Dangerfield, C.; Kay, D.; Burrage, K. Stochastic models and simulation of ion channel dynamics. *Procedia Comput. Sci.* **2010**, *1*, 1587–1596. [CrossRef]
20. Dangerfield, C.E.; Kay, D.; MacNamara, S.; Burrage, K. A boundary preserving numerical algorithm for the Wright—Fisher model with mutation. *BIT Numer. Math.* **2012**, *5*, 283–304. [CrossRef]
21. Chaleyat-Maurel, M.; Genon-Catalot, V. Filtering the Wright–Fisher diffusion. *ESAIM Probab. Stat.* **2009**, *13*, 197–217. [CrossRef]
22. Papaspiliopoulos, O.; Ruggiero, M. Optimal filtering and the dual process. *Bernoulli* **2014**, *20*, 1999–2019. [CrossRef]
23. Delbaen, F.; Shirakawa, H. An interest rate model with upper and lower bounds. *Asia-Pac. Financ. Mark.* **2002**, *9*, 191–209. [CrossRef]
24. Gourieroux, C.; Jasiak, J. Multivariate Jacobi process with application to smooth transitions. *J. Econom.* **2006**, *131*, 475–505. [CrossRef]
25. Jenkins, P.A.; Spanò, D. Exact simulation of the Wright-Fisher diffusion. *Ann. Appl. Probab.* **2017**, *27*, 1478–1509. [CrossRef]
26. Kushner, H.J. *Approximation and Weak Convergence Methods for Random Processes, with Applications to Stochastic Systems Theory*; MIT Press Series in Signal Processing, Optimization, and Control; MIT Press: Cambridge, MA, USA, 1984; Volume 6.
27. Kushner, H.J.; Yin, G.G. *Stochastic Approximation and Recursive Algorithms and Applications*, 2nd ed.; Applications of Mathematics; Springer: New York, NY, USA, 2003; Volume 35.
28. Tanabe, K.; Sagae, M. An Exact Cholesky Decomposition and the Generalized Inverse of the Variance-Covariance Matrix of the Multinomial Distribution, with Applications. *J. R. Stat. Soc. Ser. B (Methodol.)* **1992**, *54*, 211–219. [CrossRef]
29. Wright, S. *Evolution and the Genetics of Populations, Volume 2: Theory of Gene Frequencies*; Evolution and the Genetics of Populations; University of Chicago Press: Chicago, IL, USA, 1984.
30. Dunkl, C.F.; Xu, Y. *Orthogonal Polynomials of Several Variables*, 2nd ed.; Encyclopedia of Mathematics and Its Applications; Cambridge University Press: Cambridge, UK, 2014; Volume 155. [CrossRef]
31. Aletti, G.; Crimaldi, I. The rescaled Pólya urn and the Wright—Fisher process with mutation. *arXiv* **2021**, arXiv:2110.01853.
32. Griffiths, R.C.; Spanò, D. Orthogonal polynomial kernels and canonical correlations for Dirichlet measures. *Bernoulli* **2013**, *19*, 548–598. [CrossRef]

33. Griffiths, R.C.; Spanò, D. Multivariate Jacobi and Laguerre polynomials, infinite-dimensional extensions, and their probabilistic connections with multivariate Hahn and Meixner polynomials. *Bernoulli* **2011**, *17*, 1095–1125. [CrossRef]
34. Karlin, S.; Taylor, H.M. *A Second Course in Stochastic Processes*; A Subsidiary of Harcourt Brace Jovanovich; Academic Press, Inc.: New York, NY, USA; London, UK, 1981.
35. Huillet, T. On Wright–Fisher diffusion and its relatives. *J. Stat. Mech. Theory Exp.* **2007**, *2007*, P11006–P11006. [CrossRef]

Article
Mixture of Species Sampling Models

Federico Bassetti [†] and Lucia Ladelli [*,†]

Department of Mathematics, Politecnico of Milano, 20133 Milano, Italy; federico.bassetti@polimi.it
* Correspondence: lucia.ladelli@polimi.it
† These authors contributed equally to this work.

Abstract: We introduce mixtures of species sampling sequences (mSSS) and discuss how these sequences are related to various types of Bayesian models. As a particular case, we recover species sampling sequences with general (not necessarily diffuse) base measures. These models include some "spike-and-slab" non-parametric priors recently introduced to provide sparsity. Furthermore, we show how mSSS arise while considering hierarchical species sampling random probabilities (e.g., the hierarchical Dirichlet process). Extending previous results, we prove that mSSS are obtained by assigning the values of an exchangeable sequence to the classes of a latent exchangeable random partition. Using this representation, we give an explicit expression of the Exchangeable Partition Probability Function of the partition generated by an mSSS. Some special cases are discussed in detail—in particular, species sampling sequences with general base measures and a mixture of species sampling sequences with Gibbs-type latent partition. Finally, we give explicit expressions of the predictive distributions of an mSSS.

Keywords: species sampling models; exchangeable random partitions; exchangeable sequences; predictive distributions

Citation: Bassetti, F.; Ladelli, L. Mixture of Species Sampling Models. *Mathematics* **2021**, *9*, 3127. https://doi.org/10.3390/math9233127

Academic Editor: Manuel Alberto M. Ferreira

Received: 5 November 2021
Accepted: 2 December 2021
Published: 4 December 2021

Publisher's Note: MDPI stays neutral with regard to jurisdictional claims in published maps and institutional affiliations.

Copyright: © 2021 by the authors. Licensee MDPI, Basel, Switzerland. This article is an open access article distributed under the terms and conditions of the Creative Commons Attribution (CC BY) license (https://creativecommons.org/licenses/by/4.0/).

1. Introduction

Discrete random measures have been widely used in Bayesian nonparametrics. Noteworthy examples of such random measures are the Dirichlet process [1], the Pitman–Yor process [2,3], (homogeneous) normalized random measures with independent increments (see, e.g., [4–7]), Poisson–Kingman random measures [8] and stick-breaking priors [9]. All the previous random measures are of the form

$$P = \sum_{j \geq 1} p_j^{\downarrow} \delta_{Z_j}, \qquad (1)$$

where $(Z_j)_{j \geq 1}$ are i.i.d. random variables taking values in a Polish space $(\mathbb{X}, \mathcal{X})$ with common distribution H, and $(p_j^{\downarrow})_{j \geq 1}$ are random positive weights in $[0,1]$, independent of $(Z_j)_{j \geq 1}$, such that $p_1^{\downarrow} \geq p_2^{\downarrow} \geq p_3^{\downarrow} \geq \ldots$.

With a few exceptions—see, e.g., [1,4,10–14]—the *base measure* H of a random measure P in (1) is usually assumed to be diffuse, since this simplifies the derivation of various analytical results.

The diffuseness of H is assumed also to define the so-called *species sampling sequences* [15], exchangeable sequences whose directing measure is a discrete random probability of type (1). In this case, the diffuseness of H is motivated by the interpretation of species sampling sequences as sequences describing a sampling mechanism in discovering species from an unknown population. In this context, the Z_js are the possible infinite different species, and the diffuseness of H ensures that there is no redundancy in this description.

On the other hand, from a Bayesian point of view, the diffuseness of H is not always reasonable and there are situations in which a discrete (or mixed) H is indeed natural. For example, recent interest in species sampling models with a spike-and-slab base measure emerged in [16–21] in order to induce sparsity and facilitate variable selection. Other models, which are implicitly related to species sampling sequences with non-diffused base measures, are mixtures of Dirichlet processes [10] and hierarchical random measures; see, e.g., [22–25].

The combinatorial structure of species sampling sequences derived from random measure (1) with general H have been recently studied in [14].

In this paper, we discuss some relevant properties of species sampling sequences with general base measures, as well as some further generalizations, namely mixtures of species sampling sequences with general base measures (mSSS).

An mSSS is an exchangeable sequence whose directing random measure is of type (1), where $(Z_n)_{n\geq 1}$ is a sequence of exchangeable random variables and $(p_n^\downarrow)_{n\geq 1}$ are random positive weights in $[0,1]$ with $p_1^\downarrow \geq p_2^\downarrow \geq p_3^\downarrow \geq \ldots$, independent of $(Z_n)_{n\geq 1}$.

The core of the results that we prove in this paper is that all the mSSS can be obtained by assigning the values of an exchangeable sequence to the classes of a latent exchangeable random partition. We summarize the results of Section 3 in the next statement.

The following are equivalent:

1. $\xi = (\xi_n)_{n\geq 1}$ is an mSSS;
2. with probability one $(\xi_n)_{n\geq 1} = (Z_{I_n})_{n\geq 1}$, where $(I_n)_{n\geq 1}$ is a sequence of integer-valued random variables independent of the Zs such that, conditionally on $p^\downarrow := (p_1^\downarrow, p_2^\downarrow, \ldots)$, the I_n are independent and $\mathbb{P}\{I_n = i | p^\downarrow\} = p_i^\downarrow$.
3. with probability one $(\xi_n)_{n\geq 1} := (Z'_{\mathscr{C}_n(\Pi)})_{n\geq 1}$, where $(Z'_n)_{n\geq 1}$ is an exchangeable sequence with the same law of $(Z_n)_{n\geq 1}$, Π is an exchangeable partition, independent of $(Z'_n)_{n\geq 1}$, obtained by sampling from $(p_n^\downarrow)_{n\geq 1}$, and $\mathscr{C}_n(\Pi)$ is the index of the block in Π containing n.

The partition Π obtained from $p^\downarrow = (p_1^\downarrow, p_2^\downarrow, \ldots)$ is the so-called paint-box process associated with $p^\downarrow$. In general, this partition, called the latent partition, does not coincide with the partition induced by the $(\xi_n)_{n\geq 1}$. Note that also the sequence $(Z'_n)_{n\geq 1}$ is latent, in the sense that it cannot be obtained if only $(\xi_n)_{n\geq 1}$ is known. On the other hand, combining the information contained in $(Z'_n)_{n\geq 1}$ and in Π, one obtains complete knowledge of $(\xi_n)_{n\geq 1}$, and, in particular, of its clustering behavior. This last observation is essential for the development of all the other results presented in our paper.

The rest of the paper is organized as follows. Section 2 reviews some important results on species sampling models and exchangeable random partitions. Section 3 introduces mixtures of species sampling sequences and discusses how these sequences are related to various types of Bayesian models. In the same section, the stochastic representations for mixtures of species sampling sequences sketched above are proven. In Section 4, we provide an explicit expression of the *Exchangeable Partition Probability Function* (EPPF) of the partition generated by such sequences. This result is achieved considering two EPPFs arising from a suitable latent partition structure. Some special cases are further detailed. Finally, Section 5 deals with the predictive distributions of mixtures of species sampling sequences.

2. Background Materials

In this section, we briefly review some basic notions of exchangeable random partitions and species sampling models.

2.1. Exchangeable Random Partitions

A partition π_n of $[n] := \{1, \ldots, n\}$ is an unordered collection $\{\pi_{1,n}, \ldots, \pi_{k,n}\}$ of disjoint non-empty subsets (blocks) of $\{1, \ldots, n\}$ such that $\cup_{j=1}^k \pi_{j,n} = [n]$. A partition $\pi_n = \{\pi_{1,n}, \pi_{2,n}, \ldots, \pi_{k,n}\}$ has $|\pi_n| := k$ blocks (with $1 \leq |\pi_n| \leq n$) and $|\pi_{c,n}|$, with $c = 1, \ldots, k$, is the number of elements of the block c. We denote by $\mathcal{P}_n$ the collection of all partitions of $[n]$ and, given a partition, we list its blocks in ascending order of their smallest element, i.e., *in order of their appearance*. For instance, we write $[(1,3), (2,4), (5)]$ and not $[(2,4), (3,1), (5)]$.

A sequence of random partitions, $\Pi = (\Pi_n)_{n\geq 1}$, defined on a common probability space, is called a *random partition of* $\mathbb{N}$ if, for each n, the random variable Π_n takes values in $\mathcal{P}_n$ and, for $m < n$, the restriction of Π_n to $\mathcal{P}_m$ is Π_m (consistency property).

In order to define an *exchangeable* random partition, given a permutation ρ of $[n]$ and a partition π_n in $\mathcal{P}_n$, we denote by $\rho(\pi_n)$ the partition with blocks $\{\rho(j) : j \in \pi_{i,n}\}$ for $i = 1, \ldots, |\pi_n|$. A random partition of $\mathbb{N}$ is said to be *exchangeable* if Π_n has the same distribution of $\rho(\Pi_n)$ for every n and every permutation ρ of $[n]$. In other words, its law is invariant under the action of all permutations (acting on Π_n in the natural way).

The law of any exchangeable random partition on $\mathbb{N}$ is completely characterized by its *Exchangeable Partition Probability Function* (EPPF); in other words, there exists a unique symmetric function q on the integers such that, for any partition π_n in $\mathcal{P}_n$,

$$\mathbb{P}\{\Pi_n = \pi_n\} = q(|\pi_{1,n}|, \ldots, |\pi_{k,n}|) \qquad (2)$$

where k is the number of blocks in π_n. In the following, we shall write $\Pi \sim q$ to denote an exchangeable partition of $\mathbb{N}$ with EPPF q. Note that an EPPF is indeed a family of symmetric functions $q_k^n(\cdot)$ defined on $\mathcal{C}_{n,k} = \{(n_1, \ldots, n_k) \in \mathbb{N}^k : \sum_{i=1}^k n_i = n\}$. To simplify the notation, we write q instead of q_k^n. Alternatively, one can assume that q is a function on $\cup_{n \in \mathbb{N}} \cup_{k=1}^n \mathcal{C}_{n,k}$. See [26].

Given a sequence of random variables $X = (X_j)_{j \geq 1}$ taking values in some measurable space, the random partition $\Pi^*(X)$ induced by X is defined as the random partition obtained by the equivalence classes under the random equivalence relation $i(\omega) \sim j(\omega)$ if and only if $X_i(\omega) = X_j(\omega)$. One can check that a partition induced by an exchangeable random sequence is an exchangeable random partition.

Recall that, by de Finetti's theorem, a sequence $X = (X_n)_{n \geq 1}$ taking values in a Polish space $(\mathbb{X}, \mathcal{X})$ is exchangeable if and only if the X_ns, given some random probability measure Q on $\mathcal{X}$, are conditionally independent with common distribution Q. Moreover, the random probability Q, known as the *directing random measure of* X, is the almost sure limit (with respect to weak convergence) of the empirical process $\frac{1}{n}\sum_{i=1}^n \delta_{X_i}$.

Based on de Finetti's theorem, Kingman's correspondence theorem sets up a one-to-one map between the law of an exchangeable random partition on $\mathbb{N}$ (i.e., its EPPF) and the law of random ranked weights $p^\downarrow = (p_j^\downarrow)_{j \geq 1}$ satisfying $1 \geq p_1^\downarrow \geq p_2^\downarrow \geq \cdots \geq 0$ and $\sum_j p_j^\downarrow \leq 1$ (with probability one). To state the theorem, recall that a partition Π is said to be generated by a (possibly random) $p^\downarrow$, if it is generated by a sequence $(I_n)_{n \geq 1}$ of integer-valued random variables that are conditionally independent given $p^\downarrow$ with conditional distribution

$$\mathbb{P}\{I_n = i | p^\downarrow\} := \begin{cases} 1 - \sum_{j \geq 1} p_j^\downarrow & \text{if } i = -n \\ p_i^\downarrow & \text{if } i \geq 1, \end{cases} \qquad (3)$$

Note that $1 - \sum_{j \geq 1} p_j^\downarrow$ is the magnitude of the so-called "dust" component; indeed, each I_n sampled from this part, i.e., $I_n = -n$, contributes to a singleton n in the partition Π. A consequence is that if $\sum_{j \geq 1} p_j^\downarrow = 1$ a.s., the partition Π has no singleton. The partition $\Pi^*(I)$ is sometimes referred to as the $p^\downarrow$-paintbox process; see [27].

Let $\nabla := \{p_j^\downarrow \in [0,1] : p_1^\downarrow \geq p_2^\downarrow \geq \ldots, \sum_{j \geq 1} p_j^\downarrow \leq 1\}$. We are now ready to state Kingman's theorem.

Theorem 1 ([28]). *Given any exchangeable random partition Π with EPPF q, denote by $\Pi_{j,n}^\downarrow$ the blocks of the partition rearranged in decreasing order with respect to the number of elements in the blocks of Π_n. Then,*

$$\lim_n \left(\frac{|\Pi_{j,n}^\downarrow|}{n}\right)_{j \geq 1} = (p_j^\downarrow)_{j \geq 1} \quad a.s. \qquad (4)$$

for some random $p^\downarrow = (p_j^\downarrow)_{j\geq 1}$ taking values in ∇. Moreover, on a possibly enlarged probability space, there is a sequence of integer-valued random variables $I = (I_n)_{n\geq 1}$, conditionally independent given $p^\downarrow$, such that (3) holds and the partition induced by I is equal to Π a.s.

Kingman's theorem is usually stated in a slightly weaker form (see, e.g., Theorem 2.2 in [26]) and the equality between $\Pi^*(I)$ and Π is given in law. The previous "almost sure" version can be easily derived by inspecting the proof of Kingman's theorem given in [29].

A consequence of the previous theorem is that $q \mapsto \text{Law}(p^\downarrow)$ for $p^\downarrow$ in (4) defines a bijection from the set of the EPPF and the laws on ∇.

If $p^\downarrow$ is proper, i.e., $\sum_{j\geq 1} p_j^\downarrow = 1$ a.s., then Kingman's correspondence between $p^\downarrow$ and the EPPF q can be made explicit by

$$q(n_1,\ldots,n_k) = \sum_{(j_1,\ldots,j_k)\in \mathbb{N}_k} \mathbb{E}\left[\prod_{i=1}^{k}(p_{j_i}^\downarrow)^{n_i}\right]. \tag{5}$$

where $\mathbb{N}_k$ is the set of all ordered k-tuples of distinct positive integers. See Chapter 2 [26].

Given an EPPF q, one deduces the corresponding sequence of predictive distributions, which is the sequence of conditional distributions

$$P\{\Pi_{n+1} = \pi_{n+1}|\Pi_n = \pi_n\}$$

when $\Pi \sim q$. Starting with $\Pi_1 = \{1\}$, given $\Pi_n = \pi_n$ (with $|\pi_n| = k$), the conditional probability of adding a new block (containing $n+1$) to Π_n is

$$\nu_n(\pi_n) = \nu_n(|\pi_{1,n}|,\ldots,|\pi_{k,n}|) := \frac{q(|\pi_{1,n}|,\ldots,|\pi_{k,n}|,1)}{q(|\pi_{1,n}|,\ldots,|\pi_{k,n}|)}; \tag{6}$$

while the conditional probability of adding $n+1$ to the ℓ-th block of Π_n (for $\ell = 1,\ldots,k$) is

$$\omega_{n,\ell}(\pi_n) = \omega_{n,\ell}(|\pi_{1,n}|,\ldots,|\pi_{k,n}|) := \frac{q(|\pi_{1,n}|,\ldots,|\pi_{\ell,n}|+1,\ldots,|\pi_{k,n}|)}{q(|\pi_{1,n}|,\ldots,|\pi_{k,n}|)}. \tag{7}$$

2.2. Species Sampling Models

A *species sampling random probability* (SSrp) is a random probability of the form

$$P = \sum_{j\geq 1} p_j \delta_{Z_j} + (1 - \sum_{j\geq 1} p_j)H \tag{8}$$

where $(Z_j)_{j\geq 1}$ are i.i.d. random variables taking values in a Polish space $(\mathbb{X}, \mathcal{X})$ with common distribution H, and $(p_j)_{j\geq 1}$ are random positive weights in $[0,1]$, independent of $(Z_j)_{j\geq 1}$, such that $\sum_{j\geq 1} p_j \leq 1$ with probability one. These random probability measures are also known as *Type III* random probability measures; see [30].

Given the SSrp in (8), let $(p_j^\downarrow)_{j\geq 1}$ be the ranked sequence obtained from $(p_j)_{j\geq 1}$ rearranging the p_js in decreasing order. One can always write

$$P = \sum_{j\geq 1} p_j^\downarrow \delta_{\tilde{Z}_j} + (1 - \sum_{j\geq 1} p_j^\downarrow)H \tag{9}$$

where $(\tilde{Z}_j)_{j\geq 1}$ is a suitable random reordering of the original sequence $(Z_j)_{j\geq 1}$. It is easy to check that $(\tilde{Z}_j)_{j\geq 1}$ are i.i.d. random variables with law H independent of $(p_j^\downarrow)_{j\geq 1}$. Hence, H and the EPPF q associated via Kingman's correspondence with $(p_j^\downarrow)_{j\geq 1}$ completely characterize the law of P, from now on denoted by $SSrp(q, H)$.

$SSrp$ with H diffuse are also characterized as directing random measures of a particular type of exchangeable sequences, known as species sampling sequences. Let q be an EPPF and H a diffuse probability measure on a Polish space $\mathbb{X}$. An exchangeable sequence $\xi := (\xi_n)_n$ taking values in $\mathbb{X}$ is a *species sampling sequence*, $SSS(q, H)$, if the law of $(\xi_n)_n$ is characterized by the predictive system:

- (PS1) $\mathbb{P}\{\xi_1 \in dx\} = H(dx)$;
- (PS2) *the conditional distribution of* ξ_{n+1} *given* $(\xi_1, \ldots, \xi_n)$ *is*

$$\mathbb{P}\{\xi_{n+1} \in dx | \xi_1, \ldots, \xi_n\} = \sum_{c=1}^{K} \omega_{n,c} \delta_{\xi_c^*}(dx) + \nu_n H(dx),$$

where $(\xi_1^*, \ldots, \xi_K^*)$ *is the sequence of distinct observations in order of appearance*, $\omega_{n,c} = \omega_{n,c}(|\Pi_{1,n}|, \ldots, |\Pi_{K,n}|)$, $\nu_n = \nu_n(|\Pi_{1,n}|, \ldots, |\Pi_{K,n}|)$, $K = |\Pi_n|$, Π_n *is the random partition induced by* $(\xi_1, \ldots, \xi_n)$ *and* $\omega_{n,c}$ *and* ν_n *are related to the q by* (6) *and* (7).

We summarize here some results proven in [15].

Proposition 1 ([15])**.** *Let H be a diffuse probability measure; then, an exchangeable sequence $(\xi_n)_n$ is characterized by (PS1)–(PS2) if and only if its directing random measure is an $SSrp(q, H)$.*

As noted in [29], the partition induced by any exchangeable sequence taking values in $\mathbb{X}$ with directing measure $\tilde{\mu}$ depends only on the sequence $\tilde{\mu}(\tilde{x}_j)$, where $\tilde{x}_j$ are the random atoms forming the discrete component of $\tilde{\mu}$ and ordered in such a way that $\tilde{\mu}(\tilde{x}_1) \geq \tilde{\mu}(\tilde{x}_2) \geq \ldots$. Combining this observation with the previous proposition, one can see that, when H is diffuse and ξ is an $SSS(q, H)$, the partition $\Pi^*(\xi)$ is equal (a.s.) to $\Pi^*(I)$ (where I is defined as in Kingman's theorem) and $\Pi^*(\xi)$ has EPPF q. Note that [29] defines the $p^{\downarrow}$-paintbox process as any random partition $\Pi^*(\xi)$ where ξ is an exchangeable sequence with directing random measure (9) and H is a diffuse measure.

One can show (see the proof of Proposition 13 in [15]) that an $SSS(q, H)$ can be obtained by assigning the values of an i.i.d. sequence $(Z_n)_n$ with distribution H to the classes of an independent exchangeable random partition with EPPF q. More formally, for a random partition Π, let $\mathscr{C}_n(\Pi)$ be the random index denoting the block containing n, i.e.,

$$\mathscr{C}_n(\Pi) = c \text{ if } n \in \Pi_{c,n}$$

or equivalently if $n \in \Pi_{c,j}$ for some (and hence all) $j \geq n$. If $Z' = (Z'_n)_n$ is an i.i.d. sequence with law H (diffuse), Π is an exchangeable partition with $\Pi \sim q$, and Z' and Π are stochastically independent, then

$$(\xi_n)_{n \geq 1} := (Z'_{\mathscr{C}_n(\Pi)})_{n \geq 1} \tag{10}$$

is an $SSS(q, H)$. Note that the Z'_ns appearing in (10) are not the same Z_ns of (8), although they have the same law.

It is worth mentioning that the original characterization given in [15] of species sampling sequences is stronger than the one summarized here. Indeed, the original definition of SSS is given using a slightly weaker predictive assumption. For details, see Proposition 13 and the discussion following Proposition 11 in [15].

In summary, when H is diffuse, one can build a species sampling sequence $(\xi_n)_n$ by one of the following equivalent constructions:

- using the system of predictive distributions (PS1)–(PS2);
- sampling (conditionally) i.i.d. variables from (8);
- combining an i.i.d. sequence from H with an exchangeable random partition by (10).

3. Mixture of Species Sampling Models

We now discuss some possible generalizations of the notion of species sampling sequences and we show that the three constructions presented above are no more equivalent in this setting.

3.1. Definitions and Relation to Other Models

Exchangeable sequences sampled from an *SSrm* with a general base measure, also known as *generalized species sampling sequences (gSSS)*, have been introduced and studied in [14,25].

Definition 1 ($gSSS(\mathfrak{q}, H)$). $(\xi_n)_{n\geq 1}$ is a $gSSS(\mathfrak{q}, H)$ if it is an exchangeable sequence with directing random measure P, where $P \sim SSrp(\mathfrak{q}, H)$, H being any measure on $(\mathbb{X}, \mathcal{X})$ (not necessarily diffuse).

Clearly, a $gSSS(\mathfrak{q}, H)$ with H diffuse is an $SSS(\mathfrak{q}, H)$. On the contrary, if ξ is a $gSSS(\mathfrak{q}, H)$ with H non-diffuse, (PS1)–(PS2) are no longer true. Moreover, the EPPF of the random partition induced by ξ with H non-diffuse is not $\mathfrak{q}$. The relation between the partition induced by ξ and $\mathfrak{q}$ has been studied in [14].

In [25], the definition of $gSSS(\mathfrak{q}, H)$ with H not necessarily diffuse was motivated by an interest in defining the class of so-called *hierarchical species sampling models*. If $\xi_1, \xi_2, \ldots$ are exchangeable random variables with a directing random measure of hierarchical type, one has that

$$\xi_n | P_1, P_0 \overset{i.i.d.}{\sim} P_1 \quad n \geq 1$$
$$P_1 | P_0 \sim SSrp(\mathfrak{q}, P_0)$$
$$P_0 \sim SSrp(\mathfrak{q}_0, H_0).$$

In order to understand why the general definition of $gSSS(\mathfrak{q}, H)$ is useful in this context, note that, even if H_0 is diffuse and $\mathfrak{q}_0$ is proper (i.e., the $p^\downarrow$ associated with $\mathfrak{q}_0$ by Kingman's correspondence are proper), the conditional distribution of $[\xi_n]_{n\geq 1}$ given P_0 is not an SSS, since P_0 is a.s. a purely atomic probability measure on $\mathcal{X}$. Moreover, assuming that $\mathfrak{q}$ is proper, we can write

$$P_1 = \sum_j p_{j1} \delta_{Z_j}$$

where Z_j are conditionally i.i.d. with common distribution P_0, given P_0, and $(p_{j1})_j$ are associated by Kingman's correspondence with the EPPF $\mathfrak{q}$. In other words, in this case, $(\xi_n)_{n\geq 1}$ are exchangeable with directing random measure $P_1 = \sum_j p_{j1} \delta_{Z_j}$, where $(p_{j1})_j$ and $(Z_j)_j$ are independent and $(Z_j)_j$ are exchangeable with directing measure $P_0 \sim SSrp(\mathfrak{q}_0, H_0)$.

The previous observation suggests a further generalization of species sampling sequences.

Definition 2 (*mSSS*). We say that $(\xi_n)_{n\geq 1}$ is a mixture of species sampling sequences (mSSS) if it is an exchangeable sequence with directing random measure

$$P = \sum_{j\geq 1} p_j^\downarrow \delta_{Z_j} + (1 - \sum_{j\geq 1} p_j^\downarrow) \tilde{H} \tag{11}$$

where $Z := (Z_n)_{n\geq 1}$ is an exchangeable sequence with directing random measure $\tilde{H}$, $p^\downarrow$ a sequence of random weight in ∇ with EPPF $\mathfrak{q}$ such that $P\{\sum_{j\geq 1} p_j^\downarrow > 0\} > 0$, $(Z, \tilde{H})$ and $p^\downarrow$ are stochastically independent.

First of all, note that $gSSS(\mathfrak{q}, H)$ is a particular case of Definition 2, obtained from a deterministic $\tilde{H} = H$. Moreover, Definition 2 can be seen as a mixture of $gSSS$. Indeed, if $\tilde{\zeta} = (\tilde{\zeta}_n)_{n \geq 1}$ is as in Definition 2 and $\tilde{H}$ is the directing random measure of $(Z_n)_n$, then the conditional distribution of $\tilde{\zeta}$ given $\tilde{H}$ is a $gSSS(\mathfrak{q}, \tilde{H})$. This motivates the name "mixture of species sampling sequences".

It is worth noticing that one can also consider more general mixtures of SSS. The most general mixture one can take into consideration leads to a random probability measure of the form (11), where $Z := (Z_n)_{n \geq 1}$ are exchangeable random variables with directing random measure $\tilde{H}$, $p^\downarrow$ is a sequence of random weight in ∇ such that $P\{\sum_{j \geq 1} p_j^\downarrow > 0\} > 0$, where $[Z, \tilde{H}]$, and $p^\downarrow$ are not necessarily stochastically independent.

As an example of this more general situation, we describe the so-called mixtures of Dirichlet processes as defined in [10]. Recall that a Dirichlet process $P \sim \mathcal{D}ir(\alpha)$ is defined as a random probability measure characterized by the system of finite n-dimensional distributions

$$\mathbb{P}\{(P(A_1), \ldots, P(A_n)) \in \cdot\} = Dir\Big(\cdot\,; \alpha(A_1), \ldots, \alpha(A_n)\Big) \quad \forall n \geq 1, \forall A_i \in \mathcal{X}$$

where $Dir(\cdot\,; a_1, \ldots, a_n)$ is the Dirichlet measure (on the $n-1$ simplex) of parameters $a_1, \ldots, a_n$ and α is a finite σ-additive measure on $\mathcal{X}$. It is well known that a Dirichlet process is an $SSrp(\mathfrak{q}, H)$ for $H(\cdot) = \alpha(\cdot)/\alpha(\mathbb{X})$ and

$$\mathfrak{q}(n_1, \ldots, n_k) = \frac{\alpha(\mathbb{X})^k}{(\alpha(\mathbb{X}))_n} \prod_{c=1}^{k} (n_c - 1)!, \tag{12}$$

where $(x)_n = x(x+1)\ldots(x+n-1)$ is the rising factorial (or Pochhammer polynomial); see [2,31]. A mixture of Dirichlet processes is defined in [10] as a random probability measure P characterized by the n-dimensional distributions

$$\mathbb{P}\{(P(A_1), \ldots, P(A_n)) \in \cdot\} = \int_U Dir\Big(\cdot\,; \alpha_u(A_1), \ldots, \alpha_u(A_n)\Big) Q(du) \tag{13}$$

where, now, $(u, A) \mapsto \alpha_u(A)$ is a kernel measure on $U \times \mathcal{X}$ (in particular, $A \mapsto \alpha_u(A)$ is a finte σ-additive measure on $\mathcal{X}$ for every $u \in U$), $(U, \mathcal{U})$ is a (Borel) regular space (e.g., a Polish space) and Q is a probability measure on $\mathcal{U}$.

Using the fact that a Dirichlet process is the $SSrp$ described above, one can prove that any mixture of Dirichlet processes has a representation of the form (11), where $((Z_n)_{n \geq 1} \tilde{H})$ and $p^\downarrow$ are stochastically dependent. More precisely, the joint law of $(\tilde{H}, (Z_n)_{n \geq 1}, p^\downarrow)$ is characterized by the law of the (augmented) random element

$$(\tilde{H}, (Z_n)_{n \geq 1}, p^\downarrow, \tilde{u})$$

given by the following:

- $\tilde{u}$ is a random variable taking values in U with law Q;
- $\tilde{H}(\cdot) := \alpha_{\tilde{u}}(\cdot)/\alpha_{\tilde{u}}(\mathbb{X})$;
- $(Z_n)_{n \geq 1}$ are exchangeable random variables with directing random measure $\tilde{H}$;
- $p^\downarrow$ is sequence of random weight in ∇ such that $P\{\sum_{j \geq 1} p_j^\downarrow = 1\} = 1$ and the conditional distribution of $p^\downarrow$ given $\tilde{u}$ depends only on $\alpha_{\tilde{u}}(\mathbb{X})$. In particular, the (conditional) EPPF associated with the law of $p^\downarrow$ given $\tilde{u}$ has the form

$$\mathfrak{q}(n_1, \ldots, n_k | \tilde{u}) := \frac{\alpha_{\tilde{u}}(\mathbb{X})^k}{(\alpha_{\tilde{u}}(\mathbb{X}))_n} \prod_{c=1}^{k} (n_c - 1)! \tag{14}$$

Note that the marginal EPPF of the $p^\downarrow$, obtained by integrating (14) with respect to the law of $\tilde{u}$, is

$$\mathfrak{q}(n_1,\ldots,n_k) = \prod_{c=1}^{k}(n_c - 1)! \int_U \frac{\alpha_u(\mathbb{X})^k}{(\alpha_u(\mathbb{X}))_n} Q(du). \tag{15}$$

Without further assumptions, a mixture of Dirichlet processes is a mixture of SSrp with $p^\downarrow$ and $\tilde{H}(\cdot)$ possibly dependent. Nevertheless, with this representation at hand, one can easily deduce that if $(\xi_n)_{n\geq 1}$ is sampled from a mixture of Dirichlet processes under the additional hypothesis that Q is such that $\alpha_{\tilde{u}}(\mathbb{X})$ and $\alpha_{\tilde{u}}(\cdot)/\alpha_{\tilde{u}}(\mathbb{X})$ are independent, then $(\xi_n)_{n\geq 1}$ satisfies Definition 2, with $\tilde{H} = \alpha_{\tilde{u}}(\cdot)/\alpha_{\tilde{u}}(\mathbb{X})$ and $\mathfrak{q}$ given by (15).

In the rest of the paper, we focus our attention on mSSS for which $[Z, \tilde{H}]$ and $p^\downarrow$ are independent.

3.2. Representation Theorems for mSSS

In this section, we give two alternative representations for exchangeable sequences as in Definition 2, which generalize Proposition 1 in [14].

Proposition 2. *An exchangeable sequence $\xi = (\xi_n)_{n\geq 1}$ is an mSSS as in Definition 2 if and only if*

$$\xi_n = Z_{I_n} \quad a.s.$$

where $Z^+ = (Z_n)_{n\geq 1}$, $\tilde{H}$ and $p^\downarrow$ are as in Definition 2, $Z^- = (Z_n)_{n\leq -1}$ are further conditionally (given $\tilde{H}$) i.i.d. random variables with conditional distribution $\tilde{H}$, and $(I_n)_{n\geq 1}$ is a sequence of integer-valued random variables independent of the Zs and $\tilde{H}$, such that, conditionally on $p^\downarrow$, the I_n are independent and (3) holds. All the random elements are defined on a possibly enlarged probability space.

Proof. Let $\sigma_2 = [Z^+, \tilde{H}, p^\downarrow]$, where Z^+, $\tilde{H}$, $p^\downarrow$ are defined as in Definition 2 (mSSS). Set $\alpha = 1 - \sum_{j\geq 1} p_j^\downarrow$. On a possibly enlarged probability space, let $(Z')^- = (Z'_n)_{n\leq -1}$ be a sequence of random variables conditionally i.i.d. given $\tilde{H}$ with conditional distribution $\tilde{H}$ and let $I' = (I'_n)_{n\geq 1}$ be a sequence of integer-valued random variables conditionally independent given $p^\downarrow$ with conditional distribution (3) with I'_n in place of I_n. One can also assume that I' and $Z = [Z^+, (Z')^-]$ are independent given $[p^\downarrow, \tilde{H}]$; see Lemma A1 in the Appendix A. Set $\tau_1 = [I', (Z')^-]$ and define

$$(\xi'_n)_{n\geq 1} = \phi(\tau_1, \sigma_2) := (Z_{I'_n}\mathbf{1}\{I'_n \geq 1\} + Z'_{-n}\mathbf{1}\{I'_n = -n\})_{n\geq 1}.$$

Let us show that the law of $\xi' := (\xi'_n)_{n\geq 1}$ given σ_2 is the same as the law of ξ given σ_2. Take n Borel sets $A_1,\ldots,A_n$ and non-zero integer numbers $i_1,\ldots,i_n$. One has

$$\mathbb{P}\{\xi'_1 \in A_1,\ldots,\xi'_n \in A_n, I'_1 = i_1,\ldots,I'_n = i_n \big| \tilde{H}, p^\downarrow, Z\}$$
$$= \prod_{j=1}^{n} \Big[\delta_{Z_{i_j}}(A_j)p^\downarrow_{i_j}\mathbf{1}\{i_j > 0\} + \alpha\delta_{Z'_{-j}}(A_j)\mathbf{1}\{i_j = -j\}\Big].$$

Conditionally on $\tilde{H}$, the $(Z'_n)_{n\leq -1}$ are i.i.d. with law $\tilde{H}$ so that

$$\mathbb{P}\bigl\{\tilde{\zeta}'_1 \in A_1,\ldots,\tilde{\zeta}'_n \in A_n, I'_1 = i_1,\ldots,I'_n = i_n \bigm| \tilde{H}, p^\downarrow, Z^+\bigr\}$$
$$= \mathbb{E}\Bigl[\mathbb{P}\bigl\{\tilde{\zeta}'_1 \in A_1,\ldots,\tilde{\zeta}'_n \in A_n, I'_1 = i_1,\ldots,I'_n = i_n \bigm| \tilde{H}, p^\downarrow, Z\bigr\} \bigm| \tilde{H}, p^\downarrow, Z^+\Bigr]$$
$$= \mathbb{E}\Bigl[\prod_{j=1}^n \bigl[\delta_{Z_{i_j}}(A_j) p^\downarrow_{i_j} \mathbf{1}\{i_j > 0\} + \alpha \delta_{Z'_{-j}}(A_j) \mathbf{1}\{i_j = -j\}\bigr] \Bigm| \tilde{H}, p^\downarrow, Z^+\Bigr]$$
$$= \prod_{j=1}^n \mathbb{E}\Bigl[\delta_{Z_{i_j}}(A_j) p^\downarrow_{i_j} \mathbf{1}\{i_j > 0\} + \alpha \delta_{Z'_{-j}}(A_j) \mathbf{1}\{i_j = -j\} \Bigm| \tilde{H}, p^\downarrow, Z^+\Bigr]$$
$$= \prod_{j=1}^n \bigl[\delta_{Z_{i_j}}(A_j) p^\downarrow_{i_j} \mathbf{1}\{i_j > 0\} + \alpha \tilde{H}(A_j) \mathbf{1}\{i_j = -j\}\bigr].$$

Marginalizing with respect to $i_1,\ldots,i_n$,

$$\mathbb{P}\bigl\{\tilde{\zeta}'_1 \in A_1,\ldots,\tilde{\zeta}'_n \in A_n \bigm| \tilde{H}, p^\downarrow, Z^+\bigr\} = \prod_{j=1}^n P(A_j).$$

Recalling that $P = \sum_{j \geq 1} p^\downarrow_j \delta_{Z_j} + \alpha \tilde{H}$,

$$\mathbb{P}\bigl\{\tilde{\zeta}'_1 \in A_1,\ldots,\tilde{\zeta}'_n \in A_n | P\bigr\} = \prod_{j=1}^n P(A_j)$$

almost surely. Since $\mathbb{X}$ is Polish, we have proven that, given P, $(\tilde{\zeta}'_n)_{n\geq 1}$ are i.i.d. with common distribution P. In particular, we have proven that $\tilde{\zeta}' := (\tilde{\zeta}'_n)_{n\geq 1}$ given σ_2 is the same as the law of $\tilde{\zeta}$ given σ_2. This concludes the proof of the "if part", since, by the previous argument, any sequence of the form $(\tilde{\zeta}')$ is of type (mSSS). To complete the proof, it remains to conclude the "only if part". Setting $\sigma_1 = \tilde{\zeta}$, we have proven that the conditional distribution of σ_1 given σ_2 is the same as the conditional distribution of $\phi(\tau_1, \sigma_2)$ given σ_2. At this stage, Lemma A3 in the Appendix A yields that there is $\tau = [(I_n)_{n\geq 1}, (Z_n)_{n\leq -1}]$ such that $(\xi_n)_n = \phi(\tau, \sigma_2)$ a.s., i.e., $(\xi_n)_n = (Z_{I_n})_n$ a.s. In addition, $\mathcal{L}(\tau, \sigma_2) = \mathcal{L}(\tau_1, \sigma_2)$; hence, the $(Z_n)_{n\leq -1}$ are conditionally i.i.d. given $\tilde{H}$ and the I_ns are conditionally independent given $[Z^+, Z^-, \tilde{H}, p^\downarrow]$ with the conditional distribution defined by (3). $\square$

Proposition 3. *An exchangeable sequence $\xi = (\xi_n)_{n\geq 1}$ is an mSSS as defined in Definition 2 if and only if*

$$(\xi_n)_{n\geq 1} := (Z'_{\mathscr{C}_n(\Pi)})_{n\geq 1} \quad a.s. \tag{16}$$

where $Z' := (Z'_n)_{n\geq 1}$ is an exchangeable sequence with the same law of Z, Π is an exchangeable partition with EPPF q and Π and Z' are independent.

Remark 1. *Note that the Z'_ns appearing in (16) are not the same Z_ns appearing in Definition 2, although they have the same law.*

Proof of Proposition 3. If ξ is mSSS, then, by Proposition 2, we know that $\xi = (Z_{I_n})_{n\geq 1}$. Let $\Pi = \Pi^*(I)$ be the partition induced by $(I_n)_{n\geq 1}$; then, Π has EPPF q by Kingman's theorem 1. Denote by $I_1^* = I_1, I_2^*, \ldots, I_K^*$ (with $K \leq +\infty$) the distinct values of $(I_n)_{n\geq 1}$ in order of appearance, and set

$$Z'_n = Z_{I_n^*} \quad n = 1, \ldots, K.$$

When $K < +\infty$, set $I_{K+j}^* = D + j$, where $D = \max\{i : i = I_n^* \text{ for } n \leq K\}$, and define the remaining Z'_m for $m > K$ accordingly as $Z'_m = Z_{I_m^*}$. Let $\{i_1, \ldots, i_M\}$ be integers in $\mathbb{Z} \setminus \{0\}$, and denote the distinct values in $(i_1, \ldots, i_M)$ in order of appearance by $(i_1^*, \ldots, i_m^*)$. Let $A_1, \ldots, A_n$ be measurable sets in $\mathcal{X}$, if $n > m$, then

$$\mathbb{P}\{Z_1' \in A_1, \ldots, Z_n' \in A_n, I_1 = i_1, \ldots, I_M = i_M\}$$
$$= \sum_{\ell_1, \ldots, \ell_{n-m}} \mathbb{P}\Big\{Z_{i_1^*} \in A_1, \ldots, Z_{i_m^*} \in A_m, Z_{\ell_1} \in A_{m+1} \ldots,$$
$$Z_{\ell_{n-m}} \in A_n, I_1 = i_1, \ldots, I_M = i_M, I_{m+1}^* = \ell_1, \ldots, I_n^* = \ell_{n-m}\Big\}$$

where the sum runs over all the non-zero distinct integers $\ell_1, \ldots, \ell_{n-m}$ different from $i_1^*, \ldots, i_m^*$. Since I^* is a function of I and I and Z are independent, it follows that

$$\mathbb{P}\Big\{Z_{i_1^*} \in A_1, \ldots, Z_{i_m^*} \in A_m, Z_{\ell_1} \in A_{m+1} \ldots,$$
$$Z_{\ell_{n-m}} \in A_n, I_1 = i_1, \ldots, I_M = i_M, I_{m+1}^* = \ell_1, \ldots, I_n^* = \ell_{n-m}\Big\}$$
$$= \mathbb{P}\{Z_{i_1^*} \in A_1, \ldots, Z_{i_m^*} \in A_m, Z_{\ell_1} \in A_{m+1}, \ldots, Z_{\ell_{n-m}} \in A_n\}$$
$$\mathbb{P}\{I_1 = i_1, \ldots, I_M = i_M, I_{m+1}^* = \ell_1, \ldots, I_n^* = \ell_{n-m}\}$$
$$= \mathbb{P}\{Z_1 \in A_1, \ldots, Z_n \in A_n\}\mathbb{P}\{I_1 = i_1, \ldots, I_M = i_M, I_{m+1}^* = \ell_1, \ldots, I_n^* = \ell_{n-m}\}$$

where the second equality follows by exchangeability. Summing in ℓ, one obtains

$$\mathbb{P}\{Z_1' \in A_1, \ldots, Z_n' \in A_n, I_1 = i_1, \ldots, I_M = i_M\}$$
$$= \mathbb{P}\{Z_1 \in A_1, \ldots, Z_n \in A\}\mathbb{P}\{I_1 = i_1, \ldots, I_M = i_M\}.$$

For $m \geq n$, the sum is not needed and the same result follows. This shows that $(Z_n')_n$ is an exchangeable sequence with the same law of Z, and $(Z_n')_n$ and $(I_n)_{n \geq 1}$ are independent. To conclude, note that, with probability one, $I_{\mathscr{C}_n(\Pi)}^* = I_n$, and hence

$$\tilde{\xi}_n = Z_{I_n} = Z_{I_{\mathscr{C}_n(\Pi)}^*} = Z_{\mathscr{C}_n(\Pi)}'.$$

Conversely, let us assume that $\tilde{\xi}_n = Z_{\mathscr{C}_n(\Pi)}'$ and let $(p_j^\downarrow)_{j \geq 1}$ be the weights obtained from Π by (4). Let $I_1, I_2, \ldots$ be the integer-valued random variables appearing in Theorem 1 such that $\Pi = \Pi^*(I)$ a.s. It follows that $\mathscr{C}_n(\Pi) = \mathscr{C}_n(\Pi^*(I))$ and $I_{\mathscr{C}_n(\Pi)}^* = I_n$, where the I_n^* are defined as above for $n \leq K$. Setting

$$Z_m := \begin{cases} Z_k' & \text{if } I_k^* = m \\ Z_m'' & \text{if } I_k^* \neq m \ \forall \ k, \end{cases}$$

with $Z_m'', m \in \mathbb{Z}$ conditionally i.i.d. given $\tilde{H}$ with law $\tilde{H}$, independent of everything else. Arguing as above, one can check that the $(Z_m)_{m \in \mathbb{Z}, m \neq 0}$ are exchangeable random variables with the same law of Z' independent of $(I, p^\downarrow)$. To conclude, note that, in particular,

$$Z_{I_n} = Z_{I_{\mathscr{C}_n(\Pi)}^*} = Z_{\mathscr{C}_n(\Pi)}' \quad \text{a.s..}$$

The conclusion follows by Proposition 2. □

A simple consequence of the previous proposition is the following.

Corollary 1. *Let $(\tilde{\xi}_n)_{n \geq 1}$ be an mSSS as defined in Definition 2. For every $A_1, \ldots, A_n$ Borel set in $\mathbb{X}$,*

$$\mathbb{P}\{\tilde{\xi}_1 \in A_1, \ldots, \tilde{\xi}_n \in A_n\} = \sum_{\pi_n \in \mathcal{P}_n} \mathfrak{q}(|\pi_{1,n}|, \ldots, |\pi_{k,n}|) \mathbb{E}\Big[\prod_{c=1}^{|\pi_n|} \tilde{H}(\cap_{j \in \pi_{c,n}} A_j)\Big].$$

4. Random Partitions Induced by *mSSS*

Let $\tilde{\Pi} = \Pi^*(\xi)$ be the random partition induced by an exchangeable sequence ξ defined as in Definition 2, and let $\Pi^{(0)} := \Pi^*(Z')$ be the partition induced by the corresponding exchangeable sequence $(Z'_n)_n$ (see Proposition 3). Finally, let Π be the partition with EPPF q appearing in Proposition 3. As already observed, if Z' is an i.i.d. sequence from a diffuse H, then $\Pi^{(0)} = [(1), (2), (3), \ldots]$ a.s. and hence $\Pi^*(\xi) = \Pi$. The same result follows if Z' is exchangeable without ties (see Corollary 2). When $\Pi^{(0)}$ is not the trivial partition, it is clear by construction that different blocks in Π can merge in the final clustering configuration (i.e., $\Pi^*(\xi)$). In other words, two observations can share the same value because either they belong to the same block in the latent partition Π or they are in different blocks but they share the same value (from Z'). This simple observation leads us to write the EPPF of the random partition $\Pi^*(\xi)$ using the EPPF of $\Pi^{(0)}$ and of Π.

4.1. Explicit Expression of the EPPF

If $\tilde{\pi}_n = \{\tilde{\pi}_{1,n} \ldots, \tilde{\pi}_{k,n}\}$ is a partition of $[n]$ with $|\tilde{\pi}_{i,n}| = n_i$ ($i = 1, \ldots, k$) and $\boldsymbol{n} = (n_1, \ldots, n_k)$, we can easily describe all the partitions π_n more finely than $\tilde{\pi}_n$, which are compatible with $\tilde{\pi}_n$ in the merging process described above. To do this, first of all, note that any block $\tilde{\pi}_{i,n}$ can arise from the union of $1 \leq m_i \leq n_i$ blocks in the latent partition. Hence, given $\boldsymbol{n} = (n_1, \ldots, n_k)$, where $n = \sum_{i=1}^{k} n_i$, we define the set

$$\mathcal{M}(\boldsymbol{n}) = \left\{ \boldsymbol{m} = (m_1, \ldots, m_k) \in \mathbb{N}^k : 1 \leq m_i \leq n_i \right\}.$$

See Figure 1 for an example. Once a specific configuration $\boldsymbol{m}$ in $\mathcal{M}(\boldsymbol{n})$ is considered, the m_i blocks of the latent partition contributing to the block $\tilde{\pi}_{i,n}$, are characterized by the sufficient statistics $\boldsymbol{\lambda}_i = (\lambda_{i1}, \ldots, \lambda_{in_i}) \in \mathbb{N}^{n_i}$, where λ_{ij} is the number of blocks of j elements among the m_i blocks above. This leads, for $\boldsymbol{m}$ in $\mathcal{M}(\boldsymbol{n})$, to the definition of

$$\Lambda(\boldsymbol{n}, \boldsymbol{m}) := \left\{ \begin{array}{l} \boldsymbol{\lambda} = [\boldsymbol{\lambda}_1, \ldots, \boldsymbol{\lambda}_k] \text{ where } \boldsymbol{\lambda}_i = (\lambda_{i1}, \ldots, \lambda_{in_i}) \in \mathbb{N}^{n_i} : \\ \sum_{j=1}^{n_i} j \lambda_{ij} = n_i, \sum_{j=1}^{n_i} \lambda_{ij} = m_i \text{ for } i = 1, \ldots, k \end{array} \right\}.$$

In summary, the set of partitions $\tilde{\pi}_n$, which are compatible with $\tilde{\pi}_n$ in the merging process described above, can be written as

$$\mathcal{P}_{\tilde{\pi}_n} := \cup_{\boldsymbol{m} \in \mathcal{M}(\boldsymbol{n})} \cup_{\boldsymbol{\lambda} \in \Lambda(\boldsymbol{n}, \boldsymbol{m})} \mathcal{P}_{\tilde{\pi}_n}(\boldsymbol{\lambda}) \qquad (17)$$

where $\mathcal{P}_{\tilde{\pi}_n}(\boldsymbol{\lambda})$ is the set of all the partitions in $\mathcal{P}_n$ with $m_1 + \cdots + m_k =: |\boldsymbol{m}|$ blocks such that

- it is possible to determine k subset containing $m_1, \ldots, m_k$ of these blocks;
- the union of the blocks in the i-th subset coincides with the i-th block of $\tilde{\pi}_n$ for $i = 1, \ldots, k$;
- in the i-th block, there are λ_{ij} blocks with j elements, for $j = 1, \ldots, n_i$.

Given the EPPF q, let

$$\tilde{\mathfrak{q}}(\boldsymbol{\lambda}) := \mathfrak{q}(n_{11}, \ldots, n_{1m_1}, n_{21}, \ldots, n_{km_k}),$$

where $(n_{11}, \ldots, n_{1m_1}, \ldots, n_{km_k})$ is any sequence of integer numbers such that $\sum_{c=1}^{m_i} n_{ic} = \sum_j j \lambda_{ij}$ for every i and $\#\{c : n_{ic} = j\} = \lambda_{ij}$ for every i and j. Note that since the value of $\mathfrak{q}(n_{11}, \ldots, n_{1m_1}, n_{21}, \ldots, n_{km_k})$ depends only on the statistics $\boldsymbol{\lambda}$, $\tilde{\mathfrak{q}}(\boldsymbol{\lambda})$ is well-defined. See, e.g., [26].

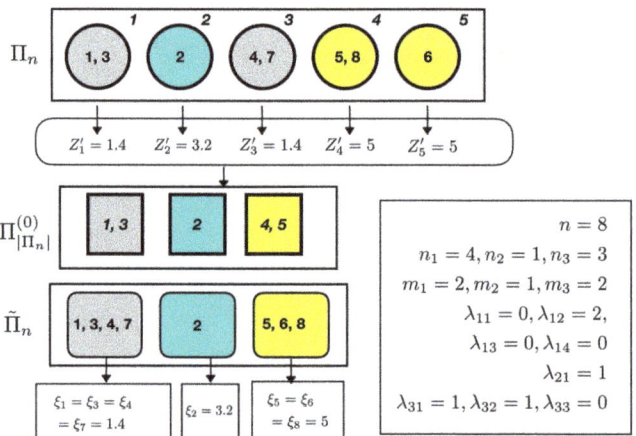

Figure 1. Pictorial representation of the latent partition structure of an mSSS. In the example, the partition induced by $(\xi_1, \ldots, \xi_n)$ for $n = 8$ is $\tilde{\Pi}_n = \{[1,3,4,7], [2], [5,6,8]\}$, and it is represented using rounded squares (left bottom). Circles at the top left represent a compatible latent partition, namely $\Pi_n = \{[1,3], [2], [4,7], [5,8], [6]\}$. The partition on $\{1, \ldots, 5\}$ induced by the latent Z'_n, i.e., $\Pi^{(0)}_{|\Pi_n|} = \{[1,3], [2], [4,5]\}$, is represented with squares in the middle of the figure. Combining Π_n and $\Pi^{(0)}_{|\Pi_n|}$, one obtains $\tilde{\Pi}_n$. The statistics n, m and λ corresponding to this particular configuration are shown in the box at the bottom right.

Finally, recall that the cardinality of $\mathcal{P}_{\tilde{\pi}_n}(\lambda)$ is

$$c(\lambda) := \prod_{i=1}^{k} \frac{(\sum_j j \lambda_{ij})!}{\prod_{j=1}^{n_i} \lambda_{ij}! (j!)^{\lambda_{ij}}} = \prod_{i=1}^{k} \frac{n_i!}{\prod_{j=1}^{n_i} \lambda_{ij}! (j!)^{\lambda_{ij}}},$$

See Equation (39) in [15].

Proposition 4. *Let $\xi = (\xi_n)_{n \geq 1}$ be an (mSSS). Denote by $\tilde{\Pi} = \Pi^*(\xi)$ the random partition induced by ξ and by $\mathsf{q}^{(0)}$ the EPPF of the partition induced by $(Z'_n)_{n \geq 1}$. If $\tilde{\pi}_n = \{\tilde{\pi}_{1,n} \ldots, \tilde{\pi}_{k,n}\}$ is a partition of $[n]$ with $|\tilde{\pi}_{i,n}| = n_i$ ($i = 1, \ldots, k$) and $\mathbf{n} = (n_1, \ldots, n_k)$, then*

$$\mathbb{P}\{\tilde{\Pi}_n = \tilde{\pi}_n\} = \sum_{m \in \mathcal{M}(n)} \mathsf{q}^{(0)}(m) \sum_{\lambda \in \Lambda(n,m)} c(\lambda) \bar{\mathsf{q}}(\lambda). \qquad (18)$$

Proof. Start by writing

$$\mathbb{P}\{\tilde{\Pi}_n = \tilde{\pi}_n\} = \mathbb{P}(\cup_{m \in \mathcal{M}(n)} \cup_{\lambda \in \Lambda(n,m)} \cup_{\pi_n \in \mathcal{P}_{\tilde{\pi}_n}(\lambda)} \{\Pi_n = \pi_n, \tilde{\Pi}_n = \tilde{\pi}_n\}), \qquad (19)$$

which gives

$$\mathbb{P}\{\tilde{\Pi}_n = \tilde{\pi}_n\} = \sum_{m \in \mathcal{M}(n)} \sum_{\lambda \in \Lambda(n,m)} \sum_{\pi_n \in \mathcal{P}_{\tilde{\pi}_n}(\lambda)} \mathbb{P}\{\tilde{\Pi}_n = \tilde{\pi}_n | \Pi_n = \pi_n\} \mathbb{P}\{\Pi_n = \pi_n\} \qquad (20)$$

Whenever $\pi_n \in \mathcal{P}_{\tilde{\pi}_n}(\lambda)$,

$$\mathbb{P}\{\Pi_n = \pi_n\} = \bar{\mathsf{q}}(\lambda).$$

Therefore, we can write (20) as

$$\mathbb{P}\{\tilde{\Pi}_n = \tilde{\pi}_n\} = \sum_{m \in \mathcal{M}(n)} \sum_{\lambda \in \Lambda(n,m)} \sum_{\pi_n \in \mathcal{P}_{\tilde{\pi}_n}(\lambda)} P\{\tilde{\Pi}_n = \tilde{\pi}_n | \Pi_n = \pi_n\} \bar{\mathsf{q}}(\lambda). \qquad (21)$$

Define now the function $M_{\tilde{\pi}_n,\pi_n}: \{1,\ldots,|m|\} \to \{1,\ldots,|\tilde{\pi}_n|\}$ as $M(j) = i$ if $\pi_{j,n}$ is in the i-th subset of blocks, i.e., if $\pi_{j,n} \subset \tilde{\pi}_{i,n}$. Recalling that k is the number of blocks in $\tilde{\pi}_n$, define now a partition $\pi(M_{\tilde{\pi}_n,\pi_n})$ on $\{1,\ldots,|m|\}$ with k block where the i-th block is

$$\{j: M_{\tilde{\pi}_n,\pi_n}(j) = i\}.$$

Recalling that $\Pi^{(0)}$ is the partition induced by the Z's, one has

$$\{\tilde{\Pi}_n = \tilde{\pi}_n, \Pi_n = \pi_n\} = \{\Pi^{(0)}_{|m|} = \pi(M_{\tilde{\pi}_n,\pi_n}), \Pi_n = \pi_n\}$$

which gives

$$P\{\tilde{\Pi}_n = \tilde{\pi}_n | \Pi_n = \pi_n\} = P\{\Pi^{(0)}_{|m|} = \pi(M_{\tilde{\Pi}_n,\pi_n}) | \Pi_n = \pi_n\}$$
$$= \mathbb{P}\{\Pi^{(0)}_{|m|} = \pi(M_{\tilde{\Pi}_n,\pi_n})\}$$

since $\Pi^{(0)}$ and Π are independent. To conclude, note that the vector of the cardinalities of the blocks in $\pi(M_{\tilde{\pi}_n,\pi_n})$ is m; hence, if $\mathfrak{q}^{(0)}$ is the EPPF of $\Pi^{(0)}$, one has $P\{\Pi^{(0)}_{|m|} = \pi(M_{\tilde{\pi}_n,\pi_n})\} = \mathfrak{q}^{(0)}(m)$. Since the cardinality of $\mathcal{P}_{\tilde{\pi}_n}(\lambda)$ is $c(\lambda)$, one obtains the thesis. □

Corollary 2. *Let $\xi = (\xi_n)_n$ be defined according to (mSSS). If $\mathbb{P}\{Z'_1 = Z'_2\} = 0$, then $\Pi^*(\xi) = \Pi$ with probability one.*

Proof. If $\mathbb{P}\{Z'_1 = Z'_2\} = 0$, by exchangeability, $\mathbb{P}\{Z'_{i_1} = Z'_{i_2} = \cdots = Z'_{i_k}\} \leq \mathbb{P}\{Z'_1 = Z'_2\} = 0$. Hence, the Z'_i's are distinct with probability one. Since $(\xi_1,\ldots,\xi_n) = (Z'_{\mathscr{C}_1(\Pi)}, \ldots, Z'_{\mathscr{C}_n(\Pi)})$ by Proposition 3, it follows that $\tilde{\Pi}_n = \Pi_n$. □

Remark 2. *Note that, as a special case, we recover the fact that if ξ is a $gSSS(\mathfrak{q},H)$ with H diffuse (i.e., it is a $SSS(\mathfrak{q},H)$), then the random partition induced by ξ is a.s. Π.*

Remark 3. *The fact that the EPPF of $\tilde{\Pi}_n$ is $\mathfrak{q}$ when $\mathbb{P}\{Z'_1 = Z'_2\} = 0$ can be deduced from (18). Indeed, if $\mathbb{P}\{Z'_1 = Z'_2\} = 0$, then the partition induced by Z' is a.s. the trivial partition $[(1),(2),(3),\ldots]$, so that $\mathfrak{q}^{(0)}(m) = 0$ for every $m \neq (1,1,\ldots,1)$. For $m = (1,1,\ldots,1)$, $\tilde{\pi}_n = \{\tilde{\pi}_{1,n} \ldots, \tilde{\pi}_{k,n}\}$ with $|\tilde{\pi}_{i,n}| = n_i$ ($i = 1,\ldots,k$), and $n = (n_1,\ldots,n_k)$, the set $\Lambda(n,m)$ reduces to the singleton $\lambda^{(1)} := [\lambda_1,\ldots,\lambda_k]$, where $\lambda_i = (0,0,\ldots,1)$ with λ_i of length n_i. Hence, $\bar{\mathfrak{q}}(\lambda^{(1)}) = \mathfrak{q}(n)$ and (18) gives $\mathbb{P}\{\tilde{\Pi}_n = \tilde{\pi}_n\} = \bar{\mathfrak{q}}(\lambda^{(1)}) = \mathfrak{q}(n)$.*

4.2. EPPF When Π Is of Gibbs Type

An important class of exchangeable random partitions is that of Gibbs-type partitions, introduced in [32] and characterized by the EPPF

$$\mathfrak{q}(n_1,\ldots,n_k) := V_{n,k} \prod_{j=1}^k (1-\sigma)_{n_j-1}, \tag{22}$$

where $(x)_n = x(x+1)\ldots(x+n-1)$, $\sigma < 1$ and $V_{n,k}$ are positive real numbers such that $V_{1,1} = 1$ and

$$(n - \sigma k)V_{n+1,k} + V_{n+1,k+1} = V_{n,k}, \quad n \geq 1, \ 1 \leq k \leq n.$$

A noteworthy example of Gibbs-type EPPF is the so-called Pitman–Yor two-parameter family. It is defined by

$$\mathfrak{q}(n_1,\ldots,n_k) := \frac{\prod_{i=1}^{k-1}(\theta + i\sigma)}{(\theta+1)_{n-1}} \prod_{c=1}^k (1-\sigma)_{n_c-1}, \tag{23}$$

where $0 \leq \sigma < 1$ and $\theta > -\sigma$; or $\sigma < 0$ and $\theta = |\sigma|m$ for some integer m; see [2,31].

In order to state the next result, we recall that

$$\sum_{\substack{(\lambda_1,\ldots,\lambda_n) \\ \sum_{j=1}^n j\lambda_j = n, \sum_{j=1}^n \lambda_j = k}} \prod_{j=1}^n [(1-\sigma)_{j-1}]^{\lambda_j} \frac{n!}{\prod_{j=1}^n \lambda_j!(j!)^{\lambda_j}} = S_\sigma(n,k) \tag{24}$$

where $S_\sigma(n,k)$ is the generalized Stirling number of the first kind; see (3.12) in [26]. In the same book, various equivalent definitions of generalized Stirling numbers are presented.

Corollary 3. *Let $\tilde{\Pi} = \Pi^*(\xi)$ be defined as in Proposition 4 with q of Gibbs type defined in (22). If $\tilde{\pi}_n = \{\tilde{\pi}_{1,n}\ldots,\tilde{\pi}_{k,n}\}$ is a partition of $[n]$ with $|\tilde{\pi}_{i,n}| = n_i$ $(i = 1,\ldots,k)$ and $\mathbf{n} = (n_1,\ldots,n_k)$, then*

$$\mathbb{P}\{\tilde{\Pi}_n = \tilde{\pi}_n\} = \sum_{m \in \mathcal{M}(n)} q^{(0)}(m) V_{n,|m|} \prod_{i=1}^k S_\sigma(n_i, m_i).$$

Proof. Combining Proposition 4 with (22), one obtains

$$\mathbb{P}\{\tilde{\Pi}_n = \tilde{\pi}_n\} = \sum_{m \in \mathcal{M}(n)} q^{(0)}(m) V_{n,|m|} \sum_{\lambda \in \Lambda(n,m)} \prod_{i=1}^k \prod_{j=1}^{n_i} [(1-\sigma)_{j-1}]^{\lambda_{i,j}} \frac{n_i!}{\prod_{j=1}^{n_i} \lambda_{ij}!(j!)^{\lambda_{ij}}}$$

$$= \sum_{m \in \mathcal{M}(n)} q^{(0)}(m) V_{n,|m|}$$

$$\times \prod_{i=1}^k \sum_{\substack{(\lambda_{i1},\ldots,\lambda_{in_i}) \\ \sum_{j=1}^{n_i} j\lambda_{ij} = n_i, \sum_{j=1}^{n_i} \lambda_{ij} = m_i}} \prod_{j=1}^{n_i} [(1-\sigma)_{j-1}]^{\lambda_{i,j}} \frac{n_i!}{\prod_{j=1}^{n_i} \lambda_{ij}!(j!)^{\lambda_{ij}}}$$

$$= \sum_{m \in \mathcal{M}(n)} q^{(0)}(m) V_{n,|m|} \prod_{i=1}^k S_\sigma(n_i, m_i).$$

□

4.3. The EPPF of a gSSS(q, H)

As a special case, we now consider the partition induced by a $gSSS(q, H)$ with general base measure H. For the rest of the section, it is useful to decompose H as

$$H(dx) = \sum_{i \geq 1} a_i \delta_{\bar{x}_i}(dx) + (1-a) H^c(dx) \tag{25}$$

where $\mathbb{X}_0 := \{\bar{x}_1, \bar{x}_2, \ldots\}$ is the collection of points with positive H probability, $a_i = H(\bar{x}_i)$, $a = H(\mathbb{X}_0) \in [0,1]$ and $H^c(\cdot) = H(\cdot \cap \mathbb{X}_0^c)/H(\mathbb{X}_0^c)$ is a diffuse probability measure on $\mathbb{X}$. The sum is assumed taken over $i \in \{1,\ldots,|\mathbb{X}_0|\}$.

We now describe $q^{(0)}$, i.e., the EPPF of the partition induced by $(Z'_n)_{n \geq 1}$. Let m in $\mathcal{M}(n)$, where $\mathbf{n} = (n_1,\ldots,n_k)$, and assume that the realization of $\Pi^{(0)}_{|m|}$ has k blocks of cardinality $m_1,\ldots,m_k$. Set $z_i = 0$ if the Z'_n corresponding to the i-th block of $\Pi^{(0)}_{|m|}$ comes from the diffuse component H^c, while $z_k = \ell$ if it is equal to $\bar{x}_\ell$. Since the blocks in $\Pi^{(0)}$ need to be associated with different values of the Z'_n, one has that necessarily $z_i = z_j = 0$ if $z_i = z_j$ for $i \neq j$. In this case, the block is a singleton, which is $m_i = m_j = 1$. On the other hand, if $m_i \geq 2$, i.e., a merging occurred, necessarily, $z_i > 0$. Note that it is also possible that $m_i = 1$ but $z_i > 0$. This motivates the definition of the set

$$\mathcal{Z}(m) := \Big\{(z_1,\ldots,z_k) \in \{0,1,\ldots,|\mathbb{X}_0|\}^k : \text{if } z_i = z_j \text{ for } i \neq j \text{ then } z_i = z_j = 0$$

$$\text{and } m_i = m_j = 1; \text{ if } m_i \geq 2 \text{ then } z_i > 0\Big\}$$

for m in $\mathcal{M}(n)$ where $n = (n_1,\ldots,n_k)$. The probability of obtaining, in an i.i.d. sample of length $|m|$ from H, exactly k ordered blocks with cardinality $m_1,\ldots,m_k$, such that observations in each block are equal and observations in distinct blocks are different, is

$$H^{\#}(m) := \sum_{(z_1,\ldots,z_k)\in \mathcal{Z}(m)} (1-a)^{\#\{j:z_j=0\}} \prod_{j:z_j>0} a_{z_j}^{m_j}.$$

By exchangeability, $H^{\#}(m)$ turns out to be $\mathfrak{q}^{(0)}(m)$. Note also that if $a=1$, $H^{\#}(m)$ reduces to

$$\sum_{(z_1,\ldots,z_k)} \prod_{j=1}^{k} a_{z_j}^{m_j}$$

where $z_1,\ldots,z_k$ runs over all distinct positive integers (less than or equal to $|\mathbb{X}_0|$ if $\mathbb{X}_0$ is finite), which is nothing else (5) for deterministic weights.

To rewrite $H^{\#}(m)$ in a different way, given $m = (m_1,\ldots,m_k)$ in $\mathcal{M}(n)$, let m^* be the vector containing all the elements $m_i > 1$ and let r be its length, with possibly $r = 0$ if $m = (1,1,\ldots,1)$, and define for $\ell \geq 0$

$$A_{m,\ell} = \sum_{j_1 \neq \cdots \neq j_{r+\ell}} a_{j_1}^{m_1^*} \ldots a_{j_r}^{m_r^*} a_{j_{r+1}} \ldots a_{j_{r+\ell}}$$

with the convention that $A_{m,0} = 1$ when $r = 0$. A simple combinatorial argument shows that

$$H^{\#}(m) = \sum_{\ell=0}^{k-r} (1-a)^{k-\ell-r} \binom{k-r}{\ell} A_{m,\ell}.$$

Proposition 4 gives immediately the next proposition.

Proposition 5. *Let ξ be a $gSSS(\mathfrak{q}, H)$ and let $\tilde{\Pi} = \Pi^*(\xi)$ be the random partition induced by ξ. If $\tilde{\pi}_n = [\tilde{\pi}_{1,n}\ldots,\tilde{\pi}_{k,n}]$ is a partition of $[n]$ with $|\tilde{\pi}_{i,n}| = n_i$ $(i=1,\ldots,k)$ and $n = (n_1,\ldots,n_k)$, then*

$$\mathbb{P}\{\tilde{\Pi}_n = \tilde{\pi}_n\} = \sum_{m \in \mathcal{M}(n)} H^{\#}(m) \sum_{\lambda \in \Lambda(n,m)} c(\lambda)\bar{\mathfrak{q}}(\lambda).$$

Remark 4. *Once again, if H is diffuse, then $H^{\#}(m) = 0$ for every $m \neq (1,1,\ldots,1)$. Hence, the above formula reduces to the familiar*

$$\mathbb{P}\{\tilde{\Pi}_n = \tilde{\pi}_n\} = \mathfrak{q}(|\tilde{\pi}_{n,1}|,\ldots,|\tilde{\pi}_{n,k}|) = \mathbb{P}\{\Pi_n = \tilde{\pi}_n\}.$$

4.4. EPPF for gSSS with Spike-and-Slab Base Measure

We now consider the special case of gSSS with a spike-and-slab base measure. A spike-and-slab measure is defined as

$$H(dx) = a\delta_{x_0}(dx) + (1-a)H^c(dx) \tag{26}$$

where $a \in (0,1)$, x_0 is a point of $\mathbb{X}$ and H^c is a diffuse measure on $\mathbb{X}$. This type of measure has been used as a base measure in the Dirichlet process by [16–20] and in the Pitman–Yor process by [21].

Here, we deduce by Proposition 5 the explicit form of the EPPF of the random partition induced by a sequence sampled from a species sampling random probability with such a base measure.

Proposition 6. Let H be as in (26), $\tilde{\Pi}$ be the random partition induced by a $gSSS(\mathfrak{q}, H)$ and Π be an exchangeable random partition with EPPF $\mathfrak{q}$. If $\pi_n = \{\pi_{1,n}, \ldots, \pi_{k,n}\}$ is a partition of $[n]$ with $|\pi_{i,n}| = n_i$ ($i = 1, \ldots, k$), then

$$\mathbb{P}\{\tilde{\Pi}_n = \pi_n\} = (1-a)^k \mathfrak{q}(n_1, \ldots, n_k)$$
$$+ (1-a)^{k-1} \sum_{i=1}^{k} \mathfrak{q}(n_1, \ldots, n_{i-1}, n_{i+1}, \ldots, n_k) \sum_{r=1}^{n_i} a^r q_n(r | n_1, \ldots, n_{i-1}, n_{i+1}, \ldots, n_k) \quad (27)$$

where, conditionally on the fact that Π_{n-n_i} has $k-1$ blocks with sizes $n_1, \ldots, n_{i-1}, n_{i+1}, \ldots, n_k$, the probability that Π_n has $k-1+r$ blocks is denoted by $q_n(r|n_1, \ldots, n_{i-1}, n_{i+1}, \ldots, n_k)$. If, in addition, $\mathfrak{q}$ is of Gibbs type (22), then

$$\mathbb{P}\{\tilde{\Pi}_n = \pi_n\} = (1-a)^k V_{n,k} \prod_{j=1}^{k}(1-\sigma)_{n_j-1}$$
$$+ (1-a)^{k-1} \sum_{i=1}^{k} \prod_{j=1, j\neq i}^{k} (1-\sigma)_{n_j-1} \sum_{r=1}^{n_i} a^r V_{n,k-1+r} S_\sigma(n_i, r).$$

Proof. In this case, $H^\#(\boldsymbol{m}) = 0$ if $m_i \geq 2$ and $m_j \geq 2$ for some $i \neq j$ because H has only one atom. Moreover, $H^\#(\boldsymbol{m})$ is clearly symmetric and

$$H^\#(1, 1, 1, \ldots, 1) = (1-a)^k + k(1-a)^{k-1} a$$

$$H^\#(m, 1, \ldots, 1) = a^m (1-a)^{k-1} \quad \text{for } m > 1.$$

By Proposition 5,

$$\mathbb{P}\{\tilde{\Pi}_n = \pi_n\} = [(1-a)^k + k(1-a)^{k-1} a] \mathfrak{q}(n_1, \ldots, n_k)$$
$$+ (1-a)^{k-1} \sum_{i=1}^{k} \sum_{m_i=2}^{n_i} a^{m_i} \sum_{\lambda \in \Lambda(\boldsymbol{m})} c(\lambda) \bar{\mathfrak{q}}(\lambda)$$
$$= [(1-a)^k + k(1-a)^{k-1}] \mathfrak{q}(n_1, \ldots, n_k) +$$
$$+ (1-a)^{k-1} \sum_{i=1}^{k} \sum_{r=2}^{n_i} a^r \sum_{\substack{\lambda \in \Lambda(\boldsymbol{m}) \text{ for } \boldsymbol{m}: \\ m_i = r,\, m_j = 1, j \neq i}} c(\lambda) \mathfrak{q}(n_1, \ldots, n_{i-1}, \mathbf{n}_r^{(i)}, n_{i+1}, \ldots, n_k)$$
$$= (1-a)^k \mathfrak{q}(n_1, \ldots, n_k)$$
$$+ (1-a)^{k-1} \sum_{i=1}^{k} \sum_{r=1}^{n_i} a^r \sum_{\substack{\lambda \in \Lambda(\boldsymbol{m}) \text{ for } \boldsymbol{m}: \\ m_i = r,\, m_j = 1, j \neq i}} c(\lambda) \mathfrak{q}(n_1, \ldots, n_{i-1}, \mathbf{n}_r^{(i)}, n_{i+1}, \ldots, n_k)$$

where $\mathbf{n}_r^{(i)}$ is any vector of r positive integers with sum n_i such that λ_{ij} of them are equal to j. In view of the definition of $c(\lambda)$, formula (27) is immediately obtained.

If $\mathfrak{q}$ is of Gibbs type, taking into account (24), then

$$q_n(r|n_1, \ldots, n_{i-1}, n_{i+1}, \ldots, n_k) = \frac{V_{n,k-1+r}}{V_{n-n_i, k-1}} S_\sigma(n_i, r)$$

and the second part of the thesis follows by simple algebra. □

Applying Proposition 6 to the Pitman–Yor EPPF defined in (23), one immediately recovers the results stated in Theorem 1 and Corollary 1 of [21].

5. Predictive Distributions

In this section, we provide some expressions for the predictive distributions of mixtures of species sampling sequences.

5.1. Some General Results

Let ξ be as in Definition 2 and let $(Z'_n)_n$ and Π_n be the sequence of exchangeable random variables and the exchangeable random partition appearing in Proposition 3. Let

$$\mathcal{G}_n = \sigma(Z'_1, \ldots, Z'_{|\Pi_n|}, \Pi_n)$$

be the σ-field generated by $(Z'_1, \ldots, Z'_{|\Pi_n|}, \Pi_n)$. By Proposition 3, one has $\xi_n = Z'_{\mathscr{C}_n(\Pi)}$ a.s.; hence, ξ_n is $\mathcal{G}_n$ measurable. Note that, in general, $\sigma(\xi_1, \ldots, \xi_n)$ can be strictly contained in $\mathcal{G}_n$. Set $\Xi_n := |\Pi_n|$ and, for any $k \geq 1$, let $\mathcal{K}_{k+1}(\cdot|\cdot)$ be a kernel corresponding to the conditional distribution of Z'_{k+1} given $Z'_1, \ldots, Z'_k$ (i.e., the $k+1$-predictive distribution of the exchangeable sequence Z'). Finally, recall that $\tilde{\Pi} = \Pi^*(\xi)$ is the partition induced by ξ and define $\xi^*_{1:\tilde{K}_n} = (\xi^*_1, \ldots, \xi^*_{\tilde{K}_n})$ as the distinct values in order of appearance of $\xi_{1:n} := (\xi_1, \ldots, \xi_n)$ with $\tilde{K}_n = |\tilde{\Pi}_n|$.

Proposition 7. *Let ξ as in Definition 2. Then,*

$$\mathbb{P}\{\xi_{n+1} \in \cdot | \mathcal{G}_n\} = \sum_{\ell=1}^{\Xi_n} \omega_{n,\ell}(\Pi_n)\delta_{Z'_\ell}(\cdot) + \nu_n(\Pi_n)\mathcal{K}_{\Xi_n+1}(\cdot|Z'_1, \ldots, Z'_{\Xi_n}) \quad (28)$$

where ν_n and $\omega_{n,\ell}$ are defined by (6) and (7). If $P\{Z'_1 = Z'_2\} = 0$, then

$$\mathbb{P}\{\xi_{n+1} \in \cdot | \xi_1, \ldots, \xi_n\} = \mathbb{P}\{\xi_{n+1} \in \cdot | \xi^*_1, \ldots, \xi^*_{\tilde{K}_n}, \tilde{\Pi}_n\}$$
$$= \sum_{\ell=1}^{\tilde{K}_n} \omega_{n,\ell}(\tilde{\Pi}_n)\delta_{\xi^*_\ell}(\cdot) + \nu_n(\tilde{\Pi}_n)\mathcal{K}_{\tilde{K}_n+1}(\cdot|\xi^*_1, \ldots, \xi^*_{\tilde{K}_n}). \quad (29)$$

Proof. Set

$$E^*_{new} = \{\Xi_{n+1} = \Xi_n + 1\}.$$

Since $\xi_n = Z'_{\mathscr{C}_n(\Pi)}$, one can write

$$\mathbb{P}\{\xi_{n+1} \in A|\mathcal{G}_n\} = \sum_{\ell=1}^{\Xi_n} \mathbb{P}\{\xi_{n+1} \in A, n+1 \in \Pi_{\ell,n}|\mathcal{G}_n\} + \mathbb{P}\{\xi_{n+1} \in A, E^*_{new}|\mathcal{G}_n\}$$
$$= \sum_{\ell=1}^{\Xi_n} \mathbb{P}\{Z'_\ell \in A, n+1 \in \Pi_{\ell,n}|\mathcal{G}_n\} + \mathbb{P}\{Z'_{\Xi_n+1} \in A, E^*_{new}|\mathcal{G}_n\}$$
$$= \sum_{\ell=1}^{\Xi_n} \delta_{Z'_\ell}(A)\mathbb{P}\{n+1 \in \Pi_{\ell,n}|\mathcal{G}_n\} + \mathbb{P}\{Z'_{\Xi_n+1} \in A, E^*_{new}|\mathcal{G}_n\}$$

Now, since Π and $(Z'_n)_n$ are independent, it follows that $\mathbb{P}\{n+1 \in \Pi_{\ell,n}|\mathcal{G}_n\} = \mathbb{P}\{n+1 \in \Pi_{\ell,n}|\Pi_n\} = \omega_{n,\ell}(\Pi_n)$ and also

$$\mathbb{P}\{Z'_{\Xi_n+1} \in A, E^*_{new}|\mathcal{G}_n\}$$
$$= \mathbb{P}\{Z'_{\Xi_n+1} \in A|Z'_1, \ldots, Z'_{\Xi_n}\}\mathbb{P}\{E^*_{new}|\Pi_n\}$$
$$= \mathcal{K}_{\Xi_n+1}(A|Z'_1, \ldots, Z'_{\Xi_n})\nu_n(\Pi_n).$$

Combining all the claims, one obtains (28). The second part of the proof follows since, if $P\{Z'_1 = Z'_2\} = 0$, the Z'_is are distinct with probability one. Since $(\xi_1, \ldots, \xi_n) = (Z'_{\mathscr{C}_1(\Pi)}, \ldots, Z'_{\mathscr{C}_n(\Pi)})$, it follows that $\tilde{\Pi}_n = \Pi_n$, $\Xi_n = \tilde{K}_n$ and $(\xi^*_1, \ldots, \xi^*_{\tilde{K}_n}) = (Z'_1, \ldots, Z'_{\Xi_n})$ with probability one and $\mathcal{G}_n = \sigma(\xi_1, \ldots, \xi_n)$. Hence, (29) follows from (28). □

Remark 5. Note that (29) can be also derived as follows. $\mathbb{P}\{Z'_1 = Z'_2\} = 0$ is equivalent to the fact that $\tilde{H}$ is almost sure diffuse. Hence, conditionally on $\tilde{H}$, we have a $SSS(\mathfrak{q}, \tilde{H})$; then, by (PS2) in Section 2.2, one has

$$\mathbb{P}\{\xi_{n+1} \in \cdot | \xi_1, \ldots, \xi_n, \tilde{H}\} = \sum_{\ell=1}^{\tilde{K}_n} \omega_{n,\ell}(\tilde{\Pi}_n) \delta_{\xi_\ell^*}(\cdot) + \nu_n(\tilde{\Pi}_n) \tilde{H}(dx). \tag{30}$$

Taking the conditional expectation of the previous equation, given $\xi_1, \ldots, \xi_n$, we obtain

$$\mathbb{P}\{\xi_{n+1} \in A | \xi_1, \ldots, \xi_n\} = \sum_{\ell=1}^{\tilde{K}_n} \omega_{n,\ell}(\tilde{\Pi}_n) \delta_{\xi_\ell^*}(A) + \nu_n(\tilde{\Pi}_n) \mathbb{E}[\tilde{H}(A) | \xi_1, \ldots, \xi_n] \tag{31}$$

and the thesis follows since one can check (arguing as in the proof of the proposition) that

$$\mathbb{E}[\tilde{H}(A)|\xi_1,\ldots,\xi_n] = \mathbb{E}[\tilde{H}(A)|Z'_1,\ldots,Z'_{\tilde{K}_n}] = \mathcal{K}_{\tilde{K}_n+1}(A|Z'_1,\ldots,Z'_{\tilde{K}_n}).$$

Assume now that the random variables Z'_j are defined on $\mathbb{X}$ by a Bayesian model with likelihood $f(z_j|u)$ and prior $Q(u)$, where f is a density with respect to a dominating measure λ and Q is a probability measure defined on a Polish space U (the space of parameters). In other words,

$$\mathbb{P}\{Z'_1 \in A_1, \ldots, Z'_k \in A_k\} = \int_U \left(\int_{A_1 \times A_2 \cdots \times A_k} \prod_{j=1}^{k} f(z_j|u) \lambda(dz_1) \ldots \lambda(dz_k) \right) Q(du).$$

Note that this means that $\tilde{H}(A) = \int_A f(z|\tilde{u})\lambda(dz)$, where $\tilde{u} \sim Q$. Bayes' theorem (see, e.g., Theorem 1.31 in [33]) gives

$$\mathbb{P}\{Z'_{k+1} \in dz_{k+1}|Z'_1,\ldots,Z'_k\} = \left(\int_U f(z_{k+1}|u) Q(du|Z'_1,\ldots,Z'_k) \right) \lambda(dz_{k+1})$$

where $Q(du|Z_1,\ldots,Z_k)$ is the usual posterior distribution, which is

$$Q(du|Z'_1,\ldots,Z'_k) := \frac{\prod_{j=1}^k f(Z'_j|u) Q(du)}{\int_U \prod_{j=1}^k f(Z'_j|v) Q(dv)}.$$

If λ is a diffuse measure, one obtains $\mathbb{P}\{Z'_1 = Z'_2\} = 0$. Hence, (29) in Proposition 7 applies and one has

$$\mathbb{P}\{\xi_{n+1} \in dx | \xi_1, \ldots, \xi_n\} = \sum_{\ell=1}^{\tilde{K}_n} \omega_{n,\ell}(\tilde{\Pi}_n) \delta_{\xi_\ell^*}(dx) \\ + \nu_n(\tilde{\Pi}_n) \left(\int_U f(x|u) Q(du|\xi_1^*,\ldots,\xi_{\tilde{K}_n}^*) \right) dx. \tag{32}$$

For example, one can apply this result to a mixture of Dirichlet processes in the sense of [10], as briefly described at the end of Section 3.1. Assume that $\alpha_{\tilde{u}}(\mathbb{X})$ and $\tilde{H}(\cdot) = \alpha_{\tilde{u}}(\cdot)/\alpha_{\tilde{u}}(\mathbb{X})$ are independent and that $\alpha_u(A)/\alpha_u(\mathbb{X}) = \int_Z f(z|u)\lambda(dz)$ for a suitable dominating diffuse measure λ.

Under these hypotheses, a sample $(\xi_n)_{n\geq 1}$ from a mixture of Dirichlet processes is an mSSS with q described in (15) and, in addition, $\mathbb{P}\{Z'_1 = Z'_2\} = 0$. Combining (15) with (6) and (7), one obtains

$$\omega_{n,\ell}(\tilde{\Pi}_n) = |\tilde{\Pi}_{n,\ell}| \frac{\int_U \frac{\alpha_u(\mathbb{X})^{\tilde{K}_n}}{(\alpha_u(\mathbb{X}))_{n+1}} Q(du)}{\int_U \frac{\alpha_u(\mathbb{X})^{\tilde{K}_n}}{(\alpha_u(\mathbb{X}))_n} Q(du)}$$

and
$$\nu_n(\tilde{\Pi}_n) = \frac{\int_U \frac{\alpha_u(\mathbb{X})^{\tilde{k}_n+1}}{(\alpha_u(\mathbb{X}))_{n+1}} Q(du)}{\int_U \frac{\alpha_u(\mathbb{X})^{\tilde{k}_n}}{(\alpha_u(\mathbb{X}))_n} Q(du)}$$

Hence, the predictive distribution of ξ_{n+1} given $(\xi_1, \ldots, \xi_n)$ is (33) for $\omega_{n,\ell}(\tilde{\Pi}_n)$ and $\nu_n(\tilde{\Pi}_n)$ given above.

Note that the same result can be deduced by combining Lemma 1 and Corollary 3.2' in [10].

Example 1 (Species Sampling NIG). *Let Z'_n be defined as a mixture of normal random variables with Normal-Inverse-Gamma prior. In other words, given $\mu_0 \in \mathbb{R}$, $k_0 > 0$, $\alpha_0 > 0$, $\beta_0 > 0$,*

$$Z_n | \tilde{\mu}, \tilde{\sigma}^2 \overset{i.i.d.}{\sim} \mathcal{N}(\tilde{\mu}, \tilde{\sigma}^2)$$
$$\tilde{\mu} | \tilde{\sigma}^2 \sim \mathcal{N}(\mu_0, \tilde{\sigma}^2/k_0)$$
$$\tilde{\sigma}^2 \sim \text{In}\Gamma(\alpha_0, \beta_0)$$

where $\mathcal{N}(\mu, \sigma^2)$ denotes a normal distribution of mean μ and variance σ^2 and $\text{In}\Gamma(\alpha, \beta)$ is the inverse gamma distribution with shape α and scale β. Let $\mathcal{T}_\nu(\cdot | \mu, \sigma^2)$ be the density of a Student-T distribution with ν degrees of freedom and (μ, σ) position/scale parameters, i.e.,

$$\mathcal{T}_\nu(x|\mu, \sigma^2) := \frac{1}{\sqrt{\sigma^2}} \frac{\Gamma(\frac{\nu+1}{2})}{\sqrt{\nu\pi}\Gamma(\frac{\nu}{2})} \left(1 + \frac{1}{\nu\sigma^2}(x-\mu)^2\right)^{\frac{-\nu+1}{2}}.$$

It is well known that, under these assumptions, $\mathcal{K}_{k+1}(A|z_1, \ldots, z_k)$ has density $\mathcal{T}_{2\alpha_k}(z|\mu_k, \sigma_k^2)$, where the parameters are updated

$$\mu_k = \frac{k_0 \mu_0 + k \bar{z}_n}{k_0 + n} \qquad \bar{z}_k = \frac{1}{k} \sum_{j=1}^k z_j \qquad \alpha_k = \alpha_0 + k/2,$$

$$\sigma_k^2 = \frac{\left(\beta_0 + \frac{1}{2}\sum_{j=1}^k (z_j - \bar{z}_k)^2 + \frac{kk_0(\bar{z}_k - \mu_0)^2}{2(n+k_0)}\right)(k_0 + k + 1)}{(\alpha_0 + k/2)(k_0 + k)}$$

Thus, in this case, if $z_1, \ldots, z_k$ are distinct real numbers and $\pi_n = [\pi_{1,n}, \ldots, \pi_{k,n}]$, one has

$$\mathbb{P}\{\xi_{n+1} \in dx | \xi_1^* = z_1, \ldots, \xi_k^* = z_k, \Pi^*(\xi) = \pi_n\} = \sum_{\ell=1}^k \omega_{n,\ell}(\pi_n) \delta_{\xi_\ell^*}(dx) \quad (33)$$
$$+ \nu_n(\pi_n) \mathcal{T}_{2\alpha_k}(x|\mu_n, \sigma_k^2) dx.$$

We show an application of (33) to a true dataset by choosing $\omega_{n,\ell}$ and ν_n according to a Pitman–Yor two-parameter family; see (23). The data are the relative changes in reported larcenies between 1991 and 1995 (relative to 1991) for the 90 most populous US counties, taken from Section 2.1 of [34]. We apply our models to both the raw data and the rounded data (approximated to the second digit) in order to obtain ties in the ξs. In the evaluation of the predictive CDFs, we fix $\mu_0 = 0$, $\alpha_0 = 0.1$, $\beta_0 = 0.1$ and $k_0 = 0.1$. In Figure 2, we report the empirical CDF of the rounded data (solid line), the predictive CDF obtained from (33) (dotted line) and the predictive CDF of a Pitman–Yor species sampling sequence (see PS2) with $H = \mathcal{T}_{2\alpha_0}(\cdot|\mu_0, \sigma_0^2)$, $\sigma_0^2 = \beta_0(k_0 + 1)/\alpha_0 k_0$ (dashed line). Similar plots are reported in Figure 3, with raw data in place of the rounded data. Note that in all the various settings, the influence of the hyper-parameters (θ, σ) is stronger in the CDF of the simple Pitman–Yor species sampling model with respect to the corresponding predictive CDF derived from (33).

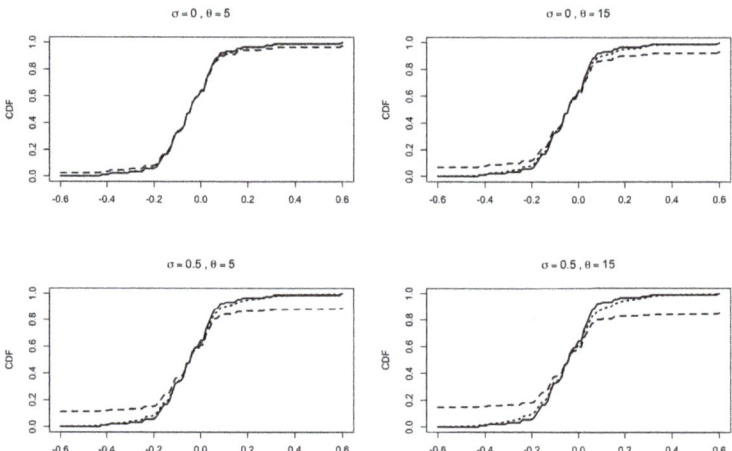

Figure 2. Predictive CDFs for the relative changes in larcenies between 1991 and 1995 (relative to 1991) for the 90 most populous US counties; data taken from Section 2.1 of [34]. Data have been rounded to the second decimal. Here, $n = 90$ and $k = 36$. Solid line: empirical CDF. Dotted line: predictive CDF from (33). Dashed line: predictive CDF from PS2 with $H = \mathcal{T}_{2\alpha_0}(\cdot|\mu_0, \sigma_0^2)$, $\sigma_0^2 = \beta_0(k_0+1)/\alpha_0 k_0$. Different plots correspond to different values of θ and σ. In all the plots, the predictive CDFs are evaluated with $\mu_0 = 0$, $\alpha_0 = 0.1$, $\beta_0 = 0.1$ and $k_0 = 0.1$.

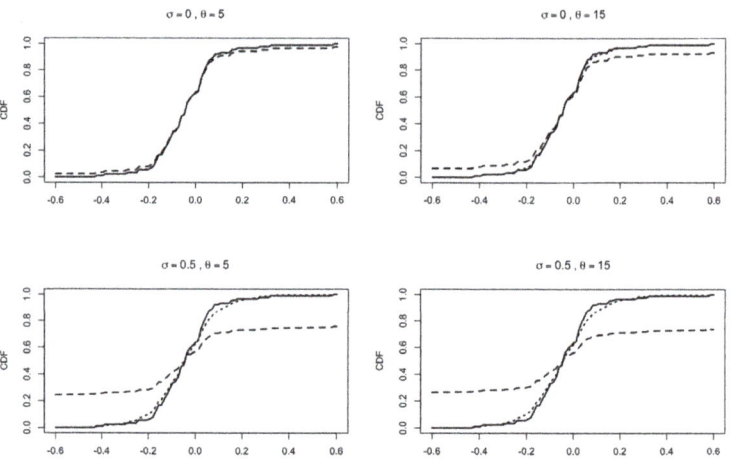

Figure 3. Predictive CDFs for the relative changes in larcenies between 1991 and 1995 (relative to 1991) for the 90 most populous US counties; data taken from Section 2.1 of [34]. Raw data, without rounding. Here, $n = 90$ and $k = 36$. Solid line: empirical CDF. Dotted line: predictive CDF from (33). Dashed line: predictive CDF from PS2 with $H = \mathcal{T}_{2\alpha_0}(\cdot|\mu_0, \sigma_0^2)$, $\sigma_0^2 = \beta_0(k_0+1)/\alpha_0 k_0$. Different plots correspond to different values of θ and σ. In all the plots, the predictive CDFs are evaluated with $\mu_0 = 0$, $\alpha_0 = 0.1$, $\beta_0 = 0.1$ and $k_0 = 0.1$.

5.2. Predictive Distributions for gSSS

We now deduce an explicit form for the predictive distribution of a *gSSS* with general base measure H given in (25).

Recall that we denote by $\tilde{\Pi}_n$ the partition induced by $\xi_{1:n}$, with $\tilde{K}_n = |\tilde{\Pi}_n|$, and by Π_n the latent partition appearing in Proposition 3. We also set

$$\zeta_i = \begin{cases} \ell & \text{if } \xi_i^* = \bar{x}_\ell \\ 0 & \text{if } \xi_i^* \in \mathbb{X}_0^c. \end{cases}$$

The variable ζ_i is a discrete random variable that takes value 0 if ξ_i^* comes from the diffuse component of H.

Let $\mathcal{P}_{\tilde{\pi}_n, z_{1:k}} \subset \mathcal{P}_{\tilde{\pi}_n}$ be the set of all the possible configurations of Π_n that are compatible with the observed partition $\tilde{\Pi}_n = \tilde{\pi}_n$ and the additional information given by $\zeta_{1:\tilde{K}_n} = z_{1:k}$, $\tilde{K}_n = k$. In order to describe this set, observe that if $z_i > 0$, then the block $\tilde{\pi}_{i,n}$ may arise from the union of more blocks in π_n, while, if $z_i = 0$, then $\tilde{\pi}_{i,n} = \pi_{\phi(i),n}$ for some ϕ. Note that it may happen that $\phi(i) \neq i$.

Recalling that the elements in $\boldsymbol{m} = (m_1, \ldots, m_k)$ in (17) are used to describe the numbers of sub-blocks into which the blocks of $\tilde{\pi}_n$ have been divided to form the latent partition π_n, it turns out that the set $\mathcal{P}_{\tilde{\pi}_n, z_{1:k}}$ has the additional constraint $m_i = 1$ whenever $z_i = 0$. These considerations yield that, starting from $\tilde{\pi}_n$ and $z_{1:k}$, the set of admissible $\boldsymbol{m}$ can also be described by resorting to the definition of $\mathcal{Z}(\boldsymbol{m})$ as follows:

$$\mathcal{M}(\boldsymbol{n}, z_{1:k}) := \{\boldsymbol{m} \in \mathcal{M}(\boldsymbol{n}) : z_{1:k} \in \mathcal{Z}(\boldsymbol{m})\}$$
$$= \{\boldsymbol{m} \in \mathbb{N}^k : m_i = 1 \text{ if } z_i = 0, 1 \leq m_i \leq n_i \text{ if } z_i > 0\}.$$

With this position, one has

$$\begin{aligned} \mathcal{P}_{\tilde{\pi}_n, z_{1:k}} &= \cup_{\boldsymbol{m} \in \mathcal{M}(\boldsymbol{n}): z_{1:k} \in \mathcal{Z}(\boldsymbol{m})} \cup_{\lambda \in \Lambda(\boldsymbol{n}, \boldsymbol{m})} \mathcal{P}_{\tilde{\pi}_n}(\lambda) \\ &= \cup_{\boldsymbol{m} \in \mathcal{M}(\boldsymbol{n}, z_{1:k})} \cup_{\lambda \in \Lambda(\boldsymbol{n}, \boldsymbol{m})} \mathcal{P}_{\tilde{\pi}_n}(\lambda), \end{aligned} \quad (34)$$

where $\Lambda(\boldsymbol{n}, \boldsymbol{m})$ and $\mathcal{P}_{\tilde{\pi}_n}(\lambda)$ have been defined in Section 4.1.

For any $\boldsymbol{m}$ in $\mathcal{M}(\boldsymbol{n}, z_{1:k})$ and any λ in $\Lambda(\boldsymbol{n}, \boldsymbol{m})$, we define

$$\lambda^{new} := [\lambda_1, \ldots, \lambda_k, 1].$$

In other words, λ^{new} corresponds to the configuration obtained from λ by adding one new element as a new block. In the following, let $\tilde{\boldsymbol{N}} = (|\tilde{\Pi}_{1,n}|, \ldots, |\tilde{\Pi}_{\tilde{K}_n, n}|)$, and let $\tilde{\boldsymbol{N}}^{i+}$ be obtained from $\tilde{\boldsymbol{N}}$ by adding 1 to its i-th component.

Proposition 8. *Let $(\xi_n)_{n \geq 1}$ be a gSSS$(\mathfrak{q}, H)$. Then, for any A in $\mathcal{X}$,*

$$P\{\xi_{n+1} \in A | \xi_{1:n}\} = \frac{1}{\mathcal{Z}_n} \left(\sum_{i=1}^{\tilde{K}_n} w_i \delta_{\xi_i^*}(A) + w_0 \bar{H}_n(A) \right) \quad \text{a.s.}$$

where

$$\bar{H}_n(A) := \left[\sum_{\ell: \bar{x}_\ell \notin \xi_{1:\tilde{K}_n}^*} a_\ell \delta_{\bar{x}_\ell}(A) + (1-a) H^c(A) \right] H(\mathbb{X} \setminus \xi_{1:\tilde{K}_n}^*)^{-1},$$

$$w_i := \sum_{\boldsymbol{m} \in \mathcal{M}(\tilde{\boldsymbol{N}}^{i+}, \zeta_{1:\tilde{K}_n})} \prod_{j: \zeta_j > 0} a_{\zeta_j}^{m_j} \sum_{\lambda \in \Lambda(\tilde{\boldsymbol{N}}^{i+}, \boldsymbol{m})} c(\lambda) \bar{\mathfrak{q}}(\lambda),$$

$$w_0 := H(\mathbb{X} \setminus \xi_{1:\tilde{K}_n}^*) \sum_{\boldsymbol{m} \in \mathcal{M}(\tilde{\boldsymbol{N}}, \zeta_{1:\tilde{K}_n})} \prod_{j: \zeta_j > 0} a_{\zeta_j}^{m_j} \sum_{\lambda \in \Lambda(\tilde{\boldsymbol{N}}, \boldsymbol{m})} c(\lambda) \bar{\mathfrak{q}}(\lambda^{new}),$$

$$\mathcal{Z}_n := \sum_{\boldsymbol{m} \in \mathcal{M}(\tilde{\boldsymbol{N}}, \zeta_{1:\tilde{K}_n})} \prod_{j: \zeta_j > 0} a_{\zeta_j}^{m_j} \sum_{\lambda \in \Lambda(\tilde{\boldsymbol{N}}, \boldsymbol{m})} c(\lambda) \bar{\mathfrak{q}}(\lambda).$$

Proof. We start by defining the following events for $i = 1, \ldots, \tilde{K}_n$:

$$E_i = \{\xi_{n+1} = \xi_i^*\}, \quad E_{new} = \{\xi_{n+1} \notin \xi_{1:\tilde{K}_n}^*\}.$$

Since conditioning on $\xi_{1:n}$ is equivalent to the condition on $[\xi_{1:\tilde{K}_n}^*, \tilde{\Pi}_n]$, one can write

$$P\{\xi_{n+1} \in A | \xi_{1:n}\} = \sum_{i=1}^{\tilde{K}_n} P\{\xi_{n+1} \in A, E_i | \xi_{1:\tilde{K}_n}^*, \tilde{\Pi}_n\} + P\{\xi_{n+1} \in A, E_{new} | \xi_{1:\tilde{K}_n}^*, \tilde{\Pi}_n\}$$

Now, set

$$E_{new}^* := \{\mathscr{C}_{n+1}(\Pi) = |\Pi_n| + 1\}$$

and

$$E_i^* = \{|\mathscr{C}_{n+1}(\Pi)| \leq |\Pi_n| \text{ and } \Pi_{\mathscr{C}_{n+1}(\Pi),n} \subset \tilde{\Pi}_{i,n}\}.$$

On $\zeta_i = 0$, one has (up to zero probability sets)

$$\{\xi_{n+1} \in A\} \cap E_i = \{\xi_i^* \in A\} \cap E_i^*$$

while, on $\zeta_i > 0$ (up to zero probability sets),

$$\{\xi_{n+1} \in A\} \cap E_i = \left(\{\xi_i^* \in A\} \cap E_i^*\right) \cup \left(\{\xi_i^* \in A\} \cap \{Z_{|\Pi_n|+1}' = \bar{x}_{\zeta_i}\} \cap E_{new}^*\right).$$

Note that (up to zero probability sets)

$$\{Z_{|\Pi_n|+1}' = \bar{x}_{\zeta_i}\} \cap E_{new}^* \cap \{\zeta_i = 0\} = \emptyset.$$

Hence,

$$\begin{aligned} \mathbb{P}\{\xi_{n+1} \in A, E_i | \xi_{1:\tilde{K}_n}^*, \tilde{\Pi}_n\} \\ = \delta_{\xi_i^*}(A) \mathbb{P}\{E_i^* \cup (\{Z_{|\Pi_n|+1}' = \bar{x}_{\zeta_i}\} \cap E_{new}^*) | \tilde{\Pi}_n, \xi_{1:\tilde{K}_n}^*\}. \end{aligned} \quad (35)$$

Similarly, using that $E_{new} \subset E_{new}^*$, one obtains

$$\begin{aligned} \mathbb{P}\{\xi_{n+1} \in A, E_{new} | \xi_{1:n}\} &= \mathbb{P}\{\xi_{n+1} \in A, E_{new}, E_{new}^* | \tilde{\Pi}_n, \xi_{1:\tilde{K}_n}^*\} \\ &= \mathbb{P}\{\xi_{n+1} \in A, E_{new} | \xi_{1:\tilde{K}_n}^*, E_{new}^*\} \mathbb{P}\{E_{new}^* | \tilde{\Pi}_n, \xi_{1:\tilde{K}_n}^*\} \\ &= H_n(A) \mathbb{P}\{E_{new}^* | \tilde{\Pi}_n, \xi_{1:\tilde{K}_n}^*\} \end{aligned} \quad (36)$$

where

$$H_n(A) = \sum_{\ell: \bar{x}_\ell \notin \xi_{1:\tilde{K}_n}^*} \bar{a}_\ell \delta_{\bar{x}_\ell}(A) + (1-a) H^c(A).$$

At this stage, note that, by construction,

$$\mathcal{L}(\xi_{1:\tilde{K}_n}^* | \Pi_n, Z_{|\Pi_n|+1}', \zeta_{1:\tilde{K}_n}, \tilde{\Pi}_n, \Pi_{n+1}) = \mathcal{L}(\xi_{1:\tilde{K}_n}^* | \zeta_{1:\tilde{K}_n})$$

where $\mathcal{L}(\xi_{1:\tilde{K}_n}^* | \zeta_{1:\tilde{K}_n})$ is characterized by

$$P(\xi_1^* \in A_1, \ldots, \xi_{\tilde{K}_n}^* \in A_{\tilde{K}_n} | \zeta_{1:\tilde{K}_n}) = \prod_{i=1}^{\tilde{K}_n} \left(H^c(A_i) \mathbf{1}\{\zeta_i = 0\} + \delta_{\bar{x}_{\zeta_i}}(A_i) \mathbf{1}\{\zeta_i > 0\}\right),$$

and then

$$\mathcal{L}(Z_{|\Pi_n|+1}', \zeta_{1:\tilde{K}_n}, \tilde{\Pi}_n, \Pi_{n+1}, \xi_{1:\tilde{K}_n}^* | \Pi_n) = \mathcal{L}(Z_{|\Pi_n|+1}', \zeta_{1:\tilde{K}_n}, \tilde{\Pi}_n, \Pi_{n+1} | \Pi_n) \mathcal{L}(\xi_{1:\tilde{K}_n}^* | \zeta_{1:\tilde{K}_n}).$$

Hence,

$$\mathcal{L}(\Pi_n, \Pi_{n+1}, Z'_{|\Pi_n|+1}, \xi^*_{1:\tilde{K}_n}, \zeta_{1:\tilde{K}_n}, \tilde{\Pi}_n) \\ = \mathcal{L}(\Pi_n)\mathcal{L}(Z'_{|\Pi_n|+1}, \zeta_{1:\tilde{K}_n}, \tilde{\Pi}_n, \Pi_{n+1}|\Pi_n)\mathcal{L}(\xi^*_{1:\tilde{K}_n}|\zeta_{1:\tilde{K}_n}) \quad (37)$$

which shows, in particular, that $[\Pi_n, \Pi_{n+1}, Z'_{|\Pi_n|+1}, \tilde{\Pi}_n]$ and $\xi^*_{1:\tilde{K}_n}$ are conditionally independent given $\zeta_{1:\tilde{K}_n}$. Since E^*_i, E^*_{new} and $\{Z'_{|\Pi_n|+1} = \bar{x}_{\zeta_i}\} \cap E^*_{new}$ depend logically only on $\Pi_{n+1}, Z'_{|\Pi_n|+1}, \tilde{\Pi}_n, \zeta_{1:\tilde{K}_n}$, one obtains

$$\mathbb{P}\{E^*_i|\tilde{\Pi}_n, \xi^*_{1:\tilde{K}_n}\} = \mathbb{P}\{E^*_i|\tilde{\Pi}_n, \zeta_{1:\tilde{K}_n}\}, \\ \mathbb{P}\{(Z'_{|\Pi_n|+1} = \bar{x}_{\zeta_i}) \cap E^*_{new}|\tilde{\Pi}_n, \xi^*_{1:\tilde{K}_n}\} = \mathbb{P}\{(Z'_{|\Pi_n|+1} = \bar{x}_{\zeta_i}) \cap E^*_{new}|\tilde{\Pi}_n, \zeta_{1:\tilde{K}_n}\} \quad (38)$$

and, finally,

$$\mathbb{P}\{E^*_{new}|\tilde{\Pi}_n, \xi^*_{1:\tilde{K}_n}\} = \mathbb{P}\{E^*_{new}|\tilde{\Pi}_n, \zeta_{1:\tilde{K}_n}\}. \quad (39)$$

Since $[\tilde{\Pi}_n, \zeta_{1:\tilde{K}_n}, \tilde{K}_n]$ are discrete random variables, we use the elementary definition of the conditional probability of events to evaluate the conditional distributions (38) and (39). Specifically, assume that $\tilde{K}_n = k$, $[\tilde{\Pi}_n, \zeta_{1:\tilde{K}_n}] = [\tilde{\pi}_n, z_{1:k}]$, $\tilde{N} = n$, and, for a given event E, write

$$\mathbb{P}\{E|\tilde{\Pi}_n = \tilde{\pi}_n, \zeta_{1:\tilde{K}_n} = z_{1:k}\} = \frac{\mathbb{P}\{E, \tilde{\Pi}_n = \tilde{\pi}_n, \zeta_{1:\tilde{K}_n} = z_{1:k}\}}{\mathbb{P}\{\tilde{\Pi}_n = \tilde{\pi}_n, \zeta_{1:\tilde{K}_n} = z_{1:k}\}}. \quad (40)$$

As for the denominator in (40), letting $\mathcal{M}^*_n = \mathcal{M}(n, z_{1:k})$ and $J = \#\{i : z_i > 0\}$, using (34), one obtains

$$\mathbb{P}\{\tilde{\Pi}_n = \tilde{\pi}_n, \zeta_{1:\tilde{K}_n} = z_{1:k}\} \\ = \sum_{\pi_n \in \mathcal{P}_{\tilde{\pi}_n, z_{1:k}}} \mathbb{P}\{\tilde{\Pi}_n = \tilde{\pi}_n, \zeta_{1:\tilde{K}_n} = z_{1:k}, \Pi_n = \pi_n\} \\ = \sum_{m \in \mathcal{M}^*_n} \sum_{\lambda \in \Lambda(n,m)} \sum_{\pi_n \in \mathcal{P}_{\tilde{\pi}_n}(\lambda)} \mathbb{P}\{\tilde{\Pi}_n = \tilde{\pi}_n, \zeta_{1:n} = z_{1:k}, \Pi_n = \pi_n\} \\ = (1-a)^{k-J} \sum_{m \in \mathcal{M}^*_n} \prod_{j:z_j>0} a^{m_j}_{z_j} \sum_{\lambda \in \Lambda(n,m)} \sum_{\pi_n \in \mathcal{P}_{\tilde{\pi}_n}(\lambda)} \bar{q}(\lambda) \\ = (1-a)^{k-J} \sum_{m \in \mathcal{M}^*_n} \prod_{j:z_j>0} a^{m_j}_{z_j} \sum_{\lambda \in \Lambda(n,m)} c(\lambda)\bar{q}(\lambda).$$

As for the numerators in (40), when $E = E^*_{new}$, we start from

$$P\{E^*_{new}, \tilde{\Pi}_n = \tilde{\pi}_n, \zeta_{1:\tilde{K}_n} = z_{1:k}, \Pi_n = \pi_n\} \\ = P\{E^*_{new}|\Pi_n = \pi_n\}P\{\tilde{\Pi}_n = \tilde{\pi}_n, \zeta_{1:\tilde{K}_n} = z_{1:k}, \Pi_n = \pi_n\} \\ = P\{E^*_{new}|\Pi_n = \pi_n\}(1-a)^{k-J} \prod_{j:z_j>0} a^{m_j}_{z_j} \bar{q}(\lambda) \\ = (1-a)^{k-J} \prod_{j:z_j>0} a^{m_j}_{z_j} \bar{q}(\lambda^{new}).$$

where, in the last equality, we used that for $\pi_n \in \mathcal{P}_{\tilde{\pi}_n}(\lambda)$, one has

$$P\{E^*_{new}|\Pi_n = \pi_n\} = \frac{\bar{q}(\lambda^{new})}{\bar{q}(\lambda)}.$$

Taking the sum over $\mathcal{P}_{\tilde{\pi}_n, z_{1:k}}$ gives

$$P\{E^*_{new}, \tilde{\Pi}_n = \tilde{\pi}_n, \zeta_{1:\tilde{K}_n} = z_{1:k}\} = (1-a)^{k-J} \sum_{m \in \mathcal{M}^*_n} \prod_{j:z_j>0} a^{m_j}_{z_j} \sum_{\lambda \in \Lambda(n,m)} c(\lambda)\bar{q}(\lambda^{new}).$$

Combining these with (39) and (40) and recalling that $\mathcal{M}_n^* = \mathcal{M}(n, z_{1:k})$, one obtains

$$\mathbb{P}\{E_{new}^* | \tilde{\Pi}_n, \zeta_{1:\tilde{K}_n}^*\} = \frac{\sum_{m \in \mathcal{M}(n, z_{1:k})} \prod_{j:z_j > 0} a_{z_j}^{m_j} \sum_{\lambda \in \Lambda(n,m)} c(\lambda) \bar{\mathfrak{q}}(\lambda^{new})}{\sum_{m \in \mathcal{M}(n, z_{1:k})} \prod_{j:z_j > 0} a_{z_j}^{m_j} \sum_{\lambda \in \Lambda(n,m)} c(\lambda) \bar{\mathfrak{q}}(\lambda)}.$$

Finally, it remains to consider (40) when $E = E_i^* \cup (\{Z'_{|\Pi_n|+1} = \bar{x}_{\zeta_i}\} \cap E_{new}^*)$. Now, observe that

$$\left(E_i^* \cup (\{Z'_{|\Pi_n|+1} = \bar{x}_{\zeta_i}\} \cap E_{new}^*)\right) \cap \{\tilde{\Pi}_n = \tilde{\pi}_n, \zeta_{1:\tilde{K}_n} = z_{1:k}\}$$
$$= \{\tilde{\Pi}_{n+1} = \tilde{\pi}_n^{i+}, \zeta_{1:\tilde{K}_n} = z_{1:k}\} = \{\tilde{\Pi}_{n+1} = \tilde{\pi}_n^{i+}, \zeta_{1:\tilde{K}_{n+1}} = z_{1:k}\}$$

where $\tilde{\pi}_n^{i+}$ denote the partition in $\mathcal{P}_{n+1}$ obtained from $\tilde{\pi}_n$ by adding $n+1$ to the i-th block of $\tilde{\pi}_n$. Note that, for the second equality, we used that, on $\tilde{\Pi}_{n+1} = \tilde{\pi}_n^{i+}$, one has $\tilde{K}_{n+1} = \tilde{K}_n = k$.

Hence, using (34) with $\tilde{\pi}_n^{i+}$ in place of $\tilde{\pi}_n$, one concludes that

$$\mathbb{P}\{\tilde{\Pi}_{n+1} = \tilde{\pi}_n^{i+}, \zeta_{1:\tilde{K}_{n+1}} = z_{1:k}\}$$
$$= \sum_{\substack{m \in \mathcal{M}(n^{i+}, z_{1:k}) \\ \lambda \in \Lambda(n,m)}} \sum_{\pi_{n+1} \in \mathcal{P}_{\tilde{\pi}_{n+1}}(\lambda)} P\{\tilde{\Pi}_{n+1} = \tilde{\pi}_n^{i+}, \Pi_{n+1} = \pi_{n+1}, \zeta_{1:\tilde{K}_{n+1}} = z_{1:k}\}$$
$$= (1-a)^{k-J} \sum_{m \in \mathcal{M}(n^{i+}, z_{1:k})} \prod_{j:z_j > 0} a_{z_j}^{m_j} \sum_{\lambda \in \Lambda(n,m)} c(\lambda) \bar{\mathfrak{q}}(\lambda)$$

where $n^{i+} = (n_1, \ldots, n_i + 1, \ldots, n_k)$. Hence, by (38)–(40), one can write

$$\mathbb{P}\{E_i^* \cup (\{Z'_{|\Pi_n|+1} = \bar{x}_{\zeta_i}\} \cap E_{new}^*) | \tilde{\Pi}_n, \zeta_{1:\tilde{K}_n}^*\}$$
$$= \frac{\sum_{m \in \mathcal{M}(n^{i+}, z_{1:k})} \prod_{j:z_j > 0} a_{z_j}^{m_j} \sum_{\lambda \in \Lambda(n,m)} c(\lambda) \bar{\mathfrak{q}}(\lambda)}{\sum_{m \in \mathcal{M}(n, z_{1:k})} \prod_{j:z_j > 0} a_{z_j}^{m_j} \sum_{\lambda \in \Lambda(n,m)} c(\lambda) \bar{\mathfrak{q}}(\lambda)}.$$

Combining these results, one obtains the thesis. □

6. Conclusions and Discussion

We have defined a new class of exchangeable sequences, called mixtures of species sampling sequences (mSSS). We have shown that these sequences include various well-known Bayesian nonparametric models. In particular, the observations of many nonparametric hierarchical models (e.g., hierarchical Dirichlet process, hierarchical Pitman–Yor process and, more generally, hierarchical species sampling models [22–25]) are mSSS. We have shown that also observations sampled from a mixture of Dirichlet processes [10] are mSSS, under some additional assumptions. Our general class also includes species sampling sequences with a general (not necessarily diffuse) base measure, which have been used in various applications, e.g., in the case of "spike-and-slab"-type nonparametric priors [16–21].

We believe that our general framework sheds light on the common structure of all the above-mentioned models, leading to a possible unified treatment of some of their important features. Our techniques provides unified proofs for various results that, up to now, have been proven with ad hoc methods.

We have proven that all the mSSS are obtained by assigning the values of an exchangeable sequence to the classes of a latent exchangeable random partition. This representation is proven in the strong sense of an almost sure equality (see Section 3) and leads to the simple and clear derivation of an explicit expression for the EPPF of an mSSS. We believe that our general proof simplifies the derivation of the EPPF of many of the above-mentioned particular cases. Moreover, our results show that the clustering and the predictive structure of various well-known models do not depend on the relation between these models and completely random measures, but are essentially a consequence of the simple combinatorial structure of these sequences. Many important differences between well-known models (such as mixtures of Dirichlet and hierarchical Dirichlet) can be explained easily by simple differences in the latent partition and the corresponding latent exchangeable sequence.

We stress that a clear understanding of the clustering structure of mSSS is fundamental for practical purposes, since these models are typically used to cluster observations. Moreover, we hope that the explicit expression for EPPFs in our general framework can lead to the development of new MCMC algorithms for sampling from the posterior distribution.

Finally, we believe that some of the results we have proven here for mSSS can be extended to the more general case of partially exchangeable arrays. In this direction, for future works, a possible generalization of mSSS is to consider partially exchangeable arrays with a mixture of species sampling random probability measures as directing measures.

Author Contributions: F.B.: Methodology, simulation, writing and editing. L.L.: Methodology, writing and editing. All authors have read and agreed to the published version of the manuscript.

Funding: This research received funding from the European Research Council (ERC) under the European Union's Horizon 2020 research and innovation programme under grant agreement No. 817257.

Institutional Review Board Statement: Not applicable.

Informed Consent Statement: Not applicable.

Data Availability Statement: Not applicable.

Acknowledgments: F. Bassetti and L. Ladelli wish to express their gratitude to Professor Eugenio Regazzini, who has been an inspiring teacher and outstanding guide in many fields of probability and statistics.

Conflicts of Interest: The authors declare no conflict of interest.

Appendix A

In what follows, $\mathcal{L}(X)$ denotes the law of a random element X. For ease of reference, we state here Lemma 5.9 and Corollary 5.11 of [35].

Lemma A1 (Extension 1). *Fix a probability kernel K between two measurable spaces S and T, and let σ be a random element defined on $(\Omega, \mathcal{F}, \mathbb{P})$ taking values in S. Then, there exists a random element η in T, defined on some extension of the original probability space Ω, such that $\mathbb{P}[\eta \in \cdot | \sigma] = K(\cdot | \sigma)$ a.s. and, moreover, η is conditionally independent given σ from any other random element on Ω.*

Lemma A2 (Extension 2). *Fix two Borel spaces S and T, a measurable mapping $f : T \to S$ and some random elements σ in S and $\tilde{\eta}$ in T with $\mathcal{L}(\sigma) = \mathcal{L}(f(\tilde{\eta}))$. Then, there is a random element η defined on some extension of the original probability space, such that $\mathcal{L}(\eta) = \mathcal{L}(\tilde{\eta})$ and $\sigma = f(\eta)$ a.s.*

We need the following variant of the previous result.

Lemma A3 (Extension 3). *Fix three Borel spaces S_1, S_2 and T_1, a measurable mapping $\phi : T_1 \times S_2 \to S_1$ and some random elements $\sigma = (\sigma_1, \sigma_2)$ in $S_1 \times S_2$ and τ_1 in T_1, all defined on a probability space $(\Omega, \mathcal{F}, P)$. Assume that the conditional law of σ_1 given σ_2 is the same as the conditional law of $\phi(\tau_1, \sigma_2)$ given σ_2 (P-almost surely). Then, there is a random element τ defined on some extension of the original probability space $(\Omega, \mathcal{F}, P)$ taking values in T_1 such that*

- $\sigma_1 = \phi(\tau, \sigma_2)$ *a.s.*
- $\mathcal{L}(\tau_1, \sigma_2) = \mathcal{L}(\tau, \sigma_2)$.

Proof. Define $f : T_1 \times S_2 =: T \to S_1 \times S_2 =: S$ by $f(a, b) = (\phi(a, b), b)$, set $\tilde{\eta} = (\tau_1, \sigma_2)$ and $\sigma = (\sigma_1, \sigma_2)$. By hypothesizing, it is clear that $\mathcal{L}(f(\tilde{\eta})) = \mathcal{L}((\phi(\tau_1, \sigma_2), \sigma_2)) = \mathcal{L}(\sigma_1, \sigma_2) =$

$\mathcal{L}(\sigma)$. Thus, by Lemma A2, one has that, on an enlargement of $(\Omega, \mathcal{F}, P)$, there exists $\eta := (\tau, \sigma_2^*)$ such that $\mathcal{L}(\eta) = \mathcal{L}(\bar{\eta})$ and $(\phi(\tau, \sigma_2^*), \sigma_2^*) = f(\eta) = \sigma = (\sigma_1, \sigma_2)$ a.s. Hence, $\sigma_2^* = \sigma_2$ a.s. but also $\phi(\tau, \sigma_2) = \phi(\tau, \sigma_2^*) = \sigma_1$ a.s. It remains to show the second part of the thesis. Since $(\tau, \sigma_2) = (\tau, \sigma_2^*) = \eta$ a.s. and $\mathcal{L}(\eta) = \mathcal{L}(\bar{\eta})$, where $\bar{\eta} = (\tau_1, \sigma_2)$, it follows that $\mathcal{L}(\tau, \sigma_2) = \mathcal{L}(\tau_1, \sigma_2)$. □

References

1. Ferguson, T.S. A Bayesian analysis of some nonparametric problems. *Ann. Stat.* **1973**, *1*, 209–230. [CrossRef]
2. Pitman, J.; Yor, M. The two-parameter Poisson-Dirichlet distribution derived from a stable subordinator. *Ann. Probab.* **1997**, *25*, 855–900. [CrossRef]
3. Perman, M.; Pitman, J.; Yor, M. Size-biased sampling of Poisson point processes and excursions. *Probab. Theory Relat. Fields* **1992**, *92*, 21–39. [CrossRef]
4. Regazzini, E.; Lijoi, A.; Prünster, I. Distributional results for means of normalized random measures with independent increments. *Ann. Stat.* **2003**, *31*, 560–585. [CrossRef]
5. James, L.F.; Lijoi, A.; Prünster, I. Posterior analysis for normalized random measures with independent increments. *Scand. J. Stat.* **2009**, *36*, 76–97. [CrossRef]
6. Lijoi, A.; Prünster, I. Models beyond the Dirichlet process. In *Bayesian Nonparametrics*; Hjort, N.L., Holmes, C., Müller, P., Walker, S., Eds.; Cambridge University Press: New York, NY, USA, 2010.
7. De Blasi, P.; Favaro, S.; Lijoi, A.; Mena, R.H.; Prunster, I.; Ruggiero, M. Are Gibbs-Type Priors the Most Natural Generalization of the Dirichlet Process? *IEEE Trans. Pattern Anal. Mach. Intell.* **2015**, *37*, 212–229. [CrossRef]
8. Pitman, J. Poisson-Kingman partitions. In *Statistics and Science: A Festschrift for Terry Speed*; IMS Lecture Notes Monograph Series; Institute of Mathematical Statistics: Beachwood, OH, USA, 2003; Volume 40, pp. 1–34.
9. Ishwaran, H.; James, L.F. Gibbs sampling methods for stick-breaking priors. *J. Am. Stat. Assoc.* **2001**, *96*, 161–173. [CrossRef]
10. Antoniak, C.E. Mixtures of Dirichlet processes with applications to Bayesian nonparametric problems. *Ann. Stat.* **1974**, *2*, 1152–1174. [CrossRef]
11. Cifarelli, D.M.; Regazzini, E. Distribution functions of means of a Dirichlet process. *Ann. Stat.* **1990**, *18*, 429–442. [CrossRef]
12. Sangalli, L.M. Some developments of the normalized random measures with independent increments. *Sankhyā* **2006**, *68*, 461–487.
13. Broderick, T.; Wilson, A.C.; Jordan, M.I. Posteriors, conjugacy, and exponential families for completely random measures. *Bernoulli* **2018**, *24*, 3181–3221. [CrossRef]
14. Bassetti, F.; Ladelli, L. Asymptotic number of clusters for species sampling sequences with non-diffuse base measure. *Stat. Probab. Lett.* **2020**, *162*, 108749. [CrossRef]
15. Pitman, J. Some developments of the Blackwell-MacQueen urn scheme. In *Statistics, Probability and Game Theory*; IMS Lecture Notes Monograph Series; Institute of Mathematical Statistics: Hayward, CA, USA, 1996; Volume 30, pp. 245–267. [CrossRef]
16. Dunson, D.B.; Herring, A.H.; Engel, S.M. Bayesian selection and clustering of polymorphisms in functionally related genes. *J. Am. Stat. Assoc.* **2008**, *103*, 534–546. [CrossRef]
17. Kim, S.; Dahl, D.B.; Vannucci, M. Spiked Dirichlet process prior for Bayesian multiple hypothesis testing in random effects models. *Bayesian Anal.* **2009**, *4*, 707–732. [CrossRef] [PubMed]
18. Suarez, A.J.; Ghosal, S. Bayesian Clustering of Functional Data Using Local Features. *Bayesian Anal.* **2016**, *11*, 71–98. [CrossRef]
19. Cui, K.; Cui, W. Spike-and-Slab Dirichlet Process Mixture Models. *Spike Slab Dirichlet Process. Mix. Model.* **2012**, *2*, 512–518. [CrossRef]
20. Barcella, W.; De Iorio, M.; Baioa, G.; Malone-Leeb, J. Variable selection in covariate dependent random partition models: An application to urinary tract infection. *Stat. Med.* **2016**, *35*, 1373–13892. [CrossRef]
21. Canale, A.; Lijoi, A.; Nipoti, B.; Prünster, I. On the Pitman–Yor process with spike and slab base measure. *Biometrika* **2017**, *104*, 681–697. [CrossRef]
22. Teh, Y.; Jordan, M.I. Hierarchical Bayesian nonparametric models with applications. In *Bayesian Nonparametrics*; Hjort, N.L., Holmes, C., Müller, P., Walker, S., Eds.; Cambridge University Press: New York, NY, USA, 2010.
23. Teh, Y.W.; Jordan, M.I.; Beal, M.J.; Blei, D.M. Hierarchical Dirichlet processes. *J. Am. Stat. Assoc.* **2006**, *101*, 1566–1581. [CrossRef]
24. Camerlenghi, F.; Lijoi, A.; Orbanz, P.; Pruenster, I. Distribution theory for hierarchical processes. *Ann. Stat.* **2019**, *1*, 67–92. [CrossRef]
25. Bassetti, F.; Casarin, R.; Rossini, L. Hierarchical Species Sampling Models. *Bayesian Anal.* **2020**, *15*, 809–838. [CrossRef]
26. Pitman, J. *Combinatorial Stochastic Processes*; Lectures from the 32nd Summer School on Probability Theory Held in Saint-Flour, 7–24 July 2002, with a Foreword by Jean Picard; Lecture Notes in Mathematics; Springer: Berlin, Germany, 2006; Volume 1875.
27. Crane, H. The ubiquitous Ewens sampling formula. *Stat. Sci.* **2016**, *31*, 1–19. [CrossRef]
28. Kingman, J.F.C. The representation of partition structures. *J. Lond. Math. Soc.* **1978**, *18*, 374–380. [CrossRef]
29. Aldous, D.J. Exchangeability and related topics. In *École d'été de Probabilités de Saint-Flour, XIII—1983*; Lecture Notes in Mathematics; Springer: Berlin, Germany, 1985; Volume 1117, pp. 1–198. [CrossRef]
30. Kallenberg, O. Canonical representations and convergence criteria for processes with interchangeable increments. *Z. Wahrscheinlichkeitstheorie Und Verw. Geb.* **1973**, *27*, 23–36. [CrossRef]
31. Pitman, J. Exchangeable and partially exchangeable random partitions. *Probab. Theory Relat. Fields* **1995**, *102*, 145–158. [CrossRef]

32. Gnedin, A.; Pitman, J. Exchangeable Gibbs partitions and Stirling triangles. *Zap. Nauchn. Sem. S.-Peterburg. Otdel. Mat. Inst. Steklov. (POMI)* **2005**, *325*, 83–102, 244–245. [CrossRef]
33. Schervish, M.J. *Theory of Statistics*; Springer Series in Statistics; Springer: New York, NY, USA, 1995. [CrossRef]
34. Marin, J.M.; Robert, C.P. *Bayesian Core: A Practical Approach to Computational Bayesian Statistics*; Springer Texts in Statistics; Springer: New York, NY, USA, 2007; pp. xiv+255.
35. Kallenberg, O. *Foundations of Modern Probability*, 3rd ed.; Probability Theory and Stochastic Modelling; Springer: New York, NY, USA, 2021; Volume 99.

Article

A Central Limit Theorem for Predictive Distributions

Patrizia Berti [1], Luca Pratelli [2] and Pietro Rigo [3,*]

[1] Dipartimento di Scienze Fisiche, Informatiche e Matematiche, Università di Modena e Reggio-Emilia, Via Campi 213/B, 41100 Modena, Italy; patrizia.berti@unimore.it
[2] Accademia Navale di Livorno, 57127 Livorno, Italy; pratel@mail.dm.unipi.it
[3] Dipartimento di Scienze Statistiche "P. Fortunati", Università di Bologna, Via delle Belle Arti 41, 40126 Bologna, Italy
* Correspondence: pietro.rigo@unibo.it

Abstract: Let S be a Borel subset of a Polish space and F the set of bounded Borel functions $f : S \to \mathbb{R}$. Let $a_n(\cdot) = P(X_{n+1} \in \cdot \mid X_1, \ldots, X_n)$ be the n-th predictive distribution corresponding to a sequence (X_n) of S-valued random variables. If (X_n) is conditionally identically distributed, there is a random probability measure μ on S such that $\int f \, da_n \xrightarrow{a.s.} \int f \, d\mu$ for all $f \in F$. Define $D_n(f) = d_n \{ \int f \, da_n - \int f \, d\mu \}$ for all $f \in F$, where $d_n > 0$ is a constant. In this note, it is shown that, under some conditions on (X_n) and with a suitable choice of d_n, the finite dimensional distributions of the process $D_n = \{ D_n(f) : f \in F \}$ stably converge to a Gaussian kernel with a known covariance structure. In addition, $E\{\varphi(D_n(f)) \mid X_1, \ldots, X_n\}$ converges in probability for all $f \in F$ and $\varphi \in C_b(\mathbb{R})$.

Keywords: bayesian predictive inference; central limit theorem; conditional identity in distribution; exchangeability; predictive distribution; stable convergence

MSC: 60B10; 60G25; 60G09; 60F05; 62F15; 62M20

1. Introduction

All random elements appearing in the sequel are defined on a common probability space, say $(\Omega, \mathcal{A}, P)$. We denote by S a Borel subset of a Polish space and by $\mathcal{B}$ the Borel σ-field on S. We let

$$\mathcal{P} = \{\text{probability measures on } \mathcal{B}\} \quad \text{and}$$
$$F = \{\text{real bounded Borel functions on } S\}.$$

Moreover, if $\lambda \in \mathcal{P}$ and $f \in F$, we write $\lambda(f)$ to denote

$$\lambda(f) = \int f \, d\lambda.$$

In other terms, depending on the context, λ is regarded as a function on $\mathcal{B}$ or a function on F. This slight abuse of notation is quite usual (see, e.g., [1,2]) and very useful for the purposes of this note.

Let

$$X = (X_1, X_2, \ldots)$$

be a sequence of S-valued random variables and

$$\mathcal{F}_0 = \{\emptyset, \Omega\} \quad \text{and} \quad \mathcal{F}_n = \sigma(X_1, \ldots, X_n).$$

The *predictive distributions* of X are the random probability measures on $(S, \mathcal{B})$ given by

$$a_n(\cdot) = P(X_{n+1} \in \cdot \mid \mathcal{F}_n) \quad \text{for all } n \geq 0.$$

Under some conditions, there is a further random probability measure μ on $(S, \mathcal{B})$ such that
$$\mu(f) \stackrel{a.s.}{=} \lim_n a_n(f) \qquad \text{for each } f \in F. \tag{1}$$
For instance, condition (1) holds if X is exchangeable. More generally, it holds if X is conditionally identically distributed (c.i.d.), as defined in Section 2. Note also that, since S is separable, condition (1) implies $a_n \to \mu$ weakly. Regarding a_n and μ as measurable functions from Ω into $\mathcal{P}$, one obtains
$$P(\{\omega \in \Omega : a_{n,\omega} \to \mu_\omega \text{ weakly}\}) = 1.$$
Assume condition (1), fix a sequence d_n of positive constants, and define
$$D_n(f) = d_n \{a_n(f) - \mu(f)\} \qquad \text{for each } f \in F.$$
This note deals with the process
$$D_n = \{D_n(f) : f \in F\}.$$
Our goal is to show that, under some conditions on X and with a suitable choice of the constants d_n, the finite-dimensional distributions of D_n stably converge, as $n \to \infty$, to a certain Gaussian limit.

To be more precise, we recall that a *kernel* on $(S, \mathcal{B})$ is a measurable map $\alpha : S \to \mathcal{P}$. This means that $\alpha(x) \in \mathcal{P}$, for each $x \in S$, and the function $x \mapsto \alpha(x)(A)$ is $\mathcal{B}$-measurable for each $A \in \mathcal{B}$. In what follows, we write
$$\alpha(x)(f) = \int f(y)\, \alpha(x)(dy) \qquad \text{for all } x \in S \text{ and } f \in F.$$
Next, as in [3], suppose the predictive distributions of X satisfy the recursive equation
$$a_{n+1} = q_n a_n + (1 - q_n)\, \alpha(X_{n+1}) \qquad \text{a.s. for all } n \geq 0, \tag{2}$$
where $q_0, q_1, \ldots \in (0, 1)$ are constants and α is a kernel on $(S, \mathcal{B})$. Moreover, let
$$\nu(\cdot) = P(X_1 \in \cdot)$$
be the marginal distribution of X_1. Under condition (2), X is c.i.d. whenever α is a regular conditional distribution for ν given a sub-σ-field $\mathcal{G} \subset \mathcal{B}$; see ([3] Section 5). Hence, we assume
$$\alpha(\cdot)(A) = E_\nu(1_A \mid \mathcal{G}), \qquad \nu\text{-a.s.,} \tag{3}$$
for all $A \in \mathcal{B}$ and some sub-σ-field $\mathcal{G} \subset \mathcal{B}$. For instance, condition (3) holds if
$$\alpha(x) = \delta_x \qquad \text{for all } x \in S$$
where δ_x denotes the unit mass at the point x (just let $\mathcal{G} = \mathcal{B}$). In addition, we assume
$$\sum_n (1 - q_n)^2 < \infty \quad \text{and} \quad \lim_n d_n \sup_{k \geq n}(1 - q_{k-1}) = 0$$
where
$$d_n = \left(\sum_{k \geq n} (1 - q_k)^2 \right)^{-1/2}.$$
In this framework, it is shown that
$$(D_n(f_1), \ldots, D_n(f_p)) \longrightarrow \mathcal{N}_p(0, \Sigma) \qquad \text{stably} \tag{4}$$

for all $p \geq 1$ and all $f_1, \ldots, f_p \in F$, where Σ is the random covariance matrix with entries

$$\sigma_{jk} = \int \alpha(x)(f_j)\, \alpha(x)(f_k)\, \mu(dx) - \mu(f_j)\, \mu(f_k).$$

We actually prove something more than (4). Let $C_b(\mathbb{R})$ denote the set of real bounded continuous functions on $\mathbb{R}$. Then, it is shown that

$$E\{\varphi(D_n(f)) \mid \mathcal{F}_n\} \xrightarrow{P} \mathcal{N}(0, \sigma^2)(\varphi) \qquad (5)$$

for all $f \in F$ and $\varphi \in C_b(\mathbb{R})$, where

$$\sigma^2 = \int \alpha(x)(f)^2\, \mu(dx) - \mu(f)^2.$$

Based on (5), it is not hard to deduce condition (4).

Before concluding the Introduction, several remarks are in order.

(i) A remarkable special case is $\alpha(x) = \delta_x$ for all $x \in S$. Indeed, Equation (2) holds with $\alpha = \delta$ in some meaningful situations, including Dirichlet sequences; see ([3] Section 4) for other examples. Thus, suppose $\alpha = \delta$. Then, the above formulae reduce to $\sigma_{jk} = \mu(f_j f_k) - \mu(f_j)\mu(f_k)$ and $\sigma^2 = \mu(f^2) - \mu(f)^2$. Moreover, if ν is non-atomic and

$$\prod_{j=0}^{n} q_j \to 0 \quad \text{and} \quad \sum_n \prod_{j=0}^{n} q_j = \infty,$$

then μ takes the form

$$\mu \stackrel{a.s.}{=} \sum_n V_n\, \delta_{Y_n}$$

where (V_n) and (Y_n) are independent sequences and (Y_n) is i.i.d. with $Y_1 \sim \nu$; see ([3] Theorem 20) and [4] for details.

(ii) Let $l^\infty(G)$ be the set of real bounded functions on G, where G is any subset of F. For instance, if $S = \mathbb{R}$, one could take $G = \{1_{(-\infty, x]} : x \in \mathbb{R}\}$. In view of (4), a natural question is whether D_n has a limit in distribution when $l^\infty(G)$ is equipped with a suitable distance. As an example, $l^\infty(G)$ could be equipped with the uniform distance (as in [1,2]) or with some weaker distance (as in [5]). Even if natural, this question is neglected in this note. We hope and plan to investigate it in a forthcoming paper.

(iii) For fixed $f \in F$, condition (4) provides some information on the convergence rate of $a_n(f)$ to $\mu(f)$. Define $L_n = u_n\, |a_n(f) - \mu(f)|$ where $u_n > 0$ is any sequence of constants. Then, condition (4) yields $L_n \xrightarrow{P} 0$ whenever $u_n/d_n \to 0$. Furthermore, $L_n \xrightarrow{P} \infty$ provided $u_n/d_n \to \infty$ and $\sigma^2 > 0$ a.s.

(iv) The condition $\lim_n d_n \sup_{k \geq n}(1 - q_{k-1}) = 0$ is just a technical assumption which guarantees that, asymptotically, there are no dominating terms. In a sense, this condition is analogous to the weak Lindeberg's condition in the classical CLT for independent summands.

(v) From a Bayesian point of view, μ can be seen as a random parameter of the data sequence X. This is quite clear if X is exchangeable, for, in this case, X is conditionally i.i.d. given μ. If X is only c.i.d., the role of μ is not as crucial, but μ still contributes to specify the probability distribution of X; see ([3] Section 2.1). Thus, in a Bayesian framework, conditions (4)–(5) may be useful to make (asymptotic) inference about μ. To this end, an alternative could be proving a limit theorem for $W_n = w_n\,(\mu_n - \mu)$, where w_n is a suitable constant and $\mu_n = (1/n)\sum_{j=1}^{n} \delta_{X_j}$ the empirical measure. However, D_n has two advantages with respect to W_n. It usually converges at a better rate and the variance of the limit distribution is smaller; see, e.g., Example 3.

(vi) Conditions (4)–(5) are our main results. They can be motivated in at least two ways. Firstly, from the theoretical perspective, conditions (4)–(5) fit into the results concerning the asymptotic behavior of conditional expectations (see, e.g., [6–8] and references therein). Secondly, from the practical perspective, conditions (4)–(5) play a role in all those fields where predictive distributions are basic objects. The main example is Bayesian predictive inference. Indeed, the predictive distributions investigated in this note have been introduced in connection with Bayesian prediction problems; see [3]. Another example is the asymptotic behavior of certain urn schemes. Related subjects, where (4)–(5) are potentially useful, are empirical processes for dependent data, Glivenko-Cantelli-type theorems and merging of opinions. Without any claim of being exhaustive, a list of references is: [3,5,9–21].

2. Preliminaries

In this note, $\mathcal{N}_p(0, C)$ denotes the Gaussian law on the Borel sets of $\mathbb{R}^p$ with mean 0 and covariance matrix C, where C is symmetric and semidefinite positive. If $p = 1$ and $c \geq 0$ is a scalar, we write $\mathcal{N}(0, c)$ instead of $\mathcal{N}_1(0, c)$ and

$$\mathcal{N}(0,c)(\varphi) = \int \varphi(x)\,\mathcal{N}(0,c)(dx)$$

for all bounded measurable $\varphi : \mathbb{R} \to \mathbb{R}$. Note that, if Σ is a random covariance matrix, $\mathcal{N}_p(0, \Sigma)$ is a random probability measure on the Borel sets of $\mathbb{R}^p$.

Let us briefly recall *stable convergence*. Let $\mathcal{A}^+ = \{H \in \mathcal{A} : P(H) > 0\}$. Fix a random probability measure K on $(S, \mathcal{B})$ and define

$$\lambda_H(A) = E\{K(A) \mid H\} \quad \text{for all } A \in \mathcal{B} \text{ and } H \in \mathcal{A}^+.$$

Each λ_H is a probability measure on $\mathcal{B}$. Then, X_n *converges stably to* K, written $X_n \to K$ stably, if

$$P(X_n \in \cdot \mid H) \longrightarrow \lambda_H \quad \text{weakly for all } H \in \mathcal{A}^+.$$

In particular, X_n converges in distribution to λ_Ω. However, stable convergence is stronger than convergence in distribution. To see this, take a further random variable $X : \Omega \to S$. Then, $X_n \xrightarrow{P} X$ if, and only if, $X_n \to \delta_X$ stably. Thus, stable convergence is strictly connected to convergence in probability. Moreover, $(X_n, X) \to K \times \delta_X$ stably whenever $X_n \to K$ stably. Therefore, if X_n converges stably, (X_n, X) still converges stably for *any* S-valued random variable X.

We next turn to conditional identity in distribution. Say that X is *conditionally identically distributed* (c.i.d.) if

$$P(X_k \in \cdot \mid \mathcal{F}_n) = P(X_{n+1} \in \cdot \mid \mathcal{F}_n) \quad \text{a.s. for all } k > n \geq 0.$$

Thus, at each time n, the future observations $(X_k : k > n)$ are identically distributed given the past. This is actually weaker than exchangeability. Indeed, X is exchangeable if, and only if, it is stationary and c.i.d.

C.i.d. sequences were introduced in [9,22] and then investigated in various papers; see, e.g., [3–5,11,23–29].

The asymptotics of c.i.d. sequences is similar to that of exchangeable ones. To see this, suppose X is c.i.d. and define the empirical measures

$$\mu_n = \frac{1}{n} \sum_{j=1}^n \delta_{X_j}.$$

Then, there is a random probability measure μ on $(S, \mathcal{B})$ such that

$$\mu(A) \stackrel{a.s.}{=} \lim_m \mu_m(A) \quad \text{for each fixed } A \in \mathcal{B}.$$

It follows that

$$E\{\mu(A) \mid \mathcal{F}_n\} = \lim_m E\{\mu_m(A) \mid \mathcal{F}_n\}$$
$$= \lim_m \frac{1}{m} \sum_{j=n+1}^m P(X_j \in A \mid \mathcal{F}_n) = P(X_{n+1} \in A \mid \mathcal{F}_n) \quad \text{a.s.}$$

for all $n \geq 0$ and $A \in \mathcal{B}$. Therefore, as in the exchangeable case, the predictive distributions can be written as

$$a_n(\cdot) = P(X_{n+1} \in \cdot \mid \mathcal{F}_n) = E\{\mu(\cdot) \mid \mathcal{F}_n\} \quad \text{a.s.}$$

Using the martingale convergence theorem, this implies

$$\mu(f) \stackrel{a.s.}{=} \lim_n E\{\mu(f) \mid \mathcal{F}_n\} = \lim_n a_n(f) \quad \text{for all } f \in F.$$

Furthermore, X is asymptotically exchangeable, in the sense that the probability distribution of the shifted sequence $(X_n, X_{n+1}, \ldots)$ converges weakly to an exchangeable probability measure on $(S^\infty, \mathcal{B}^\infty)$.

Finally, we state a technical result to be used later on.

Lemma 1. *Let (Y_n) be a sequence of real integrable random variables, adapted to the filtration $(\mathcal{F}_n)$, and*

$$Z_n = E(Y_{n+1} \mid \mathcal{F}_n).$$

Let V be a real non-negative random variable and $0 < b_1 < b_2 < \ldots$ an increasing sequence of constants, such that $b_n \uparrow \infty$ and $b_n/b_{n+1} \to 1$. Suppose (Y_n^2) is uniformly integrable, $Z_n \stackrel{a.s.}{\longrightarrow} Z$ for some random variable Z, and define

$$T_n = b_n(Z_n - Z).$$

Then,

$$E\{\varphi(T_n) \mid \mathcal{F}_n\} \stackrel{P}{\longrightarrow} \mathcal{N}(0, V)(\varphi) \quad \text{for all } \varphi \in C_b(\mathbb{R})$$

provided

$$b_n^2 \sum_{k \geq n} (Z_k - Z_{k-1})^2 \stackrel{P}{\longrightarrow} V; \quad (6)$$

$$\lim_n b_n E\left\{\sup_{k \geq n} |Z_k - Z_{k-1}|\right\} = 0; \quad (7)$$

$$\sum_{k \geq n} E\left|E(Z_{k+1} \mid \mathcal{F}_k) - Z_k\right| = o(1/b_n). \quad (8)$$

Proof. Just repeat the proof of ([10] Theorem 1) with b_n in the place of $\sqrt{n}$. □

3. Main Result

Let us go back to the notation of Section 1. Recall that $q_n \in (0, 1)$ is a constant for each $n \geq 0$ and $d_n = \left(\sum_{k \geq n}(1 - q_k)^2\right)^{-1/2}$. We aim to prove the following CLT.

Theorem 1. *Assume conditions* (2)–(3) *and*

$$\sum_n (1-q_n)^2 < \infty \quad \text{and} \quad \lim_n d_n \sup_{k \geq n}(1-q_{k-1}) = 0.$$

Then, there is a random probability measure μ on $(S, \mathcal{B})$ such that

$$\mu(f) \stackrel{a.s.}{=} \lim_n a_n(f) \quad \text{and} \quad E\{\varphi(D_n(f)) \mid \mathcal{F}_n\} \stackrel{P}{\longrightarrow} \mathcal{N}(0, \sigma^2)(\varphi)$$

for all $f \in F$ and $\varphi \in C_b(\mathbb{R})$, where

$$\sigma^2 = \int \alpha(x)(f)^2 \mu(dx) - \mu(f)^2.$$

As a consequence,

$$(D_n(f_1), \ldots, D_n(f_p)) \longrightarrow \mathcal{N}_p(0, \Sigma) \quad \text{stably}$$

for all $p \geq 1$ and all $f_1, \ldots, f_p \in F$ where the covariance matrix Σ has entries

$$\sigma_{jk} = \int \alpha(x)(f_j)\,\alpha(x)(f_k)\,\mu(dx) - \mu(f_j)\,\mu(f_k).$$

Proof. Due to conditions (2)–(3), X is c.i.d.; see ([3] Section 5). Hence, as noted in Section 2, there is a random probability measure μ on $(S, \mathcal{B})$ such that

$$a_n(f) \stackrel{a.s.}{=} E\{\mu(f) \mid \mathcal{F}_n\} \quad \text{for all } f \in F.$$

By martingale convergence, it follows that $a_n(f) \stackrel{a.s.}{\longrightarrow} \mu(f)$ for all $f \in F$.

We next prove condition (5). Fix $f \in F$ and define

$$b_n = d_n, \quad Y_n = a_n(f), \quad Z = \mu(f) \quad \text{and} \quad V = \sigma^2.$$

Then, (Y_n^2) is uniformly integrable (for f is bounded) and b_n satisfies the conditions of Lemma 1. Moreover,

$$Z_n = E(Y_{n+1} \mid \mathcal{F}_n) = E\{E(\mu(f) \mid \mathcal{F}_{n+1}) \mid \mathcal{F}_n\} = E\{\mu(f) \mid \mathcal{F}_n\} = a_n(f) \quad \text{a.s.}$$

so that $Z_n \stackrel{a.s.}{\longrightarrow} Z$. Therefore, Lemma 1 applies. Hence, to prove (5), it suffices to check conditions (6)–(8).

Let $c = \sup|f|$. Since $E(Z_{k+1} \mid \mathcal{F}_k) = Z_k$ a.s., condition (8) is trivially true. Moreover, condition (2) implies

$$\begin{aligned}
Z_k - Z_{k-1} &= a_k(f) - a_{k-1}(f) \\
&= q_{k-1} a_{k-1}(f) + (1-q_{k-1})\alpha(X_k)(f) - a_{k-1}(f) \\
&= (1-q_{k-1})\{\alpha(X_k)(f) - a_{k-1}(f)\} \quad \text{a.s. for all } k \geq 1.
\end{aligned}$$

Hence, condition (7) holds, since

$$d_n E\left\{\sup_{k \geq n}|Z_k - Z_{k-1}|\right\} \leq 2c\,d_n \sup_{k \geq n}(1-q_{k-1}) \longrightarrow 0.$$

It remains to prove condition (6), namely

$$d_n^2 \sum_{k \geq n}(1-q_{k-1})^2 \{\alpha(X_k)(f) - a_{k-1}(f)\}^2 \stackrel{P}{\longrightarrow} \sigma^2.$$

First note that, since $a_{k-1}(f)^2 \xrightarrow{a.s.} \mu(f)^2$ as $k \to \infty$, one obtains

$$d_n^2 \sum_{k \geq n} (1 - q_{k-1})^2 a_{k-1}(f)^2 = \frac{\sum_{k \geq n} (1 - q_{k-1})^2 a_{k-1}(f)^2}{\sum_{k \geq n} (1 - q_k)^2} \xrightarrow{a.s.} \mu(f)^2.$$

Next, define

$$R_k = \alpha(X_k)(f)^2 \quad \text{and} \quad M_n = d_n^2 \sum_{k \geq n} (1 - q_{k-1})^2 \{ R_k - E(R_k \mid \mathcal{F}_{k-1}) \}.$$

Then,

$$\begin{aligned} E(M_n^2) &= d_n^4 \sum_{k \geq n} (1 - q_{k-1})^4 E\{ (R_k - E(R_k \mid \mathcal{F}_{k-1}))^2 \} \\ &\leq 4 c^4 d_n^4 \sum_{k \geq n} (1 - q_{k-1})^4 \\ &\leq 4 c^4 d_n^2 \sup_{k \geq n} (1 - q_{k-1})^2 \cdot d_n^2 \sum_{k \geq n} (1 - q_{k-1})^2 \\ &\longrightarrow 0. \end{aligned}$$

Moreover,

$$E(R_k \mid \mathcal{F}_{k-1}) = E\left\{ \int \alpha(x)(f)^2 \, \mu(dx) \mid \mathcal{F}_{k-1} \right\} \xrightarrow{a.s.} \int \alpha(x)(f)^2 \, \mu(dx).$$

Therefore,

$$d_n^2 \sum_{k \geq n} (1 - q_{k-1})^2 R_k = M_n + d_n^2 \sum_{k \geq n} (1 - q_{k-1})^2 E(R_k \mid \mathcal{F}_{k-1}) \xrightarrow{P} \int \alpha(x)(f)^2 \, \mu(dx).$$

By the same argument, it follows that

$$d_n^2 \sum_{k \geq n} (1 - q_{k-1})^2 \alpha(X_k)(f) \, a_{k-1}(f) \xrightarrow{P} \mu(f) \int \alpha(x)(f) \, \mu(dx).$$

In addition, as proved in the Claim below,

$$\int \alpha(x)(f) \, \mu(dx) \stackrel{a.s.}{=} \mu(f).$$

Collecting all pieces together, one finally obtains

$$d_n^2 \sum_{k \geq n} (1 - q_{k-1})^2 \{ \alpha(X_k)(f) - a_{k-1}(f) \}^2 \xrightarrow{P} \mu(f)^2 + \int \alpha(x)(f)^2 \, \mu(dx) - 2 \mu(f)^2 = \sigma^2.$$

Hence, condition (6) holds.

This concludes the proof of (5). We next prove that (5) $\Rightarrow$ (4). Let $p \geq 1$ and $f_1, \ldots, f_p \in \mathcal{F}$. Fix $u_1, \ldots, u_p \in \mathbb{R}$ and define

$$U_n = \sum_{j=1}^p u_j D_n(f_j) \quad \text{and} \quad \sigma_u^2 = \sum_{j,k} u_j u_k \sigma_{jk}.$$

Moreover, for each $H \in \mathcal{A}^+$, define the probability measure

$$\lambda_H(A) = E\{ \mathcal{N}(0, \sigma_u^2)(A) \mid H \} \quad \text{for each Borel set } A \subset \mathbb{R}.$$

We have to show that

$$P(U_n \in \cdot \mid H) \longrightarrow \lambda_H \text{ weakly for each } H \in \mathcal{A}^+. \tag{9}$$

To this end, call ϕ_H the characteristic function of λ_H, namely

$$\phi_H(t) = E\left(\int e^{itx} \mathcal{N}(0, \sigma_u^2)(dx) \mid H\right) = E\left(e^{-t^2 \sigma_u^2/2} \mid H\right) \quad \text{for all } t \in \mathbb{R}.$$

Letting $f = \sum_{j=1}^p u_j f_j$, one obtains

$$U_n = D_n(f) \quad \text{and} \quad \sigma_u^2 = \int \alpha(x)(f)^2 \mu(dx) - \mu(f)^2.$$

Therefore, condition (5) yields

$$E\left(e^{itU_n}\right) = E\left(E\{e^{itD_n(f)} \mid \mathcal{F}_n\}\right) \longrightarrow E\left(e^{-t^2 \sigma_u^2/2}\right) = \phi_\Omega(t)$$

for each $t \in \mathbb{R}$. Hence, condition (9) holds for $H = \Omega$. Next, suppose $H \in \bigcup_n \mathcal{F}_n$ and $P(H) > 0$. Then, for large n, one obtains

$$E\left(1_H e^{itU_n}\right) = E\left(1_H E\{e^{itD_n(f)} \mid \mathcal{F}_n\}\right).$$

Hence, for each $t \in \mathbb{R}$, condition (5) still implies

$$P(H)\phi_H(t) = E\left(1_H e^{-t^2 \sigma_u^2/2}\right) = \lim_n E\left(1_H E\{e^{itD_n(f)} \mid \mathcal{F}_n\}\right) = \lim_n E\left(1_H e^{itU_n}\right).$$

Therefore, condition (9) holds whenever $H \in \bigcup_n \mathcal{F}_n$ and $P(H) > 0$. Based on this fact, by standard arguments, condition (9) easily follows for each $H \in \mathcal{A}^+$.

To conclude the proof of the Theorem, it remains only to show that:

Claim: $\int \alpha(x)(f) \mu(dx) \stackrel{a.s.}{=} \mu(f)$ for all $f \in F$.

Proof of the Claim: By (3), α is a regular conditional distribution for ν given a sub-σ-field of $\mathcal{B}$, where ν is the marginal distribution of X_1. Therefore, as proved in ([3] Lemma 6), there is a set $A \in \mathcal{B}$ such that $\nu(A) = 1$ and

$$\int \alpha(z)(f) \alpha(x)(dz) = \alpha(x)(f) \quad \text{for all } x \in A \text{ and } f \in F.$$

Since X is c.i.d. (and, thus, identically distributed) one also obtains $P(X_n \in A) = \nu(A) = 1$ for all $n \geq 1$.

Having noted these facts, fix $f \in F$. Since $a_0 = \nu$ and α is a regular conditional distribution for ν,

$$\int \alpha(x)(f) a_0(dx) = a_0(f).$$

Moreover, if $\int \alpha(x)(f) a_n(dx) = a_n(f)$ a.s. for some $n \geq 0$, then

$$\begin{aligned}\int \alpha(x)(f) a_{n+1}(dx) &= q_n \int \alpha(x)(f) a_n(dx) + (1 - q_n) \int \alpha(x)(f) \alpha(X_{n+1})(dx) \\ &= q_n a_n(f) + (1 - q_n) \alpha(X_{n+1})(f) \\ &= a_{n+1}(f) \quad \text{a.s.}\end{aligned}$$

By induction, one obtains $\int \alpha(x)(f) a_n(dx) = a_n(f)$ a.s. for each $n \geq 0$. Hence,

$$\int \alpha(x)(f) \mu(dx) = \lim_n \int \alpha(x)(f) a_n(dx) = \lim_n a_n(f) = \mu(f) \quad \text{a.s.}$$

□

We do not know whether $E\{\varphi(D_n(f)) \mid \mathcal{F}_n\}$ converges a.s. (and not only in probability) under the conditions of Theorem 1. However, it can be shown that $E\{\varphi(D_n(f)) \mid \mathcal{F}_n\}$ converges a.s. under slightly stronger conditions on q_n.

Under conditions (2)–(3), for Theorem 1 to work, it suffices that

$$\lim_n n^b (1 - q_n) = c \quad \text{for some } b > 1/2 \text{ and } c > 0. \tag{10}$$

In addition, if (10) holds, then

$$\frac{n^{b-1/2}}{d_n} \to \frac{c}{\sqrt{2b-1}}.$$

Hence, letting $D_n^* = n^{b-1/2}(a_n - \mu)$, one obtains

$$(D_n^*(f_1), \ldots, D_n^*(f_p)) \longrightarrow \mathcal{N}_p\left(0, \frac{c^2}{2b-1}\Sigma\right) \quad \text{stably,}$$

for all $p \geq 1$ and all $f_1, \ldots, f_p \in F$, provided conditions (2), (3) and (10) hold.

We close this note with some examples.

Example 1. Let

$$q_n = \frac{n + \theta_n}{n + 1 + \theta_{n+1}}$$

where (θ_n) is a bounded increasing sequence with $\theta_0 > 0$. Then, X is c.i.d. (because of (2)–(3)) but is exchangeable if and only if $\theta_n = \theta_0$ for all n. In any case, since condition (10) holds with $b = c = 1$, Theorem 1 applies and d_n can be replaced by $\sqrt{n}$. Letting $D_n^* = \sqrt{n}(a_n - \mu)$, it follows that

$$(D_n^*(f_1), \ldots, D_n^*(f_p)) \longrightarrow \mathcal{N}_p(0, \Sigma) \quad \text{stably.}$$

It is worth noting that, in the special case $\theta_n = \theta_0$ for all n, the predictive distributions of X reduce to

$$a_n = \frac{\theta_0 \nu + \sum_{i=1}^n \alpha(X_i)}{n + \theta_0}.$$

Therefore, X is a Dirichlet sequence if $\alpha = \delta$. The general case, where α is any kernel satisfying condition (3), is investigated in [30]. It turns out that X satisfies most properties of Dirichlet sequences. In particular, μ has the same distribution as

$$\mu^* = \sum_n V_n \alpha(Y_n),$$

where (V_n) and (Y_n) are independent sequences, (Y_n) is i.i.d. with $Y_1 \sim \nu$, and (V_n) has the stick breaking distribution. Nevertheless, as shown in the next example, X can behave quite differently from a Dirichlet sequence.

Example 2 (Example 1 continued). Let $\mathcal{H}$ be a countable partition of S such that $H \in \mathcal{B}$ and $\nu(H) > 0$ for all $H \in \mathcal{H}$. Define

$$\alpha(x) = \sum_{H \in \mathcal{H}} 1_H(x) \nu(\cdot \mid H) = \nu(\cdot \mid H_x) \quad \text{for all } x \in S$$

where H_x is the only element of the partition $\mathcal{H}$, such that $x \in H$. Then, α is a regular conditional distribution for ν given $\sigma(\mathcal{H})$ (i.e., condition (3) holds). If the q_n are as in Example 1 with $\theta_n = \theta_0$ for all n, one obtains

$$a_n = \frac{\theta_0 \nu + \sum_{i=1}^n \nu(\cdot \mid H_{X_i})}{n + \theta_0}.$$

Therefore,

$$a_n \ll \nu \quad \text{for all } n \geq 0. \tag{11}$$

This is a striking difference with respect to Dirichlet sequences. For instance, if ν is non-atomic, condition (11) yields
$$P(X_i = X_j \text{ for some } i \neq j) = 0$$
while $P(X_i = X_j \text{ for some } i \neq j) = 1$ if X is a Dirichlet sequence. Note also that, for each $f \in F$,
$$\sigma^2 = \int \alpha(x)(f)^2 \, \mu(dx) - \mu(f)^2 = \sum_{H \in \mathcal{H}} \nu(f \mid H)^2 \, \mu(H) - \mu(f)^2$$
while $\sigma^2 = \mu(f^2) - \mu(f)^2$ if X is a Dirichlet sequence. Other choices of α, which make X quite different from a Dirichlet sequence, are in [30].

Example 3. *A meaningful special case is $\sum_n (1 - q_n) < \infty$. In this case,*
$$\prod_{j=0}^{\infty} q_j := \lim_n \prod_{j=0}^{n} q_j$$
exists and is strictly positive. Hence, μ admits the representation
$$\mu = \nu \prod_{j=0}^{\infty} q_j + \sum_{i=1}^{\infty} \alpha(X_i) \, (1 - q_{i-1}) \prod_{j=i}^{\infty} q_j.$$
As an example, under conditions (2)–(3), Theorem 1 applies whenever
$$q_n = \exp\{-(c+n)^{-2}\} \quad \text{for some constant } c > 0.$$
With this choice of q_n, one obtains $(1 - q_n)(c + n)^2 \to 1$, so that $\sum_n (1 - q_n) < \infty$ and μ can be written as above. Note also that
$$\lim_n \frac{d_n}{(c+n)^{3/2}} = \sqrt{3}.$$
Therefore, for fixed $f \in F$, the rate of convergence of $a_n(f)$ to $\mu(f)$ is $n^{-3/2}$ and not the usual $n^{-1/2}$.

Author Contributions: Methodology, P.B., L.P. and P.R. All authors have read and agreed to the published version of the manuscript.

Funding: This research received funding from the European Research Council (ERC) under the European Union's Horizon 2020 research and innovation programme under grant agreement No 817257.

Institutional Review Board Statement: Not applicable.

Informed Consent Statement: Not applicable.

Acknowledgments: We are grateful to Giorgio Letta and Eugenio Regazzini. They not only introduced us to probability theory, they also shared with us their enthusiasm and some of their expertise.

Conflicts of Interest: The authors declare no conflict of interest.

References

1. Dudley, R.M. *Uniform Central Limit Theorems*; Cambridge University Press: Cambridge, UK, 1999.
2. van der Vaart, A.; Wellner, J.A. *Weak Convergence and Empirical Processes*; Springer: New York, NY, USA, 1996.
3. Berti, P.; Dreassi, E.; Pratelli, L.; Rigo, P. A class of models for Bayesian predictive inference. *Bernoulli* **2021**, *27*, 702–726. [CrossRef]
4. Berti, P.; Dreassi, E.; Pratelli, L.; Rigo, P. Asymptotics of certain conditionally identically distributed sequences. *Statist. Prob. Lett.* **2021**, *168*, 108923. [CrossRef]
5. Berti, P.; Pratelli, L.; Rigo, P. Limit theorems for empirical processes based on dependent data. *Electron. J. Probab.* **2012**, *17*, 1–18. [CrossRef]
6. Crimaldi, I.; Pratelli, L. Convergence results for conditional expectations. *Bernoulli* **2005**, *11*, 737–745. [CrossRef]
7. Goggin, E.M. Convergence in distribution of conditional expectations. *Ann. Probab.* **1994**, *22*, 1097–1114. [CrossRef]

8. Lan, G.; Hu, Z.C.; Sun, W. Products of conditional expectation operators: Convergence and divergence. *J. Theore. Probab.* **2021**, *34*, 1012–1028. [CrossRef]
9. Berti, P.; Pratelli, L.; Rigo, P. Limit theorems for a class of identically distributed random variables. *Ann. Probab.* **2004**, *32*, 2029–2052. [CrossRef]
10. Berti, P.; Crimaldi, I.; Pratelli, L.; Rigo, P. A central limit theorem and its applications to multicolor randomly reinforced urns. *J. Appl. Probab.* **2011**, *48*, 527–546. [CrossRef]
11. Berti, P.; Pratelli, L.; Rigo, P. Exchangeable sequences driven by an absolutely continuous random measure. *Ann. Probab.* **2013**, *41*, 2090–2102. [CrossRef]
12. Blackwell, D.; Dubins, L.E. Merging of opinions with increasing information. *Ann. Math. Statist.* **1962**, *33*, 882–886. [CrossRef]
13. Cifarelli, D.M.; Regazzini, E. De Finetti's contribution to probability and statistics. *Statist. Sci.* **1996**, *11*, 253–282. [CrossRef]
14. Cifarelli, D.M.; Dolera, E.; Regazzini, E. Frequentistic approximations to Bayesian prevision of exchangeable random elements. *Int. J. Approx. Reason.* **2016**, *78*, 138–152. [CrossRef]
15. Dolera, E.; Regazzini, E. Uniform rates of the Glivenko-Cantelli convergence and their use in approximating Bayesian inferences. *Bernoulli* **2019**, *25*, 2982–3015. [CrossRef]
16. Fortini, S.; Ladelli, L.; Regazzini, E. Exchangeability, predictive distributions and parametric models. *Sankhyā Indian J. Stat. Ser. A* **2000**, *62*, 86–109.
17. Hahn, P.R.; Martin, R.; Walker, S.G. On recursive Bayesian predictive distributions. *J. Am. Stat. Assoc.* **2018**, *113*, 1085–1093. [CrossRef]
18. Morvai, G.; Weiss, B. On universal algorithms for classifying and predicting stationary processes. *Probab. Surv.* **2021**, *18*, 77–131. [CrossRef]
19. Pitman, J. Some developments of the Blackwell-MacQueen urn scheme. *Stat. Probab. Game Theory IMS Lect. Notes Mon. Ser.* **1996**, *30*, 245–267.
20. Pitman, J. *Combinatorial Stochastic Processes*; Lectures from the XXXII Summer School in Saint-Flour; Springer: Berlin/Heidelberg, Germany, 2006.
21. Regazzini, E. Old and recent results on the relationship between predictive inference and statistical modeling either in nonparametric or parametric form. In *Bayesian Statistics 6*; Oxford University Press: Oxford, UK, 1999; pp. 571–588.
22. Kallenberg, O. Spreading and predictable sampling in exchangeable sequences and processes. *Ann. Probab.* **1988**, *16*, 508–534. [CrossRef]
23. Airoldi, E.M.; Costa, T.; Bassetti, F.; Leisen, F.; Guindani, M. Generalized species sampling priors with latent beta reinforcements. *J. Am. Stat. Assoc.* **2014**, *109*, 1466–1480. [CrossRef]
24. Bassetti, F.; Crimaldi, I.; Leisen, F. Conditionally identically distributed species sampling sequences. *Adv. Appl. Probab.* **2010**, *42*, 433–459. [CrossRef]
25. Cassese, A.; Zhu, W.; Guindani, M.; Vannucci, M. A Bayesian nonparametric spiked process prior for dynamic model selection. *Bayesian Anal.* **2019**, *14*, 553–572. [CrossRef]
26. Fong, E.; Holmes, C.; Walker, S.G. Martingale posterior distributions. *arXiv* **2021**, arXiv:2103.15671v1.
27. Fortini, S.; Petrone, S. Predictive construction of priors in Bayesian nonparametrics. *Braz. J. Probab. Statist.* **2012**, *26*, 423–449. [CrossRef]
28. Fortini, S.; Petrone, S.; Sporysheva, P. On a notion of partially conditionally identically distributed sequences. *Stoch. Proc. Appl.* **2018**, *128*, 819–846. [CrossRef]
29. Fortini, S.; Petrone, S. Quasi-Bayes properties of a procedure for sequential learning in mixture models. *J. R. Stat. Soc. B* **2020**, *82*, 1087–1114. [CrossRef]
30. Berti, P.; Dreassi, E.; Leisen, F.; Pratelli, L.; Rigo, P. Kernel based Dirichlet sequences. *arXiv* **2021**, arXiv:2106.00114.

Article

Trapping the Ultimate Success

Alexander Gnedin * and Zakaria Derbazi

School of Mathematical Sciences, Queen Mary University of London, London E1 4NS, UK; z.derbazi@qmul.ac.uk
* Correspondence: a.gnedin@qmul.ac.uk

Abstract: We introduce a betting game where the gambler aims to guess the last success epoch in a series of inhomogeneous Bernoulli trials paced randomly in time. At a given stage, the gambler may bet on either the event that no further successes occur, or the event that exactly one success is yet to occur, or may choose any proper range of future times (a trap). When a trap is chosen, the gambler wins if the last success epoch is the only one that falls in the trap. The game is closely related to the sequential decision problem of maximising the probability of stopping on the last success. We use this connection to analyse the best-choice problem with random arrivals generated by a Pólya–Lundberg process.

Keywords: best choice problem; optimal stopping time; last record; trapping strategy

MSC: 60G40

Citation: Gnedin, A.; Derbazi, Z. Trapping the Ultimate Success. *Mathematics* **2022**, *10*, 158. https://doi.org/10.3390/math10010158

Academic Editors: Emanuele Dolera and Federico Bassetti

Received: 30 October 2021
Accepted: 30 December 2021
Published: 5 January 2022

Publisher's Note: MDPI stays neutral with regard to jurisdictional claims in published maps and institutional affiliations.

Copyright: © 2022 by the authors. Licensee MDPI, Basel, Switzerland. This article is an open access article distributed under the terms and conditions of the Creative Commons Attribution (CC BY) license (https://creativecommons.org/licenses/by/4.0/).

1. Introduction

Suppose a series of inhomogeneous Bernoulli trials, with a given profile of success probabilities $p = (p_k, k \geq 1)$, is paced randomly in time by some independent point process. As the outcomes and epochs of the first $k \geq 0$ trials become known at some time t, the gambler is asked to bet on the time of the last success. The gambler is allowed to choose either a bygone action, a next action, or a proper subset of future times called *trap*. The gambler wins with bygone if no further successes occur, and with next if exactly one success occurs after time t. In the case a trapping action is chosen, the gambler wins if the last success epoch is isolated by the trap from the other success epochs.

Motivation to study this game stems from connections to the best-choice problems with random arrivals [1–9] and the random records model [10,11]. A prototype problem of this kind involves a sequence of rankable items arriving by a Poisson process with a finite horizon, where the kth arrival is relatively the best (a record) with probability $p_k = 1/k$. The optimisation task is to maximise the probability of selecting the overall best item (the last record) using a non-anticipating stopping strategy. Cowan and Zabczyk [5] showed that the optimal strategy is *myopic*, which means that the decision to stop on a particular record arrival only depends on whether the winning chance with bygone exceeds that with next. They also determined the critical cut-offs of the optimal strategy and studied some asymptotics. Similar results have been obtained for the best-choice problem with some other pacing processes [1,4,7,9]. In this context, trapping can be employed to test optimality of the myopic strategy, which fails if in some situations the action bygone outperforms next but a trapping action is better still. Simple trapping strategies are easy to evaluate and provide insight into the occurrence of records.

Regarding the pacing point process, we shall assume that it is mixed binomial [12]. This setting covers, in particular, the wide class of mixed Poisson processes. In essence, this pacing process is characterised by the *prior* distribution π of the total number of trials, and some background continuous distribution to spread the epochs of the trials in an i.i.d. manner. Without loss of generality, the distribution will be assumed uniform; hence, given the number of trials, they are scattered in time like the uniform order statistics on $[0, 1]$. We enrich the model with a natural size parameter by letting π vary within a family of power series distributions.

The most obvious instance of a trapping action amounts to leaving some fraction of time to isolate the last success. We call this trapping action the *z-strategy*, with a parameter designating the proportion of time getting skipped (as compared to the real-time cut-off in the name of the familiar '1/e-strategy' of the best choice [13,14]). The overall optimality of the class of z-strategies among all trapping actions will be explored for a fixed and a random number of trials. For the problem of stopping on the last success, the optimality of the myopic strategy will be shown to hold if the sequence of its cut-offs is decreasing and interlacing with another set of critical points of z-strategies.

Then we specialise to the best-choice problem driven by a Pólya-Lundberg pacing process, when the number of trials follows a logarithmic series distribution. In different terms, the model was introduced by Bruss and Yor [15]. Bruss and Rogers [4] recently observed that the strategy stopping at the first record after time threshold $1/e$ is not optimal. We present a more detailed analysis; in particular, we use a curious property of certain hypergeometric functions to show that the cut-offs of the myopic strategy are increasing, hence the monotone case of optimal stopping [16] does not hold. Simulation suggests, however, that the myopic strategy is very close to optimality, both in terms of the cut-offs and the winning probability. A better approximation to optimality is achieved by the strategy that stops as soon as bygone becomes more beneficial than trapping with a z-strategy.

Viewed inside a bigger picture, the log-series prior appears as the edge $\nu = 0$ instance of the random records model with negative binomial distribution $\mathrm{NB}(\nu, q)$ of the number of trials. It is known that for $\nu = 1$, corresponding to the geometric prior, all cut-offs coincide [17,18], while for integer $\nu > 1$ they are decreasing [7]. In [19], we show that for $0 < \nu < 1$ the myopic strategy is not optimal, with the pattern of cut-offs as in the log-series case treated here.

2. Setting the Scene

2.1. The Probability Model

Let π be a power series distribution

$$\pi_n = c(q) w_n q^n, \, n \geq 0, \tag{1}$$

with weights $\quad w_0 \geq 0, \quad w_n > 0 \quad \text{for } n \geq 1$

and scale parameter $q > 0$ varying within the interval of convergence of $\sum_n w_n q^n$.

The associated mixed binomial process $(N_t, \, t \in [0,1])$ is an orderly counting process with the uniform order statistics property. The process can also be seen as a time inhomogeneous pure-birth process, with a transition rate expressible through the generating function of (w_n), see [20].

Conditionally on $N_t = k$:

(i) The epochs of the trials within $[0, t]$ and $(t, 1]$ are independent;
(ii) The posterior distribution of the number of trials yet to occur is a power series distribution

$$\pi(j \,|\, t, k) := \mathbb{P}(N_1 - N_t = j | N_t = k) = f_k(x) \binom{k+j}{j} w_{k+j} x^j, \, j \geq 0, \tag{2}$$

with scale variable

$$x := (1-t)q \tag{3}$$

and a normalisation function $f_k(x)$.

(iii) $(N_{t+s/(1-t)} - N_t, \, s \in [0,1])$ is a mixed binomial process on $[0,1]$, with the number of trials distributed according to (2).

The conditioning relation (2) appears in many statistical problems related to censored or partially observable data.

In principle, instead of considering a family of distributions for (N_t) with parameter q, we could deal with one counting process on the x-scale. We prefer not to adhere to this viewpoint, as the 'real time' variable is more intuitive. Nevertheless, we will use (3) to switch back and forth between t and x, as x is better suitable for power series work.

Let $= (p_k, k \geq 1)$ be a profile of success probabilities. We assume that

$$0 \leq p_1 \leq 1, \quad 0 \leq p_k < 1 \text{ for } k > 1 \text{ and } \sum_{k=1}^{\infty} p_k = \infty.$$

The kth trial, which is occurring at index/epoch k, is a success with probability p_k, independently of other trials and the pacing process. Thus, the point process of success epochs is obtained from (N_t) by thinning out the kth point with probability $1 - p_k$. Taken by itself, the process counting the success epochs is typically intractable [10]. A notable exception is the random records model ($p_k = 1/k$) with the geometric prior π, when the process is Poisson [1].

We shall identify *state* (t, k) with the event $N_t = k$. The notation $(t, k)^\circ$ will be used to denote the event that the kth trial epoch is t and the outcome is a success. If there is at least one success, the sequence of successes $(t_i, k_i)^\circ$ increases in both components.

2.2. The Trapping Game and Stopping Problem

A single episode of the trapping game refers to the generic state (t, k). The gambler plays either next or bygone, or chooses a proper subset of the interval $(t, 1]$. The trap $[t + z(1 - t), 1]$, for $0 < z < 1$, will be called z-*strategy*; this action leaves a $(1 - z)$ portion of the remaining time to isolate the last success epoch from other successes.

Let $\mathcal{F}_t$ be the sigma-algebra generated by the epochs and outcomes of trials on $[0, t]$. Under stopping strategy τ, we mean a random variable taking values in $[0, 1]$ and adapted to the filtration $(\mathcal{F}_t, t \in [0, 1])$. The performance of τ is assessed by the probability of the event that $(\tau, N_\tau)^\circ$ is the last success state.

We call a stopping strategy Markovian if in the event $\tau \geq t$ a decision to stop or to continue in state $(t, k)^\circ$ does not depend on the trials before time t. The general theory [21] implies existence of the optimal stopping strategy and that it can be found within the class of Markovian strategies.

Conditional on $\mathcal{F}_t$, the probability that $(t, k)^\circ$ is the last success equals the winning probability with bygone, while the probability that $(t, k)^\circ$ is the penultimate success equals the winning probability with next. If for every (t, k), where bygone is at least as good as next, also every state $(t', k') \in [t, 1] \times \{k, k+1, \cdots\}$ has this property, then the optimal stopping problem is *monotone* [21].

Define the *myopic* stopping strategy τ^* to be the first record $(t, k)^\circ$, if any, such that bygone is at least as beneficial as next. In the monotone case the myopic strategy is optimal among all stopping strategies.

Suppose for each $k \geq 1$ there exists a cut-off time a_k such that the action bygone is at least as good as next precisely for $t \in [a_k, 1]$. Then τ^* coincides with the time of the first success $(t, k)^\circ$ satisfying $t \geq a_k$ (or $\tau^* = 1$ if there is no such trial). The problem is monotone, hence τ^* is optimal if the cut-offs are non-increasing, that is $a_1 \geq a_2 \geq \cdots$.

3. The Game with Fixed Number of Trials

In this section, we assess the outcomes of actions in state (t, k) conditioned on the total number of trials $n > k$. This can be interpreted as the game of an informed gambler who knows n but not the outcomes of unseen trials $k+1, \cdots, n$. The time t is not important and a comparison of bygone with next is tantamount to the discrete-time optimal stopping at the last success [22,23]. The best action will be shown to coincide with a z-strategy provided next beats bygone.

3.1. bygone vs. next

The number of successes in trials $k+1, \cdots, n$ has probability generating function

$$\lambda \mapsto \prod_{m=k+1}^{n}(1-p_m+p_m\lambda) = \left(1+\lambda \sum_{i=k+1}^{n}\frac{p_i}{1-p_i}\right)\prod_{m=k+1}^{n}(1-p_m) + O(\lambda^2).$$

From this expansion, the probability of no success is

$$s_0(k+1,n) := \prod_{m=k+1}^{n}(1-p_m),$$

and the probability of exactly one success is

$$s_1(k+1,n) := \sum_{i=k+1}^{n}\frac{p_i}{1-p_i}\prod_{m=k+1}^{n}(1-p_m) = s_0(k+1,n)\sum_{i=k+1}^{n}\frac{p_i}{1-p_i}.$$

There is an obvious recursion

$$s_1(k,n) = (1-p_k)s_1(k+1,n) + p_k s_0(k+1,n),$$

which we can write as

$$\begin{aligned} s_1(k,n) - s_1(k+1,n) &= p_k\{s_0(k+1,n) - s_1(k+1,n)\} \\ &= p_k s_0(k+1,n)\left(1 - \sum_{i=k+1}^{n}\frac{p_i}{1-p_i}\right). \end{aligned} \qquad (4)$$

Note that the sequence,

$$1 - \sum_{i=k+1}^{n}\frac{p_i}{1-p_i}, \quad 0 \leq k \leq n-1, \qquad (5)$$

has the sign pattern

$$-, \cdots, -, \geq 0, +, \cdots, +,$$

and let k^* be the index value where the sign changes from negative. It follows that:

(i) $s_1(\cdot, n)$ is unimodal with maximum at k^*;
(ii) at k^* bygone becomes at least as good as next;
(iii) k^* is non-decreasing in n.

Each $A \subset \{1, \cdots, n\}$ corresponds to a stopping strategy in the discrete time problem [22,23]. We say that A wins if the index of the last success falls in A while no other index of success does.

Lemma 1. *Among all $A \subset \{1, \cdots, n\}$, the set $A^* := \{k^*+1, \cdots, n\}$ wins with the maximal probability.*

Proof. Clearly, $n \in A$ is necessary for A to be optimal. By induction, suppose we have shown that $\{k+1, \cdots, n\} \subset A$. Including k adds to said probability

$$c\, p_k\{s_0(k+1,n) - s_1(k+1)\},$$

where $c \geq 0$ depends on $A \cap \{1, \cdots, k-1\}$ only. However, this is non-negative precisely for $k \geq k^*$. □

The next lemma improves upon Theorem 3.1 of [24] by offering a weaker condition for monotonicity.

Lemma 2. For $k^* = k^*(n)$, if $p_{k^*+1} \geq p_{n+1}$ then $\max_k s_1(k, n) \geq \max_k s_1(k, n+1)$.

Proof. It is readily checked that the maximum value of $s_1(\cdot, n+1)$ is achieved at either k^* or $k^* + 1$.

Firstly, compare the winning probability of A^* for n trials with that of $B := \{k^* + 1, \cdots, n+1\}$ for $n+1$ trials. A difference results from the event that the $(n+1)^{st}$ trial is a success, and the number of successes among trials $k^* + 1, \cdots, n$ does not exceed 1. Hence the difference of winning probabilities is

$$(s_1(k^* + 1, n) - s_0(k^* + 1, n))p_{n+1} = \left(1 - \sum_{i=k^*+1}^{n} \frac{p_i}{1 - p_i}\right) s_0(k^* + 1, n) \geq 0.$$

Secondly, compare A^* with the other possible maximiser, $C := \{k^* + 2, \cdots, n, n+1\}$. The difference of winning probabilities of A^* in the setting with n trials and C with $(n+1)$ trials has four components:

(a) $p_{k^*+1} s_0(k^* + 2, n)(1 - p_{n+1})$, equal the probability that $(k^* + 1)^{st}$ trial is a success, A^* wins while C loses,
(b) $(1 - p_{k^*+1}) s_1(k^* + 2, n) p_{n+1}$, equal the probability that $(k^* + 1)^{st}$ trial is a failure, A^* wins while C loses,
(c) $p_{k^*+1} s_1(k^* + 2, n)(1 - p_{n+1})$, equal the probability that $(k^* + 1)^{st}$ trial is a success, A^* loses while C wins,
(d) $(1 - p_{k^*+1}) s_0(k^* + 2, n) p_{n+1}$, equal the probability that $(k^* + 1)^{st}$ trial is a failure, A^* loses while C wins.

After simplification, (a) + (b) − (c) − (d) becomes

$$\left(1 - \sum_{i=k^*+2}^{n} \frac{p_i}{1 - p_i}\right)(p_{k^*+1} - p_{n+1}),$$

which has the same sign as $p_{k^*+1} - p_{n+1}$ because the first factor is non-negative by the optimality of A^*. □

3.2. z-Strategies

For n fixed, the winning probability of a z-strategy in state (t, k) does not depend on t and is given by a Bernstein polynomial in $z \in [0, 1]$,

$$S_1(k, n; z) := \sum_{j=0}^{n-k-1} \binom{n-k}{j} z^j (1-z)^{n-k-j} s_1(k+j+1, n). \tag{6}$$

In particular, $S_1(k, n; 0) = s_1(k+1, n)$ is the probability to win with next. Similarly,

$$S_0(k, n; z) := \sum_{j=0}^{n-k} \binom{n-k}{j} z^j (1-z)^{n-k-j} s_0(k+j+1, n)$$

is the probability that none of the successes occurs in the time interval $(t + z(1-t), 1]$, so $S_0(k, n; 0) = s_0(k+1, n)$ equals the probability to win with bygone.

Note that $s_0(k+1, n) = S_0(k, n; 0)$ and $s_1(k+1, n) = S_1(k, n; 0)$. From (i) and (ii) above

$$k \geq k^* \iff S_0(k, n; 0) \geq S_1(k, n; 0) \implies S_1(k, n; 0) = \max_z S_1(k, n; z). \tag{7}$$

This is also valid for the maximum taken over *all* trapping actions.

From the unimodality of $s_1(\cdot, n)$ and the shape-preserving properties of the Bernstein polynomials (see [25], Theorem 3.3), it follows that (6) is unimodal. Thus, either the maximum is at 0 and `next` beats all z-strategies, or there exists a unique optimal z-strategy.

Next result stating that the optimum can be understood in a strong sense is a continuous-time counterpart of Lemma 1.

Theorem 1. *If $S_0(k,n;0) < S_1(k,n;0)$ then the optimal trapping action is a z-strategy with threshold determined as the unique maximiser of $S_1(k,n;\cdot)$.*

Proof. By a change of variables we reduce the claim to the case $(t,k) = (0,0)$. There is certainly a final interval that belongs to the optimal trap, because close to the end of the time, the probability of two or more successes is of order $o(1-t)$. Now, suppose $[z,1]$ belongs to the trap and we are assessing if the length element $[z-dz,z]$ is worth including. The change of the winning probability due to the inclusion is a multiple of

$$\sum_{j=1}^{n} \binom{n-1}{j-1} z^{j-1}(1-z)^{n-j} p_j \{s_0(j+1,n) - s_1(j+1,n)\} \, n\, h + o(h) = \qquad (8)$$

$$(1-z)^n \sum_{j=1}^{n} \binom{n-1}{j-1} \left(\frac{z}{1-z}\right)^j p_k \{s_0(j+1,n) - s_1(j+1,n)\} \, n\, h + o(h),$$

with some positive factor depending on the structure of the trap within $[0, z-h]$. By (4), in the variable $z/(1-z)$ the polynomial $\sum(\cdots)$ has at most one variation of sign in the coefficients. Applying Descartes' rule of signs, we see that the polynomial has at most one positive root. This implies that the optimal trap is a final interval with the cut-off coinciding with the root, or $[0,1]$ (action next) if there are no roots.

It remains to check that the root, if any, coincides with the maximiser of

$$S_1(0,n;z) = \sum_{j=0}^{n} \binom{n}{j} z^j (1-z)^{n-j} s_1(j+1,n).$$

Indeed, we have for the derivative using (4)

$$D_z S_1(0,n;z) =$$

$$\sum_{j=1}^{n} \binom{n-1}{j-1} n z^{j-1} (1-z)^{n-j} s_1(j+1,n) - \sum_{j=0}^{n-1} \binom{n-1}{j} n z^j (1-z)^{n-j-1} s_1(j+1,n)$$

$$= \sum_{k=1}^{n}(\cdots) - \sum_{k=1}^{n} \binom{n-1}{k-1} n z^{k-1}(1-z)^{n-k} s_1(k,n)$$

$$= \sum_{k=1}^{n} \binom{n-1}{k-1} n z^{k-1}(1-z)^{n-k} \{s_1(k+1,n) - s_1(k,n)\}$$

$$= \sum_{k=1}^{n} \binom{n-1}{k-1} n z^{k-1}(1-z)^{n-k} p_k \{s_1(k+1,n) - s_0(k+1,n)\},$$

which is the negative of the polynomial in (8). This provides the desired conclusion. □

3.3. Examples

The best-choice problem is related to the profile $p_k = 1/k$. The associated Bernstein polynomials satisfy

$$S_1(k,n;z) \to -z \log z, \quad n \to \infty,$$

where the convergence is uniform. Both maximiser and the maximum value converge to $1/e$ as $n \to \infty$

The case $k=0$ was studied in much detail [13,14,17,26]. The winning probability of z-strategy can be alternatively written as a Taylor polynomial

$$S_1(0,n;z) = 1 - z - \sum_{j=2}^{n} \frac{(1-z)^j}{j(j-1)},$$

which decreases pointwise to $z \mapsto -z \log z$ as n increases (see Figure 1). The maximisers increase monotonically to $1/e$ and also $\max_z S_1(0,n;z) \downarrow 1/e$. These facts underlie the minimax property that the $1/e$-strategy ensures winning probability of at least $1/e$ for every $n \geq 1$.

The nice monotonicity properties do not extend to $k > 0$, the minimax value is below $1/e$ and the $1/e$-strategy is not minimax. This is already seen in the case $k=1$, where the Bernstein polynomials become

$$\begin{aligned} S_1(1,n;z) &= \frac{n-1}{n} - \sum_{j=2}^{n-1} \frac{(n-j)(1-z)^j}{nj(j-1)} \\ &= S_1(0,n;z) + \sum_{j=1}^{n-1} \frac{(1-z)^{j+1}}{nj} - \frac{(1-z)}{n}. \end{aligned}$$

The first formula is derived by conditioning on the highest rank j of trials that occur before the threshold of z-strategy.

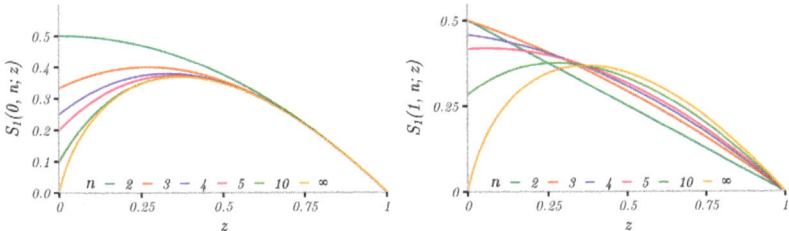

Figure 1. The winning probability $S_1(k,n;z)$ of z-strategy in the best-choice problem for $k=0$ and 1.

The more general profile

$$p_k = \frac{\theta}{\theta + k - 1}, \quad k \geq 1, \tag{9}$$

with parameter $\theta > 0$, plays a central role in the combinatorial structures related to the Ewens sampling formula for random partitions [27]. The term *Karamata–Stirling law* was coined in [28] for the distribution of the number of successes with these probabilities. The number of successes in trials $k+1, \cdots, n$ has probability generating function

$$\lambda \mapsto \frac{(k+\theta\lambda)_{n-k}}{(k+\theta)_{n-k}}.$$

As $n \to \infty$, $S_1(k,n;z) \to -\theta z^\theta \log z$. The maximum values still converge to $1/e$ but the maximisers approach $e^{-1/\theta}$. The shapes vary considerably with θ, see Figure 2. For θ large, the minimax winning probability is close to zero.

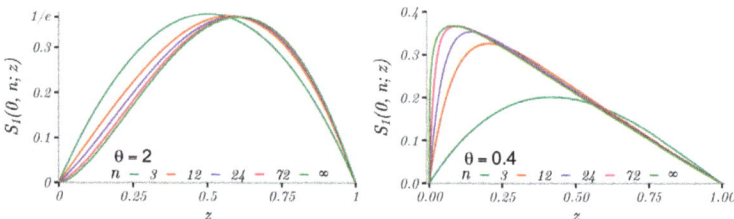

Figure 2. Bernstein polynomials for $p_k = \theta/(\theta + k - 1)$.

4. Random Number of Trials: z-Strategies

We proceed with the continuous time setting, assuming p and π are given. In state (t, k), the probability of isolating the last success by means of a z-strategy is a convex mixture of the Bernstein polynomials:

$$\mathcal{S}_1(t,k;z) := \sum_{j=1}^{\infty} \pi(j|t,k) \sum_{i=0}^{j-1} \binom{j}{i} z^i (1-z)^{j-1} s_1(k+i+1, k+j). \tag{10}$$

The $z = 0$ instance,

$$\mathcal{S}_1(t,k;0) = \sum_{j=1}^{\infty} \pi(j|t,k) s_1(k+1, k+j),$$

is the probability to win with next, and $\mathcal{S}_1(t,k;1) = 0$. Similarly, the probability that none of the successes are trapped by the z-strategy is:

$$\mathcal{S}_0(t,k;z) := \sum_{j=0}^{\infty} \pi(j|t,k) \sum_{i=0}^{j-1} \binom{j}{i} z^i (1-z)^{j-1} s_0(k+i+1, k+j),$$

and $\mathcal{S}_0(t,k;0)$ is the probability to win with bygone.

Being a convex mixture of unimodal functions, $\mathcal{S}_1(t,k;\cdot)$ itself need not be unimodal. Accordingly, the optimal trap need not be a final interval. It may rather include a few disjoint intervals akin to 'islands' in the discrete time best-choice problems [29].

Concavity is a simple condition to ensure unimodality. We say that $s_1(\cdot, n)$ is concave if for every $n \geq 1$ the second difference in the first variable is non-positive.

Theorem 2. *Suppose $s_1(\cdot, n)$ is concave. Then $\mathcal{S}_1(t,k;\cdot)$ is unimodal with maximum at some z^*. If $z^* \in (0,1)$ then for $z = z^*$ the z-strategy is optimal among all trapping actions, and if $z^* = 0$ then* next *outperforms every trapping action.*

Proof. By the shape-preserving properties of Bernstein polynomials [25], the internal sum in (10) is a concave function in z, therefore the mixture $\mathcal{S}_1(t,k;\cdot)$ is also concave hence unimodal. The maximum is attained at 0 if $D_z \mathcal{S}_1(t,k;0) \leq 0$, and $z^* > 0$ otherwise. The overall optimality follows from the unimodality as in Theorem 1. □

The concavity is easy to express in terms of p explicitly. The second difference in the variable k of the probability generating function

$$\lambda \mapsto \prod_{j=k}^{n} (1 - p_j + \lambda p_j)$$

becomes

$$\{(1-p_k+\lambda p_k)(1-p_{k+1}+\lambda p_{k+1}) - 2(1-p_{k+1}+\lambda p_{k+1}) + 1\}\prod_{j=k+2}^{n}(1-p_j+\lambda p_j).$$

Computing D_λ at $\lambda = 0$ yields the second difference of $s_1(\cdot, n)$

$$(p_k - 2p_k p_{k+1} - p_{k+1}) + (p_k p_{k+1} - p_k + p_{k+1})\sum_{j=k+2}^{n}\frac{p_j}{1-p_j}. \quad (11)$$

From this, a sufficient condition for the concavity of $s_1(\cdot, n)$ is

$$p_k - 2p_k p_{k+1} - p_{k+1} \leq 0, \quad p_k p_{k+1} - p_k + p_{k+1} \leq 0, \quad k \geq 1. \quad (12)$$

Notably, (12) ensures unimodality for arbitrary π and only involves two consecutive success probabilities. The price to pay for the simplicity is that the condition is restrictive, as seen in Figure 3.

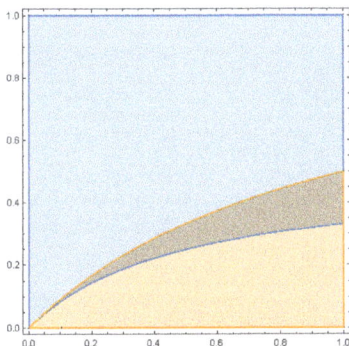

Figure 3. The concavity condition (12) holds for profiles p with (p_k, p_{k+1}) squeezed between the parabolas.

For the profile (9), straight calculation shows that (11) is non-positive, hence $s_1(\cdot, n)$ is concave, iff

$$\frac{1}{2} \leq \theta \leq 1.$$

This is only a half range, but it includes two most important for application cases $\theta = 1$ and $\theta = 1/2$.

5. Tests for the Monotone Case of Optimal Stopping

Using (2) and (3), we can cast the winning probabilities with actions bygone, next and a z-strategy as:

$$\begin{aligned}
S_0(t,k;0) &= f_k(x)P_k(x), \\
S_1(t,k;0) &= f_k(x)Q_k(x), \\
S_1(t,k;z) &= f_k(x)R_k(x,z),
\end{aligned} \quad (13)$$

where $x = q(1-t)$ and

$$P_k(x) := \sum_{j=0}^{\infty} \binom{k+j}{j} w_{k+j} x^j s_0(k+1, k+j),$$

$$Q_k(x) := \sum_{j=1}^{\infty} \binom{k+j}{j} w_{k+j} x^j s_1(k+1, k+j),$$

$$R_k(x, z) := \sum_{j=1}^{\infty} \binom{k+j}{j} w_{k+j} x^j \sum_{i=0}^{j-1} \binom{j}{i} z^i (1-z)^{j-i} s_1(k+i+1, k+j).$$

Thus, $Q_k(x) = R_k(x, 0)$. We are looking next at some critical points for the trapping game and the optimal stopping problem.

Lemma 3. *Equation $P_k(x) = Q_k(x)$ has at most one root $\alpha_k > 0$, for every $k \geq 1$.*

Proof. Coefficients of the series $P_k(x) - Q_k(x)$ have at most one change of sign from $+$ to $-$, hence Descartes' rule of signs for power series [30] entails that there is at most one positive root. □

We set $\alpha_k = \infty$ if the root does not exist. Define the cut-off

$$a_k = \left(1 - \frac{\alpha_k}{q}\right)_+.$$

This is the earliest time when bygone becomes at least as good as next. Keep in mind that if the sequence (α_k) is monotone, then (a_k) is also monotone but with the monotonicity direction reversed. The monotone case of optimal stopping holds for every q, hence τ^* is optimal, if $\alpha_k \uparrow$.

Example 1. *In the paradigmatic case $p_k = 1/k$ and the geometric prior with $w_n = 1$, we have*

$$s_0(k+1, n) = \frac{k}{n}, \quad s_1(k+1, n) = \frac{k}{n} \sum_{j=k+1}^{n} \frac{1}{j-1},$$

and explicitly computable power series

$$P_k(x) = \frac{1}{(1-x)^k}, \quad Q_k(x) = \frac{-\log(1-x)}{(1-x)^k}.$$

The equation $P_k(x) = Q_k(x)$ yields identical roots $\alpha_k = 1 - 1/e$ and coinciding cut-offs $a_k = (1 - (1 - e^{-1})/q)_+$. Thus, τ^* stops at the first success trial after a time threshold. See [1,7,17–19] for details on this remarkable case.

Lemma 4. *Equation $D_z R_k(x, 0) = 0$ has at most one root $\beta_k > 0$, for every $k \geq 0$. If the root exists, then $\beta_k \leq \alpha_{k+1}$.*

Proof. We follow the argument in Lemma 3. The derivative at $z = 0$ is

$$D_z R_k(x, 0) = p_{k+1} \sum_{j=1}^{\infty} \binom{k+j}{j} w_{k+j} j x^j \{s_0(k+2, k+j) - s_1(k+2, k+j)\},$$

which has at most one change of sign for $x \geq 0$, and then from $+$ to $-$. Furthermore,

$$\begin{aligned} D_z R_k(x,0) &\geq p_{k+1} \sum_{j=1}^{\infty} \binom{k+j}{j} w_{k+j} x^j \{s_0(k+2,k+j) - s_1(k+2,k+j)\} \\ &= p_{k+1} \{P_{k+1}(x) - Q_{k+1}(x)\}. \end{aligned}$$

This follows by comparing the series and noting that the weights at positive terms in D_z are higher. □

If there is no finite root, we set $\beta_k = \infty$. Let

$$b_k := \left(1 - \frac{\beta_k}{q}\right)_+.$$

We have $D_z R_k(q(1-t),0) < 0$ for $t \in (b_k, 1]$, and $b_k \geq a_{k+1}$ by Lemma 4. Thus, b_k is the earliest time when the action next at index k cannot be improved by a z-strategy with small enough z.

To summarise the above: for $t < a_k$ action next is better than bygone, and tor $t < b_k$ a trapping strategy is better than next.

Theorem 3. *The optimal stopping problem belongs to the monotone case (for every admissible q) if and only if $\alpha_1 \leq \alpha_2 \leq \cdots$. In that case we have the interlacing pattern of roots*

$$\cdots \leq \alpha_k \leq \beta_k \leq \alpha_{k+1} \leq \beta_{k+1} \leq \cdots. \tag{14}$$

Proof. We argue in probabilistic terms. The bivariate sequence of success epochs $(t,k)^\circ$ is an increasing Markov chain. The monotone case of optimal stopping occurs iff the set of states where bygone outperforms next is closed, which holds iff this is an upper subset with respect to the partial order in $[0,1] \times \{1,2,\cdots\}$. The latter property amounts to the monotonicity condition $\alpha_k \uparrow$.

By Lemma 3, the inequality $\alpha_k \leq \beta_{k+1}$ always holds. In the monotone case, if in some state $(t,k)^\circ$ the actions bygone and next are equally good, then trapping cannot improve upon these by optimality of the myopic strategy. In the analytic terms, the above translates as the inequality $\beta_k \leq \alpha_k$. □

6. The Best-Choice Problem under the Log-Series Prior

In this section we consider the random records model with the classic profile $p_k = 1/k$, and a pacing process with the logarithmic series prior

$$\pi_n = c(q) \frac{q^n}{n}, \quad n \geq 1, \tag{15}$$

(so $\pi_0 = 0$), where $0 < q < 1$ and $c(q) = |\log(1-q)|^{-1}$. See [31] for Poisson mixture representations of π. The function $S_1(t,k;\cdot)$ is concave, hence by Theorem 2 it is sufficient to consider z-strategies.

Let T_1 be the time of the first trial.

Lemma 5. *Under the logarithmic series prior (15) the pacing process has the following features:*

(i) *The time of the first trial T_1 has probability density function*

$$t \mapsto \frac{c(q) q}{1 - (1-t)q}, \quad t \in [0,1].$$

(ii) $(N_t, t \in [0,1])$ is a Pólya-Lundberg birth process with transition rates

$$\mathbb{P}(N_{t+dt} - N_t = 1 \mid N_t = k) = \begin{cases} \dfrac{c((1-t)q)\,q}{1-(1-t)q}, & k = 0, \\ \dfrac{k}{t+q^{-1}-1}, & k \geq 1. \end{cases}$$

(iii) Given $N_t = k$, the posterior distribution $\pi(\cdot \mid t, k)$ of $N_1 - N_t$ is $\mathrm{NB}(k, (1-t)q)$. In particular, conditionally on $T_1 = t_1$, the posterior distribution is geometric with the 'failure' probability $(1-t_1)q$.

Proof. Assertion (i) follows from

$$\mathbb{P}(T_1 > t) = \mathbb{P}(N_t = 0) = \sum_{n=1}^{\infty} \frac{c(q) q^n (1-t)^n}{n},$$

and (iii) from the identity

$$\binom{k+j}{j} \frac{x^j}{k+j} = \binom{k+j-1}{j} \frac{x^j}{k}$$

underlying the formula for $\pi(j \mid t, k)$ in terms of $x = (1-t)q$. □

In view of part (ii), we will use $\mathrm{NB}(0, q)$ to denote the log-series prior (15).

6.1. Hypergeometrics

The power series of interest can be expressed via the Gaussian hypergeometric function

$$F(a, b; c; x) := \sum_{j=0}^{\infty} \frac{(a)_j (b)_j}{(c)_j} \frac{x^j}{j!}.$$

Recall the differentiation formula

$$D_x F(a, b; c, x) = \frac{ab}{c} F(a+1, b+1; c+1, x),$$

the parameter transformation formula

$$F(a, b; c; x) = (1-x)^{c-a-b} F(c-a, c-b; c; x),$$

and Euler's integral representation for $c > b > 0$

$$F(a, b; c; x) = \frac{\Gamma(c)}{\Gamma(b)\Gamma(c-b)} \int_0^1 \frac{y^{b-1}(1-y)^{c-b-1} dy}{(1-xy)^a}.$$

The probability generating function for the number of successes following state (t, k), for $k \geq 1$, is given by a hypergeometric function:

$$\lambda \mapsto (1-x)^k \sum_{j=0}^{\infty} \binom{k+j-1}{j} x^j \frac{(k+\lambda)_j}{(k+1)_j} =$$

$$(1-x)^k \sum_{j=0}^{\infty} \frac{(k)_j (k+\lambda)_j}{(k+1)_j} \frac{x^j}{j!} =$$

$$(1-x)^k F(k+\lambda, k; k+1; x).$$

Expanding at $\lambda = 0$ we identify two basic power series as:

$$P_k(x) = k^{-1} F(k,k;k+1;x),$$
$$Q_k(x) = k^{-1} D_a F(k,k;k+1;x),$$

where as before $x = (1-t)q \in [0,1]$ and D_a is the derivative in the first parameter. The differentiation formula implies backward recursions:

$$D_x P_k(x) = k P_{k+1}(x),$$
$$D_x Q_k(x) = P_{k+1}(x) + k Q_{k+1}(x). \quad (16)$$

The normalisation function for probabilities (14) is $f_k(x) = k(1-x)^k$ for $k \geq 1$, and $f_0(x) = |\log(1-x)|^{-1}$. Applying the transformation formula yields $P_k(x) = (1-x)^{1-k} F(1,1;k+1,x)$, hence, we may write the winning probability with bygone as the series

$$S_0(t,k;0) = (1-x) \sum_{j=0}^{\infty} \frac{j! \, x^j}{(k+1)_j}, \quad x = (1-t)q.$$

It is readily seen that, as k increases, this function decreases to $1-x$. This result was already observed in [18] using a probabilistic argument. The convergence to $1-x$ relates to the fact that for large k, the point process of record epochs approaches a Poisson process.

For $R_k(x,z)$, we derive an integral formula. Consider first the case $k \geq 1$. The probability generating function of the number of record epochs following (t,k) and falling in the final interval $[t+z(1-t),1]$ has probability generating function

$$\lambda \mapsto (1-x)^k \sum_{j=0}^{\infty} \binom{k+j-1}{j} x^j \sum_{i=0}^{j} \binom{j}{i} z^i (1-z)^{j-i} \frac{(k+i+\lambda)_{j-i}}{(k+i+1)_{j-i}} =$$

$$(1-x)^k \sum_{i=0}^{\infty} \binom{k+i-1}{i} (xz)^i F(k+i+\lambda, k+i, k+i+1; x-xz) =$$

$$k(1-x)^k \sum_{i=0}^{\infty} \binom{k+i}{i} (xz)^i \int_0^1 \frac{y^{k+i-1} dy}{(1-xy+xyz)^{k+i+\lambda}} =$$

$$k(1-x)^k \int_0^1 \frac{y^{k-1}(1-xy+xyz)^{1-\lambda} dy}{(1-xy)^{k+1}}.$$

Differentiating at $\lambda = 0$ yields $S_1(k,t;z)$, which is the same as $k(1-x)^k R_k(x,z)$ for $x = (1-t)q$, whence

$$R_k(x,z) = \int_0^1 \frac{y^{k-1}(1-xy+xyz)|\log(1-xy+xyz)| dy}{(1-xy)^{k+1}}. \quad (17)$$

For $k = 0$, a similar calculation with log-series weights $NB(0,x)$ gives

$$R_0(x,z) = \int_0^1 \frac{(1-xy+xyz)\log(1-xy+xyz)}{y(1-xy)} dy.$$

6.2. The Myopic Strategy

The positive root obtained by equating

$$P_1(x) = \frac{|\log(1-x)|}{x} \quad \text{and} \quad Q_1(x) = \frac{|\log(1-x)|^2}{2x}$$

is $\alpha_1 = 1 - e^{-2} = 0.864665 \cdots$. On the other hand, solving $D_z R_1(x,0) = 0$ yields a smaller value $\beta_1 = 0.756004 \cdots$, hence the interlacing condition of Theorem 3 fails for $k = 1$. Translating in terms of the best-choice problem, this means that τ^* stops at the first trial if this occurs before $a_1 = (1 - \alpha_1/q)_+$, but a z-strategy will be more beneficial for a bigger range of times $t \leq b_1 = (1 - \beta_1/q)_+$. Therefore, at least for $q > \beta_1$, it is not optimal to stop at the first trial before b_1 and the myopic strategy can be beaten.

The root $\alpha_2 := 0.755984 \cdots$ is found by equating

$$P_2(x) = \frac{2(x - L + xL)}{(1-x)x^2} \quad \text{and} \quad Q_2(x) = \frac{-2x + 2L - L^2 + xL^2}{(1-x)x^2},$$

where for shorthand $L := -\log(1-x)$. Formulas become more complicated for larger k.

We see that $\alpha_1 > \alpha_2$, which suggests monotonicity of the whole sequence. To show this, pass to the quotient and re-define the root α_k as a unique solution on $[0,1)$ to

$$\frac{Q_k(x)}{P_k(x)} = 1 \quad \iff \quad \frac{D_a F(k,k;k+1;x)}{F(k,k;k+1;x)} = 1, \tag{18}$$

where D_a acts in the first parameter. As x increases from 0 to 1, this logarithmic derivative runs from 0 to ∞.

Lemma 6. *The logarithmic derivative* (18) *increases in k, hence the sequence of roots α_k is strictly decreasing.*

Proof. Euler's integral specialises as:

$$F(k+\lambda, k; k+1; x) = k \int_0^1 \frac{y^{k-1}}{(1-xy)^{k+\lambda}} dy.$$

Expanding in parameter at $\lambda = 0$ gives the integral representations

$$P_k(x) = \int_0^1 \frac{y^{k-1}}{(1-xy)^k} dy, \quad Q_k(x) = \int_0^1 \frac{y^{k-1}|\log(1-xy)|}{(1-xy)^k} dy.$$

From these formulas,

$$Q_k(x) P_{k+1}(x) = \int_0^1 \frac{y^{k-1}|\log(1-xy)|}{(1-xy)^k} dy \int_0^1 \frac{z^k}{(1-xz)^{k+1}} dz$$

$$= \int_0^1 \int_0^1 \frac{y^{k-1}z^{k-1}|\log(1-xy)|}{(1-xy)^k(1-xz)^k} \frac{z}{(1-xz)} dydz.$$

By the same kind of argument, a similar formula is obtained for $Q_{k+1}(x)P_k(x)$. Splitting the integration domain, and using symmetries of the integrand yields for $x \in [0,1)$

$$Q_k(x)P_{k+1}(x) - Q_{k+1}(x)P_k(x) =$$

$$\int_0^1 \int_0^1 \frac{y^{k-1}z^{k-1}|\log(1-xy)|}{(1-xy)^{k+1}(1-xz)^{k+1}} (z-y) dydz =$$

$$\iint_{0<y<z<1} \frac{y^{k-1}z^{k-1}}{(1-xy)^{k+1}(1-xz)^{k+1}} \log\left(\frac{1-xz}{1-xy}\right)(z-y) dydz < 0,$$

which implies the asserted monotonicity. □

Figure 4 shows some shapes of $f_k(x)P_k(x)$ and $f_k(x)Q_k(x)$ for $k = 1, 2, 3$.

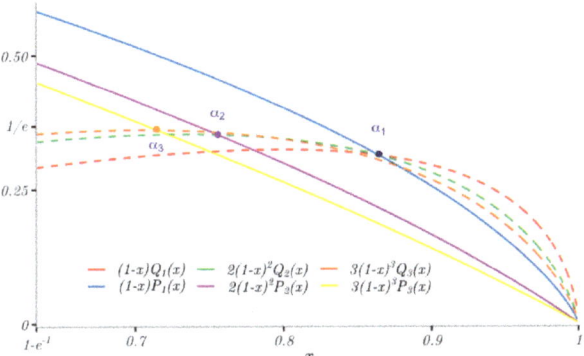

Figure 4. next and bygone curves for $k = 1, 2, 3$.

The log-series distribution weights satisfy $w_{n+1}/w_n \uparrow 1$. Comparison with the geometric distribution, as in [19], in combination with the lemma give $a_k \downarrow (1 - 1/e)$ as $k \to \infty$. The same limit has been shown for analogous roots in the best-choice problem with the negative binomial prior NB(ν, q) for integer $\nu \geq 1$; however, the monotonicity direction in that setting is different [7].

To summarise findings of this section, we have:

Theorem 4. *The monotone case of optimal stopping does not hold. The myopic strategy τ^* is not optimal and has the following features:*

(i) *for $q > 1 - 1/e$, the cut-offs of τ^* satisfy $a_k \uparrow 1 - (1 - 1/e)/q$;*
(ii) *for $t \geq (1 - (1 - 1/e)/q)_+$, bygone is the optimal action for every $(t, k)°$;*
(iii) *for times as in (ii), the myopic strategy coincides with the optimal stopping strategy (in the event $\tau^* \geq t$).*

6.3. Optimality and Bounds

For state (t, k) and $x = q(1 - t)$, define the *continuation value* $V_k(x)$ to be the maximum probability of the best choice, as achievable by stopping strategies starting in the state. By the optimality principle, the overall optimal stopping strategy, starting from $(0, 0)$, stops at the first record $(t, k)°$ satisfying $k(1 - x)^k P_k(x) \geq V_k(x)$.

Given $N_t = k$, let T_{k+1} be the next trial epoch (or 1 in the event $N_1 = k$). Similar to the argument in Lemma 5, we find that the random variable $(1 - T_{k+1})/(1 - t)$ has density

$$y \mapsto \frac{kx(1-x)^k}{(1-x+xy)^{k+1}}, \ y \in (0, 1].$$

By the $(k+1)^{\text{st}}$ trial, the optimal stopping strategy stops if this is a record and bygone is more beneficial than the optimal continuation, hence integrating out T_{k+1} we obtain

$$V_k(x) = \int_0^1 \left[\frac{1}{k+1} \max\{(1-y)^{k+1} P_{k+1}(y), V_{k+1}(y)\} + \frac{k}{k+1} V_{k+1}(y) \right] \frac{kx(1-x)^k dy}{(1-x+xy)^{k+1}}.$$

This has the equivalent differential form for $k \geq 1$,

$$(1-x) D_x V_k(x) = \frac{k}{k+1} \left((1-x)^{k+1} P_{k+1}(x) - V_{k+1}(x) \right)_+ + k\{V_{k+1}(x) - V_k(x)\}. \quad (19)$$

For the special instance $k = 0$, integrating out the variable T_1 gives

$$V_0(x) = \int_0^1 \max((1-y)P_1(y), V_1(y)) \frac{dy}{(1-x+xy)|\log(1-x)|},$$

or, in the differential form with initial conditions $V_0(0) = 1$ and $V_k(0) = 0$, for $k \geq 1$

$$(1-x)|\log(1-x)| D_x V_0(x) = \max\{(1-x)P_1(x), V_1(x)\} - V_0(x). \qquad (20)$$

By Corollary 4, the continuation value coincides with the winning probability of next in a segment of the range; therefore:

$$V_k(x) = k(1-x)^k Q_k(x), \quad \text{for } 0 \leq x \leq 1 - 1/e, \ k \geq 0. \qquad (21)$$

As a check, for $k \geq 1$ let $\widehat{V}_k(x) := k^{-1}(1-x)^{-k} V_k(x)$. With this change of variable, (19) simplifies as

$$D_x \widehat{V}_k(x) = (P_{k+1}(x) - \widehat{V}_{k+1}(x))_+ + (k+1)\widehat{V}_{k+1}(x).$$

For x in the range where $P_{k+1}(x) - \widehat{V}_{k+1}(x) \geq 0$, this becomes the recursion (16).

Outside the range covered by (21), Equations (19) and (20) should be complemented by a '$k = \infty$' boundary condition

$$\lim_{k \to \infty} V_k(x) = \begin{cases} 1/e, & \text{for } 1 - 1/e \leq x \leq 1, \\ -(1-x)\log(1-x), & \text{for } 0 \leq x \leq 1 - 1/e. \end{cases}$$

Figure 5 shows stop, continuation and z-strategy curves for $k = 1, 2$ and 3. The numerical simulation suggests that the equation $k(1-x)^k P_k(x) = V_k(x), k \geq 1$ has a unique solution γ_k and that the critical points increase with k, so the optimal stopping strategy is similar to the myopic. These critical points have lower bounds δ_k defined as the solution to $k(1-x)^k P_k(x) = I_k(x)$ and upper bounds ρ_k defined as the critical points where bygone is the same as the z-strategy.

To approximate the continuation value in the range $1 - 1/e < x < 1$, we simulated some easier computable bounds

$$k(1-x)^k Q_k \leq k(1-x)^k \max_z R_k(x, z) \leq V_k(x) < I_k(x).$$

The upper *information* bound $I_k(x)$ (see Figure 6) is the winning probability of an informed gambler who in state (t, k) (with $x = q(1-t)$) knows the total number of trials N_1, as in Section 3. Two lower bounds stem from the comparison with the myopic and z-strategies. The points β_k computed for $k \leq 10$ all satisfy $\beta_k < \alpha_k$, and so the first relation turns equality for $0 \leq x \leq \beta_k$. Therefore, the critical points satisfy

$$\delta_k < \gamma_k < \rho_k \leq \alpha_k.$$

The results of computation are presented in Figure 5 and Tables 1–4. The data show excellent performance of the strategy that by the first trial chooses between stopping and proceeding with a z-strategy.

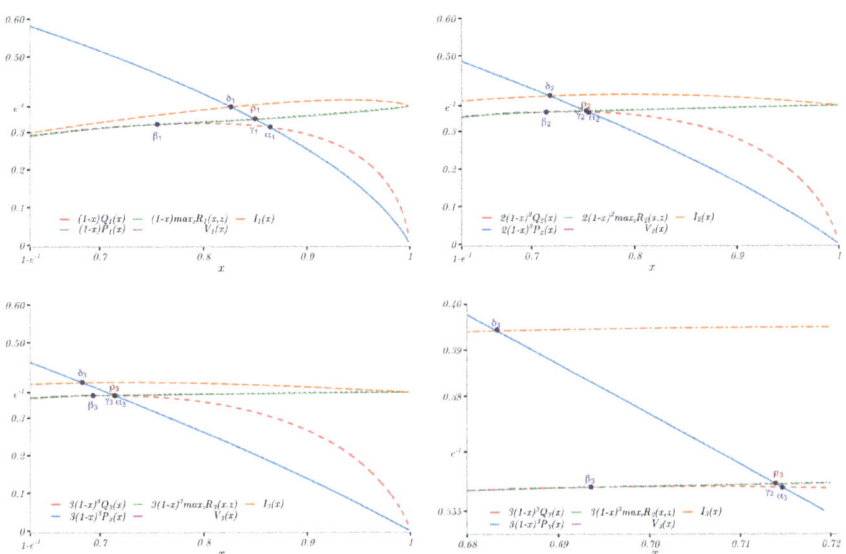

Figure 5. Stop, continuation, z-strategy values and bounds; $k = 1, 2, 3$ and zoomed-in view for $k = 3$.

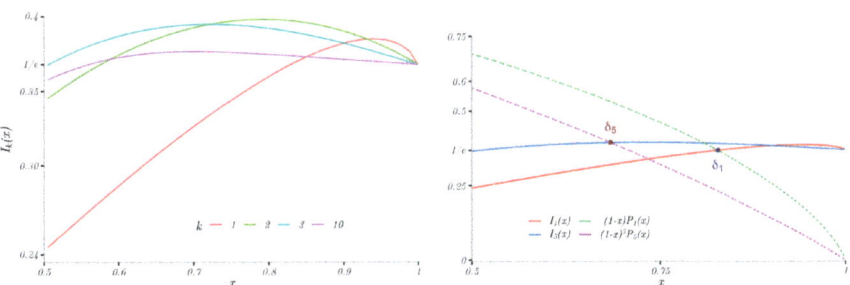

Figure 6. Information bounds on the optimal strategy $I_k(x)$.

Table 1. Critical points: α_k: Solution to $P_k(x) = Q_k(x)$, β_k: Solution to $D_z R_k(x, z) = 0$, γ_k: Solution to $k(1 - x)^k P_k(x) = V_k(x)$, δ_k: Solution to $k(1 - x)^k P_k(x) = I_k(x)$, ρ_k: Solution to $P_k(x) = \max_z R_k(x, z)$.

k	β_k	δ_k	γ_k	ρ_k	α_k
1	0.756004	0.826893	0.849635	0.850335	0.864665
2	0.714616	0.718332	0.753621	0.753727	0.755984
3	0.693549	0.683295	0.713957	0.713995	0.714596
4	0.680931	0.668986	0.693275	0.693311	0.693529
5	0.672567	0.661520	0.680687	0.680814	0.680911
6	0.666632	0.656902	0.672194	0.672499	0.672547
7	0.662206	0.653656	0.665900	0.666584	0.666611
8	0.658782	0.651188	0.661005	0.662169	0.662186
9	0.656055	0.649234	0.657108	0.658751	0.658761
10	0.653833	0.647653	0.653911	0.656028	0.656034

Table 2. Winning probability and bounds for $k = 1$.

x	$(1-x)P_1(x)$	$(1-x)Q_1(x)$	$(1-x)\max_z R_1(x,z)$	$V_1(x)$	$I_1(x)$
0.60	0.6109	0.2799	0.2799	0.2799	0.2864
0.65	0.5653	0.2967	0.2967	0.2967	0.3069
0.70	0.5160	0.3106	0.3106	0.3106	0.3262
0.75	0.4621	0.3203	0.3203	0.3204	0.3439
0.80	0.4024	0.3238	0.3269	0.3275	0.3597
0.85	0.3348	0.3176	0.3342	0.3354	0.3728
0.90	0.2558	0.2945	0.3428	0.3446	0.3821
0.95	0.1577	0.2362	0.3532	0.3555	0.3848
0.995	0.0266	0.0705	0.3659	0.3667	0.3731

Table 3. Winning probability and bounds for $k = 2$.

x	$2(1-x)^2 P_2(x)$	$2(1-x)^2 Q_2(x)$	$2(1-x)^2 \max_z R_2(x,z)$	$V_2(x)$	$I_2(x)$
0.60	0.5189	0.3297	0.3297	0.3297	0.3743
0.65	0.4682	0.3429	0.3429	0.3429	0.3850
0.70	0.4149	0.3509	0.3509	0.3509	0.3926
0.75	0.3586	0.3521	0.3541	0.3543	0.3970
0.80	0.2988	0.3440	0.3570	0.3575	0.3981
0.85	0.2348	0.3227	0.3600	0.3608	0.3960
0.90	0.1654	0.2809	0.3630	0.3643	0.3903
0.95	0.0887	0.2018	0.3659	0.3674	0.3811
0.995	0.0098	0.0428	0.3678	0.3679	0.3694

Table 4. Winning probability and bounds for $k = 3$.

x	$3(1-x)^3 P_3(x)$	$3(1-x)^3 Q_3(x)$	$3(1-x)^3 \max_z R_3(x,z)$	$V_3(x)$	$I_3(x)$
0.60	0.4811	0.3460	0.3460	0.3460	0.3869
0.65	0.4296	0.3562	0.3562	0.3562	0.3923
0.70	0.3762	0.3603	0.3604	0.3605	0.3947
0.75	0.3207	0.3568	0.3620	0.3622	0.3946
0.80	0.2629	0.3431	0.3635	0.3640	0.3923
0.85	0.2026	0.3155	0.3649	0.3660	0.3881
0.90	0.1391	0.2674	0.3663	0.3679	0.3824
0.95	0.0719	0.1846	0.3673	0.3685	0.3755
0.995	0.0075	0.0359	0.3679	0.3679	0.3687

Author Contributions: Methodology, A.G.; validation, A.G. and Z.D.; formal analysis, A.G. and Z.D.; writing—original draft preparation, A.G.; writing—review and editing, A.G. and Z.D.; visualization, A.G. and Z.D.; supervision, A.G. All authors have read and agreed to the published version of the manuscript.

Funding: This research received funding from the European Research Council (ERC) under the European Union's Horizon 2020 research and innovation programme under grant agreement No 817257.

Institutional Review Board Statement: Not applicable.

Informed Consent Statement: Not applicable.

Data Availability Statement: Data sharing not applicable.

Conflicts of Interest: The authors declare no conflict of interest.

References

1. Browne, S. *Records, Mixed Poisson Processes and Optimal Selection: An Intensity Approach*; Working Paper; Columbia University: New York, NY, USA, 1994.
2. Bruss, F.T. On an optimal selection problem of Cowan and Zabczyk. *J. Appl. Probab.* **1987**, *24*, 918–928. [CrossRef]
3. Bruss, F.T.; Samuels, S.M. A unified approach to a class of optimal selection problems with an unknown number of options. *Ann. Probab.* **1987**, *15*, 824–830. [CrossRef]
4. Bruss, F.T.; Rogers, L.C.G. The $1/e$-strategy is sub-optimal for the problem of best choice under no information. *Stoch. Process. Their Appl.* **2021**, Special issue: In Memory of Professor Larry Shepp, in press. [CrossRef]
5. Cowan, R.; Zabczyk, J. An optimal selection problem associated with the Poisson process. *Theory Probab. Appl.* **1978**, *23*, 584–592. [CrossRef]
6. Berezovsky, B.A.; Gnedin, A.V. *The Best Choice Problem*; Akad. Nauk: Moscow, Russia, 1984. (In Russian)
7. Kurushima, A.; Ano, K. A Poisson arrival selection problem for Gamma prior intensity with natural number parameter. *Sci. Math. Jpn.* **2003**, *57*, 217–231.
8. Stewart, T.J. The secretary problem with an unknown number of options. *Oper. Res.* **1981**, *29*, 130–145. [CrossRef]
9. Tamaki, M.; Wang, Q. A random arrival time best-choice problem with uniform prior on the number of arrivals. In *Optimization and Optimal Control*; Chinchuluun, A., Enkhbat, R., Tseveendorj, I., Pardalos, P.M., Eds.; Springer: New York, NY, USA, 2010; pp. 499–510.
10. Browne, S.; Bunge, J. Random record processes and state dependent thinning. *Stoch. Process. Their Appl.* **1995**, *55*, 131–142. [CrossRef]
11. Bunge, G.; Goldie, C.M. Record sequences and their applications. In *Handbook of Statistics*; Shanbhag, D.N., Rao, C.R., Eds.; Elsevier: Amsterdam, The Netherlands, 2001; Volume 19, pp. 277–308.
12. Kallenberg, O. *Random Measures, Theory and Applications*; Springer: Cham, Switzerland, 2017.
13. Bruss, F.T. A unified approach to a class of best choice problems with an unknown number of options. *Ann. Probab.* **1984**, *12*, 882–889. [CrossRef]
14. Gnedin, A. The best choice problem with random arrivals: How to beat the $1/e$-strategy. *Stoch. Process. Their Appl.* **2021**, in press. [CrossRef]
15. Bruss, F.T.; Yor, M. Stochastic processes with proportional increments and the last-arrival problem. *Stoch. Process. Their Appl.* **2012**, *122*, 3239–3261. [CrossRef]
16. Ferguson, T.S. Optimal Stopping and Applications. 2008. Available online: https://www.math.ucla.edu/~tom/Stopping/Contents.html (accessed on 10 April 2021).
17. Bruss, F.T.; Samuels, S.M. Conditions for quasi-stationarity of the Bayes rule in selection problems with an unknown number of rankable options. *Ann. Probab.* **1990**, *18*, 877–886. [CrossRef]
18. Bruss, F.T.; Rogers, L.C.G. Embedding optimal selection problems in a Poisson process. *Stoch. Process. Their Appl.* **1991**, *38*, 1384–1391. [CrossRef]
19. Gnedin, A.; Derbazi, Z. On the Last-Success Optimal Stopping Problem. in progress.
20. Puri, P.S. On the characterization of point processes with the order statistic property without the moment condition. *J. Appl. Probab.* **1982**, *19*, 39–51. [CrossRef]
21. Chow, Y.S.; Robbins, H.; Siegmund, D. *The Theory of Optimal Stopping*; Dover: New York, USA, 1991.
22. Bruss, F.T. Sum the odds to one and stop. *Ann. Probab.* **2000**, *28*, 1384–1391. [CrossRef]
23. Grau Ribas, J.M. A note on last-success-problem. *Theory Probab. Math. Stat.* **2020**, *103*, 155-165. [CrossRef]
24. Bruss, F.T. Odds-theorem and monotonicity. *Math. Appl.* **2019**, *47*, 25–43. [CrossRef]
25. DeVore, R.A.; Lorentz, G.G. *Constructive Approximation*; Springer: Berlin, Germany, 1993.
26. Bruss, F.T. Invariant record processes and applications to best choice modelling. *Stoch. Process. Their Appl.* **1988**, *30*, 303–316. [CrossRef]
27. Arratia, R.; Barbour, A.D.; Tavaré, S. *Logarithmic Combinatorial Structures: A Probabilistic Approach*; European Mathematical Society: Berlin, Germany, 2003.
28. Bingham, N.H. Tauberian theorems for Jakimovski and Karamata-Stirling methods. *Mathematika* **1988**, *35*, 216–224. [CrossRef]
29. Presman, E.; Sonin, I. The best choice problem for a random number of objects. *Theory Probab. Appl.* **1972**, *17*, 657–668. [CrossRef]
30. Curtiss, D.R. Recent extensions of Descartes' rule of signs. *Ann. Math.* **1918**, *19*, 251–278. [CrossRef]
31. Johnson, N.L.; Kemp, A.W.; Kotz, S. *Univariate Discrete Distributions*, 3rd ed.; John Wiley & Sons: Hoboken, NJ, USA, 2005.

Article

Partial Exchangeability for Contingency Tables

Persi Diaconis

Department of Mathematics and Statistics, Sequoia Hall, Stanford University, Stanford, CA 94305, USA; diaconis@math.stanford.edu

Abstract: A parameter free version of classical models for contingency tables is developed along the lines of de Finetti's notions of partial exchangeability.

Keywords: algebraic statistics; contingency tables; de Finetti representation theorem; Markov basis; partial exchangeability

1. Introduction

Consider cross-classified data: $X_1, X_2, \ldots, X_n$, where $X_a = (i_a, j_a)$, $i_a \in [I]$, $j_a \in [J]$ (for $[I] = \{1, 2, \ldots, I\}$). Such data are often presented as an $I \times J$ contingency table $T = (t_{ij})$ where t_{ij} is the number of times (i, j) happens. Suppose that $X_1, \ldots, X_n$ are exchangeable and extendible. Then, de Finetti's theorem says:

Theorem 1. *For exchangeable $\{X_i\}_{i=1}^{\infty}$ taking values in $[I] \times [J]$*

$$P[X_1 = (i_1, j_1), \ldots, X_n = (i_n, j_n)] = \int_{\Delta_{I \times J}} \prod_{i,j} p_{ij}^{t_{ij}} \mu(dp),$$

where $\Delta_{I \times J} = \{p_{ij} \geq 0, \sum_{i,j} p_{ij} = 1\}$. The representing measure μ is unique.

A popular model for cross classified data is

$$p_{ij} = \theta_i \eta_j.$$

Here is a Bayesian, parameter free, description.

Theorem 2. *For exchangeable $\{X_i\}_{i=1}^{\infty}$ taking values in $[I] \times [J]$, a necessary and sufficient condition for the mixing measure μ in Theorem 1 to be supported on $\Delta_I \times \Delta_J$ (with $\Delta_I = \{p_1, \ldots, p_I : p_i \geq 0, \sum_i p_i = 1\}$), so*

$$P[X_1 = (i_1, j_1), \ldots, X_n = (i_n, j_n)] = \int_{\Delta_I \times \Delta_J} \prod \theta_i^{t_{i*}} \eta_j^{t_{*j}} \mu(d\theta, d\eta),$$

is that

$$P[X_1 = (i_1, j_1), X_2 = (i_2, j_2), X_3 = (i_3, j_3), \ldots, X_n = (i_n, j_n)] =$$
$$P[X_1 = (i_1, j_2), X_2 = (i_2, j_1), X_3 = (i_3, j_3), \ldots, X_n = (i_n, j_n)]. \quad (1)$$

Condition (1) is to hold for any $n \geq 2$ and any (i_a, j_a) $1 \leq a \leq n$.

Proof. Condition (1) implies for all n and $h \geq 1$ (surpressing P a.s. throughout)

$$P[X_1 = (i_1, j_1), X_2 = (i_2, j_2) | X_n = (i_n, j_n), \ldots, X_{n+h} = (i_{n+h}, j_{n+h})] =$$
$$P[X_1 = (i_1, j_2), X_2 = (i_2, j_1) | X_n = (i_n, j_n), \ldots, X_{n+h} = (i_{n+h}, j_{n+h})]. \quad (2)$$

Let $h \uparrow \infty$ and then $n \uparrow \infty$. Let $\mathcal{T}$ be the tail field of $\{X_i\}_{i=1}^{\infty}$. Then, Doob's increasing and decreasing martingale theorems show

$$P[X_1 = (i_1, j_1), X_2 = (i_2, j_2)|\mathcal{T}] = P[X_1 = (i_1, j_2), X_2 = (i_2, j_1)|\mathcal{T}].$$

However, a standard form of de Finetti's theorem says that, given $\mathcal{T}$, the $\{X_i\}_{i=1}^{\infty}$ are i.i.d. with $P[X_1 = (i,j)] = p_{ij}$. Thus

$$p_{ij} p_{i'j'} = p_{ij'} p_{i'j} \quad \text{for all } i, i', j, j'. \tag{3}$$

Finally, observe that (3) implies (writing $p_{i*} := \sum_j p_{ij}$, $p_{*j} := \sum_i p_{ij}$)

$$p_{i*} p_{*j} = \sum_{h,l} p_{ih} p_{lj} = \sum_{hl} p_{ij} p_{hl} = p_{ij}.$$

□

We remark the following points.

1. If $X_i = (Y_i, Z_i)$ condition (2) is equivalent to

$$\mathcal{L}((Y_1, Z_1), (Y_2, Z_2), ..., (Y_n, Z_n)) = \mathcal{L}((Y_1, Z_{\sigma(1)}), ..., (Y_n, Z_{\sigma(n)}))$$

    for all n and $\sigma \in S_n$ (S_n is the symmetric group over $1, 2, \ldots, n$). Since $\{(Y_i, Z_i)\}_{i=1}^{n}$ are exchangeable this is equivalent to saying the law is invariant under $S_n \times S_n$.

2. The mixing measure $\mu(d\theta, d\eta)$ allows general dependence between the row parameters θ and column parameters η. Classical Bayesian analysis of contingency tables often chooses μ so that θ and η are independent. A parameter free version is that under P, the row sums t_{i*} and column sums t_{*j} are independent. It is natural to weaken this to "close to independent" along the lines of [1] or [2]. See also [3].

3. Theorems 1 and 2 have been stated for discrete state spaces. By a standard discretization argument, they hold for quite general spaces. For example:

Theorem 3. *Let $X_i = (Y_i, Z_i)$ be exchangeable with $Y_i \in \mathcal{Y}$, $Z_i \in \mathcal{Z}$, complete separable metric spaces, $1 \leq i < \infty$. Suppose*

$$P[X_1 \in (A_1, B_1), X_2 \in (A_2, B_2), ..., X_n \in (A_n, B_n)] = $$
$$P[X_1 \in (A_1, B_2), X_2 \in (A_2, B_1), ..., X_n \in (A_n, B_n)]$$

for all measurable A_i, B_i and all n. Then,

$$P(X_1 \in (A_1, B_1), ..., X_n \in (A_n, B_n)) = \int_{\mathcal{P}(\mathcal{Y}) \times \mathcal{P}(\mathcal{Z})} \prod_1^n \theta(A_i) \eta(B_i) \mu(d\theta, d\eta),$$

with $\mathcal{P}(\mathcal{Y})$, $\mathcal{P}(\mathcal{Z})$ the probabilities on the Borel sets of $\mathcal{Y}$, $\mathcal{Z}$. The mixing measure μ is unique.

4. Theorem 2 is closely related to de Finetti's work in [1,4].
5. De Finetti's law of large numbers holds as well, in Theorem 3

$$\frac{1}{n} \sum \delta_{X_i}(A \times B) \to \mu(\theta(A), \eta(B)).$$

One object of this paper is to develop similar parameter free de Finetti theorems for widely used log-linear models for discrete data. Section 2 begins by relating this to an ongoing conversation with Eugenio Regazzini. Section 3 provides needed background on discrete exponential families and algebraic statistics. Sections 4 and 5 apply those tools to give de Finetti style partially exchangeable theorems for some widely used hierarchical and graphical models for contingency tables. Section 6 shows how these exponential family tools can be used for other Bayesian tasks: building "de Finetti priors" for "almost exchangeability" and running the "exchange" algorithm for doubly intractable Bayesian computation. Some philosophy and open problems are in the final section.

2. Some History

I was lucky enough to be able to speak at Eugenio Regazzini's 60TH birthday celebration, in Milan, in 2006. My talk began this way:

≪ Hello, my name is Persi and I have a problem. ≫

For those of you not aware of the many "10 step-programs" (alcoholics anonymous, gamblers anonymous, ...) they all begin this way, with the participants admitting to having a problem. In my case the problem was this:

(a) After 50 years of thinking about it, I think that the subjectivist approach to probability, induction and statistics is the only thing that works;
(b) At the same time, I have done a lot of work inventing and analyzing various schemes for generating random samples for things like contingency tables with given row and column sums; graphs with given degree sequences; ...; Markov Chain Monte Carlo. These are used for things like permutation tests and Fisher's exact test.

There is a lot of nice mathematics and hard work in (b) but such tests violate the likelihood principle and lead to poor scientific practice. Hence my problem (I still have it): (a) and (b) are incompatible.

There has been some progress. I now see how some of the tools developed for (b) can be usefully employed for natural tasks suggested by (a). Not so many people care about such inferential questions in these 'big data' days. However, there are also lots of small datasets where the inferential details matter. There are still useful questions for people like Eugenio (and me).

3. Background on Exponential Families and Algebraic Statistics

The following development is closely based on [5], which should be considered for examples, proofs and more details.

Let $\mathcal{X}$ be a finite set. Consider the exponential family:

$$p_\theta(x) = \frac{1}{Z(\theta)} e^{\theta \cdot T(x)} \quad \theta \in \mathbb{R}^d, x \in \mathcal{X}. \tag{4}$$

Here, $Z(\theta)$ is a normalizing constant and $T : \mathcal{X} \to \mathbb{N}^d - \{0\}$. If $X_1, X_2, ..., X_n$ are independent and identically distributed from (4), the statistic $t = T(X_1) + \cdots + T(X_n)$ is sufficient for θ. Let

$$\mathcal{Y}_t = \{(x_1, ..., x_n) : T(x_1) + \cdots + T(x_n) = t\}.$$

Under (4), the distribution of $X_1, ..., X_n$ given t is uniform on $\mathcal{Y}_t$. It is usual to write

$$t = \sum_{i=1}^n T(X_i) = \sum_{\mathcal{X}} \sigma(x) T(x) \quad \text{with } \sigma(x) = \#\{i : T(X_i) = T(x)\}.$$

Let

$$\mathcal{F}_t = \{f : \mathcal{X} \to \mathbb{N} : \sum f(x) T(x) = t\}.$$

Example 1. *For contingency tables $\mathcal{X} = \{(i,j) : 1 \le i \le I, 1 \le j \le J\}$. The usual model for independence has $T(i,j) \in \mathbb{N}^{I+J}$ a vector of length $I+J$ with two non zero entries equal 1. The 1's in $T(i,j)$ are in the ith place and position j of the last j places. The sufficient statistic t contains the row and column sums of the contingency table associated to the first n observations. The set $\mathcal{F}_t$ is the set of an $I \times J$ tables with these row and column sums.*

A Markov chain on this $\mathcal{F}_t$ can be based on the following moves: pick $i \neq i'$, $j \neq j'$ and change the entries in the current f by adding ± 1 in pattern

$$\begin{array}{cc} & j \quad j' \\ i & +\ - \\ i' & -\ + \end{array} \quad \text{or} \quad \begin{array}{cc} -\ + \\ +\ - \end{array}$$

This does not change the row sums and it does not change the column sums. If told to go negative, just pick new i, i', j, j'. This gives a connected, aperiodic Markov chain on $\mathcal{F}_t$ with a uniform stationary distribution. See [6].

Returning to the general case, an analog of $\begin{array}{c}+\ -\\ -\ +\end{array}$ moves is given by the following:

Definition 1 (Markov basis). *A Markov basis is a set of functions $f_1, f_2, ..., f_L$ from $\mathcal{X}$ to $\mathbb{Z}$ such that*

$$\sum_{\mathcal{X}} f_i(x) T(x) = 0 \quad 1 \leq i \leq L \tag{5}$$

and that for any t and $f, f' \in \mathcal{F}_t$ there are $(t_1, f_{i_1}), ..., (t_A, f_{i_A})$ with $t_i = \pm 1$, such that

$$f' = f + \sum_{j=1}^{A} t_j f_{i_j} \quad \text{and} \quad f + \sum_{j=1}^{a} t_j f_{i_j} \geq 0, \text{ for } 1 \leq a \leq A. \tag{6}$$

This allows the construction of a Markov chain on $\mathcal{F}_t$: from f, pick $I \in \{1, 2, ..., L\}$ and $t = \pm 1$ at random and consider $f + t f_I$. If this is positive, move there. If not, stay at f. Assumptions (5) and (6) ensure that this Markov chain is symmetric and ergodic with a uniform stationary distribution. Below, I will use a Markov basis to formulate a de Finetti theorem to characterize mixtures of the model (4).

One of the main contributions of [5] is a method of effectively constructing Markov bases using polynomial algebra. For each $x \in \mathcal{X}$, introduce an indeterminate, also called x. Consider the ring of polynomials $k[\mathcal{X}]$ in these indeterminates where k is a field, e.g., the complex numbers. A function $g : \mathcal{X} \to \mathbb{N}$ is represented as a monomial $\mathcal{X}^g = \prod_\mathcal{X} x^{g(x)}$. The function $T : \mathcal{X} \to \mathbb{N}^d$ gives a homomorphism

$$\varphi_T : k[\mathcal{X}] \longrightarrow k[t_1, ..., t_d]$$
$$x \longmapsto t_1^{T_1(x)} t_2^{T_2(x)} \cdots t_d^{T_d(x)},$$

extended linearly and multiplicatively ($\varphi_T(x+y) = \varphi_T(x) + \varphi_T(y)$ and $\varphi_T(x^2) = \varphi_T(x)^2$ and so on). The basic object of interest is the kernel of φ_T:

$$I_T = \{p \in k[\mathcal{X}] : \varphi_T(p) = 0\}.$$

This is an ideal in $k[\mathcal{X}]$. A key result of [5] is that a generating set for I_T is equivalent to a Markov basis. To state this, observe that any $f : \mathcal{X} \to \mathbb{Z}$ can be written $f = f_+ - f_-$ with $f_+(x) = \max(f(x), 0)$ and $f_-(x) = \max(-f(x), 0)$. Observe $\sum f(x) T(x) = 0$ iff $\mathcal{X}^{f_+} - \mathcal{X}^{f_-} \in I_T$. The key result is

Theorem 4. *A collection of functions $f_1, f_2, ..., f_L$ is a Markov basis if and only if the set*

$$\mathcal{X}^{f_{i+}} - \mathcal{X}^{f_{i-}} \quad 1 \leq i \leq L$$

generates the ideal I_T.

Now, the Hilbert Basis Theorem shows that ideals in $k[\mathcal{X}]$ have finite bases and modern computer algebra packages give an effective way of finding bases.

I do not want (or need) to develop this further. See [5] or the book by Sullivant [7] or Aoki et al. [8]. There is even a Journal of Algebraic Statistics.

I hope that the above gives a flavor for what I mean by "working in (b) is hard honest work". Most of the applications are for standard frequentist tasks. In the following sections, I will give Bayesian applications.

4. Log Linear Model for Contingency Tables

Log linear models for multiway contingency tables are a healthy part of the modern statistics. The index set is $\mathscr{X} = \prod_{\gamma \in \Gamma} I_\gamma$ with Γ indexing categories and I_γ the levels of γ. Let $p(x)$ be the probability of falling into cell $x \in \mathscr{X}$. A log linear model can be specified by writing:

$$\log p(x) = \sum_{a \subseteq \Gamma} \varphi_a(x).$$

The sum ranges over subsets a of Γ and $\varphi_a(x)$ means a function that only depends on x through the coordinates in a. Thus, $\varphi_\varnothing(x)$ is a constant and $\varphi_\Gamma(x)$ is allowed to depend on all coordinates. Specifying $\varphi_a = 0$ for some class of sets a determines a model. Background and extensive references are in [9]. If the a with $\varphi_a \neq 0$ permitted form a simplicial complex $\mathcal{C}$ (so $a \in \mathcal{C}$ and $\varnothing \neq a' \subseteq a \Rightarrow a' \in \mathcal{C}$) the model is called *hierarchical*. If $\mathcal{C}$ consists of the cliques in a graph, the model is called *graphical*. If the graph is chordal (every cycle of length ≥ 4 contains a chord) the graphical model is called *decomposable*.

Example 2 (3 way contingency tables). *The graphical models for three way tables are:*

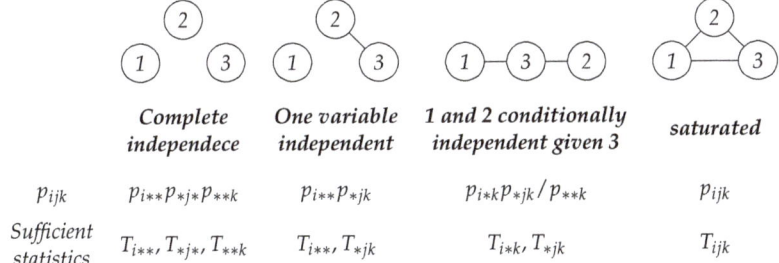

	Complete independece	One variable independent	1 and 2 conditionally independent given 3	saturated
	p_{ijk}	$p_{i**}p_{*j*}p_{**k}$	$p_{i**}p_{*jk}$	$p_{i*k}p_{*jk}/p_{**k}$	p_{ijk}
Sufficient statistics	$T_{i**}, T_{*j*}, T_{**k}$	T_{i**}, T_{*jk}	T_{i*k}, T_{*jk}	T_{ijk}

The simplest hierarchical model that is not graphical is *No Three Way Interaction Model*. This can be specified by saying 'the odds rate of any pair of variables does not depend on the third'. Thus,

$$\frac{p_{ijk}p_{i'j'k}}{p_{ij'k}p_{i'jk}} \quad \text{is constant in } k \text{ for fixed } i, i', j, j'. \tag{7}$$

As one motivation, recall that for two variables, the independence model is specified by

$$p_{ij} = \theta_i \eta_j.$$

For three variables, suppose there are parameters $\theta_{ij}, \eta_{jk}, \psi_{ik}$ satisfying:

$$p_{ijk} = \theta_{ij}\eta_{jk}\psi_{ik} \quad \text{for all } i, j, k. \tag{8}$$

It is easy to see that (8) entails (7) hence 'no three way interaction'. Cross multiplying (7) entails

$$p_{ijk}p_{i'j'k}p_{ij'k'}p_{i'jk'} = p_{ijk'}p_{i'j'k'}p_{ijk}p_{i'jk}. \tag{9}$$

This is the form we will work with for the de Finetti theorems below.

For background, history and examples (and some nice theorems) see ([10], Section 8.2), [11,12], Simpsons 'paradox' [13] is based on understanding the no three way interaction model. Further discussion is in Section 5 below.

5. From Markov Bases to de Finetti Theorems

Suppose $\mathscr{X}$ is a finite set, $T : \mathscr{X} \to \mathbb{N}^d - \{0\}$ is a statistic and $\{f_i\}_{i=1}^L$ is a Markov basis as in Section 3. The following development shows how to translate this into de Finetti theorems for the contingency table examples of Section 4. The first argument abstracts the argument used for Theorem 2 above.

Lemma 1 (Key Lemma). *Let $\mathscr{X}$ be a finite set and $\{X_i\}_{i=1}^\infty$ an exchangeable sequence of $\mathscr{X}$-valued random variables. Suppose for all $n > m$*

$$P[X_1 = x_1, ..., X_m = x_m, X_{m+1} = x_{m+1}, ..., X_n = x_n] =$$
$$P[X_1 = y_1, ..., X_m = y_m, X_{m+1} = x_{m+1}, ..., X_n = x_n]. \quad (10)$$

In (10), $x_1, ..., x_m, y_1, ..., y_m$ are fixed and $x_{m+1}, ..., x_n$ are arbitrary. Then, if $\mathcal{T}$ is the tail field of $\{X_i\}_{i=1}^\infty$ and $p(x) = P[X_1 = x | \mathcal{T}]$,

$$\prod_{i=1}^m p(x_i) = \prod_{i=1}^m p(y_i). \quad (11)$$

Proof. From (10) and exchangeability

$$P[X_1 = x_1, ..., X_m = x_m, X_{n+1} = x_{n+1}, ..., X_{n+h} = x_{n+h}] =$$
$$P[X_1 = y_1, ..., X_m = y_m, X_{n+1} = x_{n+1}, ..., X_{n+h} = x_{n+h}]$$

so

$$P[X_1 = x_1, ..., X_m = x_m | X_{n+1} = x_{n+1}, ..., X_{n+h} = x_{n+h}] =$$
$$P[X_1 = y_1, ..., X_m = y_m | X_{n+1} = x_{n+1}, ..., X_{n+h} = x_{n+h}].$$

Let $h \uparrow \infty$ and then $n \uparrow \infty$, use Doob's upward and then downward martingale convergence theorems to see:

$$P[X_1 = x_1, ..., X_m = x_m | \mathcal{T}] = P[X_1 = y_1, ..., X_m = y_m | \mathcal{T}].$$

Now, de Finetti's theorem implies (11). □

Remark 1. *The Key Lemma shows that the $p(x)$ satisfy certain relations. Using choices of $\{x_i\}, \{y_i\}$ derived from a Markov basis will show that $p(x)$ satisfy the required independence properties. Suppose that $\sum_{\mathscr{X}} f(x)T(x) = 0$, $\sum_{\mathscr{X}} f(x) = 0$ and $f \in \{0, \pm 1\}$. Let $S_+ = \{x : f(x) = 1\}$, $S_- = \{y : f(y) = -1\}$. Say $|S_+| = |S_-| = m$. Enumerate $S_+ = \{x_1, ..., x_m\}$, $S_- = \{y_1, ..., y_m\}$. Assumptions (10) and conclusion (11) will give our theorems.*

Example 3 (Independence in a two way table). *Let $\mathscr{X} = [I] \times [J]$. A minimal basis for the independence model is given by $f_{i,j,i',j'}$:*

	j	j'
i	$+$	$-$
i'	$-$	$+$

(all other entries = 0).

The condition of the Key Lemma becomes:

$$P[X_1 = (i,j), X_2 = (i',j'), X_3 = (i_3, j_3), ..., X_n = (i_n, j_n)] =$$

$$P[X_1 = (i,j'), X_2 = (i',j), X_3 = (i_3,j_3), ..., X_n = (i_n,j_n)].$$

Passing to the limit gives
$$p_{ij}p_{i'j'} = p_{ij'}p_{i'j}$$
and so
$$p_{i*}p_{*j} = \sum_{i'j'} p_{ij'}p_{i'j} = p_{ij}.$$

This is precisely Theorem 2 of the Introduction. □

Example 4 (Complete independence in a three way table). *The sufficient statistics are $T_{i**}, T_{*j*}, T_{**k}$. From [5], there are two kinds of moves in a minimal basis. Up to symmetries, these are:*

Passing to the limit, this entails:

$$p_{ijk}p_{ij'k} = p_{ij'k}p_{ijk} \text{ and } p_{ijk}p_{i'j'k'} = p_{ij'k}p_{ijk'}.$$

These may be said as 'the product of any $p_{ijk}, p_{i'jk}$ remains unchanged if the middle coordinates are exchanged'. By symmetry, this remains true if the two first or last coordinates are exchanged. As above, this entails

$$p_{i**}p_{*j*}p_{**k} = p_{ijk}.$$

These observations can be rephrased into a statement that looks more similar to the classical de Finetti theorem; using symmetry:

Theorem 5. *Let $\{X_i\}_{i=1}^{\infty}$ be exchangeable, taking values in $[I] \times [J] \times [K]$. Then*

$$P[X_1 = (i_1,j_1,k_1), ..., X_n = (i_n,j_n,k_n)] =$$
$$P[X_1 = (\sigma(i_1),\zeta(j_1),\eta(k_1)), ..., X_n = (\sigma(i_n),\zeta(j_n),\eta(k_n))]$$

for all n, $\{(i_a,j_a,k_a)\}_{a=1}^n$ and $(\sigma,\zeta,\eta) \in S_I \times S_J \times S_K$ is necessary and sufficient for there to exist a unique μ on $\Delta_I \times \Delta_J \times \Delta_K$ with

$$P[X_a = (i_a,j_a,k_a), 1 \le a \le n] = \int_{\Delta_I \times \Delta_J \times \Delta_K} \prod_{a=1}^n p_{i_a}q_{j_a}r_{k_a}\mu(dp,dq,dr).$$

□

Example 5 (One variable independent of the other two). *Suppose, without loss, that the graph is*

$$(1) \quad (2)\text{—}(3)$$

Identify the pairs (j,k) with $\{1,2,...,L\}$ with $L = JK$. The problem reduces to Example 4. A minimal basis consists of (again, up to relabeling)

	l	l'
i	+	−
i'	−	+

We may conclude

Theorem 6. Let $\{X_i\}_{i=1}^{\infty}$ be exchangeable, taking values in $[I] \times [J] \times [K]$. Then

$$P[X_1 = (i_1, j_1, k_1), ..., X_n = (i_n, j_n, k_n)] =$$
$$P[X_1 = (\sigma(i_1), \zeta(j_1, k_1)), ..., X_n = (\sigma(i_n), \zeta(j_n, k_n))]$$

for all n, $\{(i_a, j_a, k_a)\}_{a=1}^{n}$ and $(\sigma, \zeta) \in S_I \times S_{J \times K}$ is necessary and sufficient for there to exist a unique μ on $\Delta_I \times \Delta_{JK}$ with

$$P[X_a = (i_a, j_a, k_a), 1 \leq a \leq n] = \int_{\Delta_I \times \Delta_{JK}} \prod_{a=1}^{n} p_a q_a \mu(dp, dq).$$

□

Example 6 (Conditional independence). *Suppose variable i and j are conditionally independent given k.*

$$\text{(1)}-\text{(3)}-\text{(2)}$$

Rewrite the parameter condition of section four as

$$p_{**k} p_{ijk} = p_{i*k} p_{*jk} \quad \text{for all } i, j, k$$

The sufficient statistics are $\{T_{i*k}\}_{i,k}$, $\{T_{*jk}\}_{jk}$. From [5], a minimal generating set is

	j_k	j'_k
i_k	+	−
i'_k	−	+

$K \times \dfrac{I(I-1)}{2} \times \dfrac{J(J-1)}{2}$ moves in all.

From this, the Key Lemma shows (for all i, j, k)

$$p_{ijk} p_{i'j'k} = p_{ij'k} p_{i'jk}.$$

This entails:

$$p_{i*k} p_{*jk} = \sum_{i',j'} p_{ij'k} p_{i'jk} = \sum_{i'j'} p_{ijk} p_{i'j'k} = p_{ijk} p_{**k}.$$

Again, phrasing the condition (10) in terms of symmetry.

Theorem 7. Let $\{X_i\}_{i=1}^{\infty}$ be exchangeable, taking values in $[I] \times [J] \times [K]$. Then,

$$P[X_1 = (i_i, j_i, k_i), ..., X_n = (i_n, j_n, k_n)] =$$
$$P[X_1 = (\sigma^{k_1}(i_1), \zeta^{k_1}(j_1), k_1), ..., X_n = (\sigma^{k_n}(i_n), \zeta^{k_n}(j_n), k_n)] \quad (12)$$

for all n, $\{(i_a, j_a, k_a)\}_{a=1}^{n}$ and $\sigma^k, \zeta^k \in S_I \times S_J, 1 \leq k \leq K$, is necessary and sufficient for there to exist a unique family $\mu \times \prod_{b=1}^{k} \mu_{b,r}$ on $\Delta_K \times (\Delta_I \times \Delta_J)^K$

$$P[X_a = (i_a, j_a, k_a), 1 \leq a \leq n] =$$

$$\int_{\Delta_K \times (\Delta_I \times \Delta_J)^K} \prod_{a=1}^{n} r_{k_a} p_{i_a}^{k_a} q_{j_a}^{k_a} \prod_{b=1}^{k} \mu_{b,r}(p^{i_b} q^{i_b}) \mu(dr). \quad (13)$$

Both (12) and (13) have a simple interpretation. For (12), $\{X_i\}_{i=1}^{n}$ are exchangeable 3-vectors. For any k and specified sequence of values $\{(i_a, j_a, k)\}_{a=1}^{n}$ the chance of observing these values is unchanged under permuting the (i_a, j_a, k), by permutations $\sigma^k \in S_I, \zeta^k \in S_J$. Here σ^k, ζ^k are allowed to depend on k.

On the right of (13), the mixing measure may be understood as follows. There is a probability μ on Δ_K. Pick $r = (r_1, ..., r_K) \in \Delta_K$. Given this r, pick (p^k, q^k) from $\mu_{k,r}$ on the kth copy of $\Delta_I \times \Delta_J$. These choices are allowed to depend on r but are independent, conditional on r, $1 \leq k \leq K$.

All of this simply says that, conditional on the tail field,

$$P[X_a = (i, j, k)|\mathcal{T}] = P[X_a = (i, *, k)|\mathcal{T}]P(X_a = (*, j, k)|\mathcal{T}].$$

The first two coordinates are conditionally independent given the third.

Example 7 (No three way interaction). *The model is described in Section 4. The sufficient statistics are $\{T_{ij*}\}, \{T_{i*k}\}, \{T_{*jk}\}$. Minimal Markov bases have proved intractable. See [5] or [8]. For any fixed I, J, K, the computer can produce a Markov basis but these can have a huge number of terms. See [7,8] and their references for a surprisingly rich development.*

There is a pleasant surprise. Markov bases are required to connect the associated Markov chain. There is a natural subset, the first moves anyone considers, and and these are enough for a satisfactory de Finetti theorem (!).

Described informally, for an $I \times J \times K$ array, pick a pair of parallel planes, say the k, k' planes in the three dimensional array, and consider moves depicted as

	j	j'		j	j'
i	$+$	$-$	i	$+$	$-$
i'	$-$	$+$	i'	$-$	$+$
	k			k'	

*These moves preserve all line sums (the sufficient statistics). They are **not** sufficient to connect any two datasets with the same sufficient statistics. Using the prescription in the Key Lemma, suppose:*

$$P[X_1 = (i, j, k), X_2 = (i', j', k), X_3 = (i, j', k'), X_4 = (i', j, k'),$$
$$X_a = (i_a, j_a, k_a)\ 5 \leq a \leq n] =$$
$$P[X_1 = (i, j', k), X_2 = (i', j, k), X_3 = (i, j, k'), X_4 = (i', j', k'),$$
$$X_a = (i_a, j_a, k_a)\ 5 \leq a \leq n]. \quad (14)$$

Passing to the limit gives

$$p_{ijk}p_{i'j'k}p_{ij'k'}p_{i'jk'} = p_{ij'k}p_{i'jk}p_{ijk'}p_{i'j'k'}. \quad (15)$$

This is exactly the no three way interaction condition. Or, equivalently:

$$\frac{p_{ijk}p_{i'j'k}}{p_{ij'k}p_{i'jk}} = \frac{p_{ijk'}p_{i'j'k'}}{p_{ij'k'}p_{i'jk'}}.$$

The odds ratios are constant on the kth and k'^{th} planes (of course, they depend on i, j, i', j'). These considerations imply:

Theorem 8. *Let $\{X_i\}_{i=1}^{\infty}$ be exchangeable, taking values in $[I] \times [J] \times [K]$. Then, condition (14) is necessary and sufficient for the existence of a unique probability μ on Δ_{IJK}, supported on the no three way interaction variety (15) satisfying*

$$P[X_a = (i_a, j_a, k_a), 1 \leq a \leq n] = \int_{\Delta_{IJK}} \prod p_{ijk}^{\eta_{ijk}} \mu(dp_{ijk}).$$

We remark on the following points.

1. It follows from theorems in [12] and [11] that, if all $p_{ijk} > 0$, condition (15) is equivalent to the unique representation,
$$p_{ijk} = r\alpha_{jk}\beta_{ki}\gamma_{ij}, \qquad (16)$$
where r, α, β, γ have positive entries and satisfy
$$\sum_k \alpha_{jk} = \sum_i \beta_{ki} = \sum_j \gamma_{ij} = 1 \text{ for all } i,j,k$$
and
$$r \sum_{i,j,k} \alpha_{jk}\beta_{ki}\gamma_{ij} = 1.$$
The integral representation in the theorem can be stated in this parametrization. The condition $p_{ijk} > 0$ is equivalent to $P(X_1 = (i,j,k)) > 0$ on observables.
2. Condition (14) does not have an obvious symmetry interpretation.
3. Conditions (14) and (15) are stated via varying the third variable when i,j,i',j' are fixed. Because of (16), if they hold in this form, they hold for any two variables fixed as the third varies.
4. It is possible to go on, but, as John Darroch put it, 'the extensions to higher order interactions... are not likely to be of practical interest'. The most natural development—the generalization to decomposable models—is being developed by Paula Gablenz.
5. There are many extensions of the Key Lemma above. These allow a similar development for more general log linear models and exponential families.

6. Discussion and Conclusions

The tools of algebraic statistics have been harnessed above to develop partial exchangeability for standard contingency table models. I have used them for two further Bayesian tasks: approximate exchangeability and the problem of 'doubly intractable priors'. As both are developed in papers, I will be brief.

Approximate exchangeability. Consider n men and m women along with a binary outcome. If the men are judged exchangeable (for fixed outcomes for the women) and vice versa, and, if both sequences are extendable, de Finetti [1] shows that there is a unique prior on the unit square $[0,1]^2$ such that, for any outcomes $t_1,...,t_n, \sigma_1,...,\sigma_m$ in $\{0,1\}$
$$P[X_1 = t_1,...,X_n = t_n, Y_1 = \sigma_1,...,Y_m = \sigma_m] = $$
$$\int_{[0,1]^2} p^S(1-p)^{n-S}\theta^T(1-\theta)^{m-T}\mu(dp,d\theta),$$
with $S = \sum_{i=1}^n t_i$, $T = \sum_{j=1}^m \sigma_j$.

If, for the outcome of interest, $\{X_i, Y_j\}$ were almost fully exchangeable (so the men/women difference is judged practically irrelevant) the prior μ would be concentrated near the diagonal of $[0,1]^2$. De Finetti suggested implementing this by considering priors of the form
$$\mu(dp,d\theta) = Z^{-1}e^{-A(p-\theta)^2}dpd\theta$$
for A large.

In joint work with Sergio Bacallado and Susan Holmes [3], multivariate versions of such priors are developed. These are required to concentrate near sub-manifolds of cubes or products of simplices; think about 'approximate no three way interaction'. We used the tools of algebraic statistics to suggest appropriate many variable polynomials which vanish on submanifold of interest. Many ad hoc choices were involved. Sampling from such priors or posteriors is a fresh research area. See [2,14,15].

Doubly intractable priors. Consider an exponential family as in Section 3:
$$p_\theta(x) = \frac{1}{Z(\theta)}e^{\theta \cdot T(x)}.$$

Here $x \in \mathscr{X}$ a finite set, $T : \mathscr{X} \to \mathbb{R}^d$ and $\theta \in \mathbb{R}^d$. In many real examples, the normalizing constant $Z(\theta)$ will be unknown and unknowable. For a Bayesian treatment, let $\Pi(d\theta)$ be a prior distribution on $\mathbb{R}^d$. For example, the conjugate prior.

If $X_1, X_2, ..., X_n$ is as i.i.d. sample from p_θ, T is a sufficient statistic and the posterior has the form
$$Z(Z^{-1}(\theta))^n e^{\theta F} \Pi(d\theta),$$
with $F = \sum_{i=1}^{n} T(X_i)$ and Z another normalizing constant. The problem is that $Z^{-1}(\theta)$ depends on θ and is unknown!

The exchange algorithm and many variants offer a useful solution. See [16,17].

In practical implementations, there is an intermediary step requiring a sample form $p_{\theta'}^T$, the measure induced by p_θ^n under $\sum_i^n T(x_i) : \mathscr{X}^n \to \mathbb{R}$. This is a discrete sampling task and Markov basis techniques have been proved useful. See [16].

A philosophical comment. The task undertaken above, finding believable Bayesian interpretations for widely used log linear models, goes somewhat against the grain of standard statistical practice. I do not think anyone takes a reasonably complex, high dimensional hierarchical model seriously. They are mostly used as a part of exploratory data analysis; this is not to deny their usefulness. Making any sense of a high dimensional dataset is a difficult task. Practitioners search through huge collections of models in an automated way. Usually, any reflection suggests the underlying data is nothing like a sample from a well specified population. Nonetheless, models are compared using product likelihood criteria. It is a far far cry from being based on anyone's reasoned opinion.

I have written elsewhere about finding Bayesian justification for important statistical tasks such as graphical methods or exploratory data analysis [18]. These seem like tasks similar to 'how do you form a prior'. Different from the focus of even the most liberal Bayesian thinking.

The sufficiency approach. There is a different approach to extending de Finetti's theorem. This uses 'sufficiency'. Consider exchangeable $\{X_i\}_{i=1}^{\infty}$. For each n, suppose $T_n : \mathscr{X}^n \to \mathscr{Y}$ is a function. The $\{T_n\}$ have to fit together according to simple rules satisfied in all of the examples above. Call $\{X_i\}$ *partially exchangeable* with respect to T_n if $P[X_1 = x_1, \ldots, X_n = x_n | T_n = t_n]$ is uniform. Then, Diaconis and Freedman [19] show that a version of de Finetti's theorem holds. The law of $\{X_i\}$ is a mixture of i.i.d. laws indexed by extremal laws. In dozens of examples, these extremal laws can be identified with standard exponential families. This last step remains to be carried out in the generality of Section 3 above. What is required is a version of the Koopman–Pitman–Darmois theorem for discrete random variables. This is developed in [19] when $\mathscr{X} \subseteq \mathbb{N}$ and $T_n(X_1, \ldots, X_n) = X_1 + \cdots + X_n$. Passing to interpretation, this version of partial exchangeability has the following form:

if $T_n(x_1, \ldots, x_n) = T_n(y_1, \ldots, y_n)$,
$$\text{then } P[X_1 = x_1, \ldots, X_n = x_n] = P[X_1 = y_1, \ldots, X_n = y_n].$$

This is neat mathematics (and allows a very general theoretical development). However, it does not seem as easy to think about in natural examples. Exchangeability via symmetry is much easier. The development above is a half-way house between symmetry and sufficiency. A close relative of the sufficiency approach is the topic of 'extremal models' as developed by Martin-Löf and Lauritzen. See [20] and its references. Moreover, Refs. [21,22] are recent extensions aimed at contingency tables.

Classical Bayesian contingency table analysis. There is a healthy development of parametric analysis for the examples of Section 5. This is based on natural conjugate priors. It includes nice theory and R packages to actually carry out calculations in real problems. Three papers that I like are [23–26]. The many wonderful contributions by I.J. Good are still very much worth consulting. See [27] for a survey. Section 5 provides 'observable characterizations' of the models. The problem of providing 'observable characterizations' of the associated conjugate priors (along the lines of [28]) remains open.

Funding: This research received funding from the European Research Council (ERC) under the European Union's Horizon 2020 research and innovation programme under grant agreement No 817257 and funding from NSF grant No 1954042.

Institutional Review Board Statement: Not applicable.

Informed Consent Statement: Not applicable.

Data Availability Statement: Not applicable.

Acknowledgments: The author would like to thank Paula Gablenz, Sourav Chatterjee and Emanuele Dolera for help throughout.

Conflicts of Interest: The author declares no conflict of interest.

References

1. de Finetti, B. On the condition of partial exchangeability. *Stud. Inductive Log. Probab.* **1980**, *2*, 193–205.
2. Bruno, A. On the notion of partial exchangeability (Italian). In *Giornale dell'Istituto Italiano degli Attuari*; English Translation in: de Finetti, *Probability, Induction and Statistics*; International Statistical Institute: Leidschenveen, The Netherlands, 1964 ; Volume 27, Chapter 10; pp. 174–196.
3. Baccalado, S.; Diaconis, P.; Holmes, S. De Finetti priors using Markov chain Monte Carlo computations. *J. Stat. Comput.* **2015**, *25*, 797–808. [CrossRef] [PubMed]
4. de Finetti, B. *Probability, Induction and Statistics: The Art of Guessing*; Wiley: Hoboken, NJ, USA, 1972.
5. Diaconis, P.; Sturmfels, B. Algebraic algorithms for sampling from conditional distributions. *Ann. Stat.* **1998**, *26*, 363–397. [CrossRef]
6. Diaconis, P.; Gangolli, A. Rectangular arrays with fixed margins. In *Discrete Probability and Algorithms*; Springer: New York, NY, USA, 1995 ; Volume 72, pp. 15–41.
7. Sullivant, S. *Algebraic Statistics*; AMS: Providence, RI, USA, 2018 .
8. Aoki, S.; Hara, H.; Takemura, A. *Markov Bases in Algebraic Statistics*; Springer: Berlin/Heidelberg, Germany, 2012.
9. Lauritzen, S.L. *Graphical Models*, 2nd ed.; Oxford University Press: Oxford, UK, 2004 .
10. Agresti, A. *Categorical Data Analysis*, 2nd ed.; Wiley: Hoboken, NJ, USA, 2002.
11. Birch, M.W. Maximum likelihood in three-way contingency tables. *J. R. Stat. Soc. Ser. B* **1963**, *25*, 220–233. [CrossRef]
12. Darroch, J.N. Interactions in multi-factor contingency tables. *J. R. Stat. Soc. Ser.* **1962**, *24*, 251–263. [CrossRef]
13. Simpson, E.H. The interpretation of interaction in contingency tables. *J. R. Stat. Soc. Ser.* **1951**, *13*, 238–241. [CrossRef]
14. Diaconis, P.; Holmes, S.; Shahshahani, M. Sampling From a Manifold. In *Advances in Modern Statistical Theory and Applications: A Festschrift in Honor of Morris L. Eaton* ; IMS Statistics Collections: Beachwood, OH, USA, 2013 ; pp. 102–125.
15. Gerencsér, B.; Ottolini, A. Rates of convergence for Gibbs sampling in the analysis of almost exchangeable data. *arXiv* **2020**, arXiv:2010.15539v2.
16. Diaconis, P.; Wang, G. Bayesian goodness of fit tests: A conversation for David Mumford. *Ann. Math. Sci. Appl.* **2018**, *3*, 287–308. [CrossRef]
17. Wang, G. On the Theoretical Properties of the Exchange Algorithm. *arXiv* **2021**, arXiv:2005.09235v4.
18. Diaconis, P. Theories of data analysis: From magical thinking through classical statistics. In *Exploring Data Tables, Trends, and Shapes*; Hoaglin, D.C., Mosteller, F., Tukey, J.W., Eds.; Wiley: Hoboken, NJ, USA, 1985; pp. 1–36.
19. Diaconis, P.; Freedman, D. Partial Exchangeability and Sufficiency. In *Statistics:Applications and New Directions*; Ghosh, K., Roy, F. Eds.; Indian Statistical Institute: Calcutta, India, 1984; pp. 205–236
20. Lauritzen, S.L. General Exponential Models for Discrete Observations. *Scand. J. Stat.* **1975**, *2*, 23–33.
21. Lauritzen, S.L.; Rinaldo, A.; Sadeghi, K. Random Networks, Graphical Models, and Exchangeability. *arXiv* **2017**, arXiv:1701.08420v2.
22. Lauritzen, S.L.; Rinaldo, A.; Sadeghi, K. On exchangeability in network models. *J. Algebr. Stat.* **2019**, *10*, 85–114. [CrossRef]
23. Albert, J.H.; Gupta, A.K. Mixtures of Dirichlet distributions and estimation in contingency tables. *Ann. Stat.* **1982**, *10*, 1261–1268. [CrossRef]
24. Murray, I.; Ghahramani, Z.; MacKay, D.J.C. MCMC for doubly-intractable distributions. In Proceedings of the 22nd Conference in Uncertainty in Artificial Intelligence (UAI '06), Cambridge, MA, USA, 13–16 July 2006.
25. Letac, G.; Massam, H. Bayes factors and the geometry of discrete hierarchical loglinear models. *Ann. Stat.* **2012**, *40*, 861–890. [CrossRef]
26. Tarantola, C.; Ntzoufras, I. Bayesian Analysis of Graphical Models of Marginal Independence for Three Way Contingency Tables. In *Quaderni di Dipartimento from University of Pavia*; No 172; Department of Economics and Quantitative Methods, University of Pavia: Pavia, Italy, 2012. Available online: http://dem-web.unipv.it/web/docs/dipeco/quad/ps/RePEc/pav/wpaper/q172.pdf (accessed on 30 December 2021).
27. Diaconis, P.; Efron, B. Testing for independence in a two-way table: New interpretations of the Chi-Square statistic. *Ann. Stat.* **1985**, *13*, 845–874. [CrossRef]
28. Diaconis, P.; Ylvisaker, D. Quantifying prior opinion. In *Bayesian Statistics, II*; Bernardo, J., DeGroot, M., Lindley, D., Smith, A.F.M., Eds.; North-Holland: Amsterdam, The Netherlands, 1985 ; pp. 133–156.

Article

Single-Block Recursive Poisson–Dirichlet Fragmentations of Normalized Generalized Gamma Processes

Lancelot F. James

Department of Information Systems, Business Statistics and Operations Management, The Hong Kong University of Science and Technology, Clear Water Bay, Kowloon, Hong Kong; lancelot@ust.hk

Abstract: Dong, Goldschmidt and Martin (2006) (DGM) showed that, for $0 < \alpha < 1$, and $\theta > -\alpha$, the repeated application of independent single-block fragmentation operators based on mass partitions following a two-parameter Poisson–Dirichlet distribution with parameters $(\alpha, 1-\alpha)$ to a mass partition having a Poisson–Dirichlet distribution with parameters (α, θ) leads to a remarkable nested family of Poisson—Dirichlet distributed mass partitions with parameters $(\alpha, \theta + r)$ for $r = 0, 1, 2, \ldots$. Furthermore, these generate a Markovian sequence of α-diversities following Mittag-Leffler distributions, whose ratios lead to independent Beta-distributed variables. These Markov chains are referred to as Mittag-Leffler Markov chains and arise in the broader literature involving Pólya urn and random tree/graph growth models. Here we obtain explicit descriptions of properties of these processes when conditioned on a mixed Poisson process when it equates to an integer n, which has interpretations in a species sampling context. This is equivalent to obtaining properties of the fragmentation operations of (DGM) when applied to mass partitions formed by the normalized jumps of a generalized gamma subordinator and its generalizations. We focus primarily on the case where $n = 0, 1$.

Keywords: fragmentations of mass partitions; generalized gamma process; Mittag-Leffler Markov Chains; Poisson—Dirichlet distributions; species sampling

Citation: James, L.F. Single-Block Recursive Poisson–Dirichlet Fragmentations of Normalized Generalized Gamma Processes. *Mathematics* **2022**, *10*, 561. https://doi.org/10.3390/math10040561

Academic Editors: Emanuele Dolera and Federico Bassetti

Received: 1 January 2022
Accepted: 9 February 2022
Published: 11 February 2022

Publisher's Note: MDPI stays neutral with regard to jurisdictional claims in published maps and institutional affiliations.

Copyright: © 2022 by the author. Licensee MDPI, Basel, Switzerland. This article is an open access article distributed under the terms and conditions of the Creative Commons Attribution (CC BY) license (https://creativecommons.org/licenses/by/4.0/).

1. Introduction

Let $\mathbf{Z} = (Z_r, r \geq 0)$ denote a Markov chain characterized by a stationary transition density $Z_r|Z_{r-1} = z$ given for $y > z$ and $0 < \alpha < 1$:

$$\mathbb{P}(Z_r \in dy|Z_{r-1} = z)/dy = \frac{\alpha(y-z)^{\frac{1-\alpha}{\alpha}-1} y g_\alpha(y)}{\Gamma(\frac{1-\alpha}{\alpha}) g_\alpha(z)}, \quad (1)$$

where $g_\alpha(s) := f_\alpha(s^{-\frac{1}{\alpha}}) s^{-\frac{1}{\alpha}-1}/\alpha$ is the density of a variable $T_\alpha^{-\alpha}$, with a Mittag-Leffler distribution, $T_\alpha := T_{\alpha,0}$ is a positive stable variable with density denoted as $f_\alpha(t)$, and Laplace transform $\mathbb{E}[e^{-\lambda T_\alpha}] = e^{-\lambda^\alpha}$. More generally, as in [1–4], for $\theta > -\alpha$, let $T_{\alpha,\theta}$ denote a variable with density $f_{\alpha,\theta}(t) = t^{-\theta} f_\alpha(t)/\mathbb{E}[T_\alpha^{-\theta}]$; then, $T_{\alpha,\theta}^{-\alpha}$ is said to have a generalized Mittag-Leffler distribution with parameters (α, θ) and distribution denoted as $\mathrm{ML}(\alpha, \theta)$. In the cases where $Z_0 = T_{\alpha,\theta}^{-\alpha} \sim \mathrm{ML}(\alpha, \theta)$, the marginal distributions of each Z_r are $\mathrm{ML}(\alpha, \theta + r)$. Furthermore, there is a sequence of random variables $(B_j, j \geq 1)$ defined for each integer j as $B_j = Z_{j-1}/Z_j$; hence, there is the exact point-wise relation $Z_{j-1} = Z_j \times B_j$, where, remarkably, the B_j are independent Beta$(\frac{\theta+\alpha+j-1}{\alpha}, \frac{1-\alpha}{\alpha})$ variables, and $(B_1, \ldots, B_j)$ is independent of Z_j, for $j = 1, 2, \ldots$. Note further that by setting $Z_r = T_{\alpha,\theta+r}^{-\alpha}$, there is the point-wise equality $T_{\alpha,\theta} = T_{\alpha,\theta+r} \times \prod_{j=1}^{r} B_j^{-\frac{1}{\alpha}}$, where all the variables on the right-hand side are independent. In these cases, the sequence may be referred to as a Mittag-Leffler Markov chain with law denoted as $\mathbf{Z} \sim \mathrm{MLMC}(\alpha, \theta)$, as in [5] and, subsequently, [6]. The Markov chain is described prominently in various generalities, that is, ranges of α and θ, in [5–9]. See for example [5,6,10–15] for more references concerning Pólya urn and random tree/graph growth models.

Now, let $PD(\alpha, \theta)$ denote a two-parameter Poisson–Dirichlet distribution over the space of mass partitions summing to 1, say $\mathcal{P}_\infty := \{\mathbf{s} = (s_1, s_2, \ldots) : s_1 \geq s_2 \geq \cdots \geq 0$ and $\sum_{i=1}^\infty s_i = 1\}$, as described in [3,4,16]. Let $(P_\ell) := ((P_\ell), \ell \geq 1) \sim PD(\alpha, \theta)$ correspond in distribution to the ranked lengths of excursion of a generalized Bessel bridge on $[0, 1]$, as described and defined in [1,4]. In particular, $PD(1/2, 0)$ and $PD(1/2, 1/2)$ correspond to excursion lengths of standard Brownian motion and Brownian bridge, on $[0,1]$, respectively. As noted in [6], the single-block $PD(\alpha, 1-\alpha)$ fragmentation results for $PD(\alpha, \theta)$ mass partitions by [17], which we shall describe in more detail in Section 1.2, allow one to couple a version of $\mathbf{Z} \sim MLMC(\alpha, \theta)$ with a nested family of mass partitions $((P_{\ell,r}), r \geq 0)$, where each $(P_{\ell,r}) := ((P_{\ell,r}), \ell \geq 1)$ takes its values in $\mathcal{P}_\infty$, initial $(P_{\ell,0}) \sim PD(\alpha, \theta)$ has α-diversity $Z_0 = T_{\alpha,\theta}^{-\alpha}$, and each successive $(P_{\ell,r}) \sim PD(\alpha, \theta + r)$ has α-diversity $Z_r = T_{\alpha,\theta+r}^{-\alpha}$. The distribution of this family is denoted as $((P_{\ell,r}), Z_r; r \geq 0) \sim MLMC_{frag}(\alpha, \theta)$.

Recall from [2] that for $(P_{\ell,0}) \sim PD(\alpha, 0)$, $(P_{\ell,0})|T_\alpha = t$ has distribution $PD(\alpha|t)$, and for a probability measure ν on $(0, \infty)$, one may generate the general class of Poisson–Kingman distributions generated by an α-stable subordinator with mixing ν, by forming $PK_\alpha(\nu) = \int_0^\infty PD(\alpha|t)\nu(dt)$. Some prominent examples of interest in this work are $PD(\alpha, \theta) = \int_0^\infty PD(\alpha|t)f_{\alpha,\theta}(t)dt$ and $\mathbb{P}_\alpha^{[n]}(\lambda) = \int_0^\infty PD(\alpha|t)f_\alpha^{[n]}(t|\lambda)dt$, where $f_\alpha^{[n]}(t|\lambda) \propto t^n e^{-\lambda t} f_\alpha(t)$. Hence, $\mathbb{P}_\alpha^{[0]}(\lambda)$ corresponds to the law of the ranked normalized jumps of a generalized gamma subordinator, say $(\tau_\alpha(y); y \geq 0)$, where $\tau_\alpha(\lambda^\alpha)/\lambda$ has density $f_\alpha^{[0]}(t|\lambda) = e^{-\lambda t} e^{\lambda^\alpha} f_\alpha(t)$. In [6], we obtained some general distributional properties of $((P_{\ell,r}), Z_r; r \geq 0)$ formed by repeated application of the fragmentation operations in [17] to the case where $(P_{\ell,0}) \sim PK_\alpha(\nu)$. Furthermore, letting $(\mathbf{e}_\ell)$ denote a sequence of iid $\mathrm{Exp}(1)$ variables forming the arrival times, say $(\Gamma_\ell = \sum_{j=1}^\ell \mathbf{e}_j; \ell \geq 1)$, of a standard Poisson process, we ([6], Section 4.3) focused in more detail on the special case of $((P_{\ell,r}), Z_r; r \geq 0)|N_{T_{\alpha,\theta}^{-\alpha}}(\lambda) = j$ for $j = 0, 1, 2, \ldots$, when $((P_{\ell,r}), Z_r; r \geq 0) \sim MLMC_{frag}(\alpha, \theta)$ and $(N_{T_{\alpha,\theta}^{-\alpha}}(t) = \sum_{\ell=1}^\infty \mathbb{I}_{\{\Gamma_\ell/T_{\alpha,\theta}^{-\alpha} \leq t\}}, t \geq 0)$ is a mixed Poisson process with random intensity depending on $T_{\alpha,\theta}^{-\alpha}$. That is to say, $(P_{\ell,0})|N_{T_{\alpha,\theta}^{-\alpha}}(\lambda) = j$ corresponds in distribution to $(P_{\ell,0}(\lambda))$ following a $PK_\alpha(\nu)$ distribution, where ν corresponds to the distribution of $T_{\alpha,\theta}^{-\alpha}|N_{T_{\alpha,\theta}^{-\alpha}}(\lambda) = j$.

In this work, we obtain results for the case where $((P_{\ell,r}), Z_r; r \geq 0)$ is such that $(P_{\ell,0}) \sim \mathbb{P}_\alpha^{[n]}(\lambda)$, which is when $(P_{\ell,0})$ corresponds to the ranked normallized jumps of a generalized gamma process, $(\tau_\alpha(y); y \geq 0)$, and its size-biased generalizations. Interestingly, our results equate in distribution to the following setup involving $((P_{\ell,r}), Z_r; r \geq 0) \sim MLMC_{frag}(\alpha, 0)$. Let N_{T_α} be a mixed Poisson process defined by replacing $T_{\alpha,\theta}^{-\alpha}$ in $N_{T_{\alpha,\theta}^{-\alpha}}$ with T_α. Using the mixed Poisson framework in the manuscript of Pitman [18] (see also [6,19] for more details), we obtain some explicit distributional properties of $((P_{\ell,r}), Z_r; r \geq 0)|N_{T_\alpha}(\lambda) = n$ and corresponding variables $(B_1, \ldots, B_r, T_{\alpha,r})|N_{T_\alpha}(\lambda) = n$ for $n = 0, 1, 2, \ldots$, when $((P_{\ell,r}), Z_r; r \geq 0) \sim MLMC_{frag}(\alpha, 0)$. That is when $(P_{\ell,0}) \sim PD(\alpha, 0)$. The equivalence in distribution to the fragmentation operations of [17] applied in the generalized gamma cases may be deduced from [18], who shows that when $(P_{\ell,0}) \sim PD(\alpha, 0)$, $(P_{\ell,0})|N_{T_\alpha} = n$ corresponds to the distribution of $(P_{\ell,0}(\lambda)) \sim \mathbb{P}_\alpha^{[n]}(\lambda)$. We shall primarily focus on the case of $n = 0, 1$, corresponding to the generalized gamma density and its sized biased distribution, which yields the most explicit results. The fragmentation operations (6) applied to $((P_{\ell,0})) \sim P_\alpha^{[1]}(\lambda)$ allow one to recover the entire range of $PD(\alpha, \theta)$ distributions for $\theta > -\alpha$, by gamma randomization, whereas the case for $((P_{\ell,0})) \sim P_\alpha^{[0]}(\lambda)$ only applies to $\theta \geq 0$. We note that descriptions of our results for $n = 0, 1$, albeit less refined ones, appear in the unpublished manuscript ([9], Section 6). See also [20] for an application of $P_\alpha^{[0]}(\lambda)$ for randomized λ.

We close this section by recalling the definition of the first size-biased pick from a random mass partition $(P_\ell) \in \mathcal{P}_\infty$ (see [2,3,16]). Specifically, $\tilde{P}_1$ is referred to as the first size-biased pick from (P_ℓ), if it satisfies, for $k = 1, 2, \ldots,$

$$\mathbb{P}(\tilde{P}_1 = P_k | (P_\ell)) = P_k. \tag{2}$$

Hereafter, let $(P_\ell)_1 := (P_\ell) \setminus \tilde{P}_1$ denote the remainder, such that $(P_\ell) = \text{Rank}(((P_\ell)_1, \tilde{P}_1))$, where $\text{Rank}(\cdot)$ denotes the operation corresponding to ranked re-arrangement. From [1], $\tilde{P}_1$ may be interpreted as the length of excursion (i.e., one of the (P_ℓ)), first discovered by dropping a uniformly distributed random variable onto the interval $[0, 1]$. The fragmentation operation of [17] may be interpreted as shattering/fragmenting that interval by the excursion lengths of a process on $[0, 1]$, with distribution $\text{PD}(\alpha, 1 - \alpha)$ and then re-ranking. For clarity and comparison, we first recall some details of the more well-known Markovian size-biased deletion operation leading to stick-breaking representations, as described in [1–3], and more related notions arising in a Bayesian nonparametric context in the $\text{PD}(\alpha, \theta)$ setting, in the next section.

Remark 1. *Although we acknowledge the influence and contributions of the manuscript [18], the pertinent distributional results we use from that work are re-derived at the beginning of Section 2. Otherwise, the interpretation of N_{T_α} from that work is briefly mentioned in Section 1.3.*

1.1. PD(α, θ) Markovian Sequences Obtained from Successive Size-Biased Deletion

Following [1], we may define $\text{SBD}(\cdot)$ to be a *size-biased deletion* operator on $\mathcal{P}_\infty$, as $\text{SBD}((P_\ell)) := \text{Rank}(((P_\ell)_1 / (1 - \tilde{P}_1)))$, where it can be recalled from (2) that $(P_\ell) = \text{Rank}(((P_\ell)_1, \tilde{P}_1))$. Now, let $(\text{SBD}^{(j)}(\cdot), j \geq 1)$ be a collection of such operators. From [1], as per the description in ([4], Proposition 34, p. 881), it follows that for $(P_{\ell,0}) := (\hat{P}_{\ell,0}) \sim \text{PD}(\alpha, \theta)$, $\text{SBD}^{(1)}((\hat{P}_{\ell,0})) := (\hat{P}_{\ell,1}) \sim \text{PD}(\alpha, \theta + \alpha)$ and is independent of the first size-biased pick $\tilde{P}_1 := V_1 \sim \text{Beta}(1 - \alpha, \theta + \alpha)$, and hence, for $r = 2, \ldots,$

$$(\hat{P}_{\ell,r}) := \text{SBD}^{(r)}((\hat{P}_{\ell,r-1})) = \text{SBD}^{(r)} \circ \cdots \circ \text{SBD}^{(1)}((\hat{P}_{\ell,0})) \sim \text{PD}(\alpha, \theta + r\alpha). \tag{3}$$

This leads to a nested Markovian family of mass partitions $((\hat{P}_{\ell,r}), r \geq 0)$, where $(P_{\ell,0}) := (\hat{P}_{\ell,0}) \sim \text{PD}(\alpha, \theta)$ with inverse local time at time 1, $T_{\alpha,\theta}$ (see ([3], Equation (4.20), p. 83)), and for each r, $(\hat{P}_{\ell,r}) \sim \text{PD}(\alpha, \theta + r\alpha)$ with inverse local time at time 1, $T_{\alpha,\theta+r\alpha}$. Furthermore, $(T_{\alpha,\theta+r\alpha}, r \geq 0)$ form a Markov chain with pointwise equality $T_{\alpha,\theta+(j-1)\alpha} = T_{\alpha,\theta+j\alpha}/(1 - V_j)$, where V_j are independent $\text{Beta}(1 - \alpha, \theta + j\alpha)$ variables and are the respective first size-biased picks from $(\hat{P}_{\ell,j-1})$ for $j \geq 1$. Furthermore, $(V_1, \ldots, V_r)$ is independent of $T_{\alpha,\theta+r\alpha}$ and, more generally, $(\hat{P}_{\ell,r})$ for $r = 1, 2, \ldots$.

From this, one obtains the size-biased re-arrangement of a $\text{PD}(\alpha, \theta)$ mass partition, say $(\tilde{P}_\ell) \sim \text{GEM}(\alpha, \theta)$, satisfying $\tilde{P}_1 = V_1 \sim \text{Beta}(1 - \alpha, \theta + \alpha)$, and for $\ell \geq 2$, $\tilde{P}_\ell = V_\ell \prod_{j=1}^{\ell-1} (1 - V_j)$. Refs. [3,21] discuss the $\text{GEM}(\alpha, \theta)$ distribution and these other concepts in a species sampling and Bayesian context. We mention the roles of corresponding random distribution functions as priors in a Bayesian non-parametric context. Let (U_ℓ) denote a sequence of iid Uniform$[0, 1]$ variables independent of $(P_\ell) \sim \text{PD}(\alpha, \theta)$; then, the random distribution $F_{\alpha,\theta}(y) = \sum_{\ell=1}^\infty P_\ell \mathbb{I}_{\{U_\ell \leq y\}}$ is said to follow a Pitman–Yor distribution with parameters (α, θ), (see [21,22]). $F_{\alpha,\theta}$ is a two-parameter extension of the Dirichlet process [23] (which corresponds to $F_{0,\theta}$) and has been applied extensively as a more flexible prior in a Bayesian context, but it also arises in a variety of areas involving combinatorial stochastic processes [3,21]. An attractive feature of $F_{\alpha,\theta}$ is that it may be represented as $F_{\alpha,\theta}(y) = \sum_{\ell=1}^\infty \tilde{P}_\ell \mathbb{I}_{\{\tilde{U}_\ell \leq y\}}$, where $(\tilde{U}_\ell)$ are the iid Uniform$[0, 1]$ concomitants of the $(\tilde{P}_\ell)$, as exploited in [22] (see also [21]). This constitutes the stick-breaking representation of $F_{\alpha,\theta}$. Furthermore, we can describe $\tilde{P}_1$ as folllows: let $X_1 | F_{\alpha,\theta}$ have distribution $F_{\alpha,\theta}$, and denote the first value drawn from $F_{\alpha,\theta}$; then, $\tilde{P}_1$ is the mass in (P_ℓ) corresponding to that atom of $F_{\alpha,\theta}$. The size-biased deletion operation described above, as in (3), leads to the following decomposition of $F_{\alpha,\theta}$:

$$F_{\alpha,\theta}(y) = (1 - \tilde{P}_1) F_{\alpha,\theta+\alpha}(y) + \tilde{P}_1 \mathbb{I}_{\{\tilde{U}_1 \leq y\}} \qquad (4)$$

where $(\tilde{P}_1, \tilde{U}_1)$ are independent of $F_{\alpha,\theta+\alpha}(y) \stackrel{d}{=} \sum_{k=1}^{\infty} \hat{P}_{k,1} \mathbb{I}_{\{U_{k,1} \leq y\}}$, where $(\hat{P}_{\ell,1}) \sim \text{PD}(\alpha, \theta + \alpha)$, and independent of this, where $(U_{\ell,1}) \stackrel{iid}{\sim} \text{Uniform}[0,1]$. See [1,4,24] and references therein for various interpretations of (4).

1.2. DGM Fragmentation

The single-block $\text{PD}(\alpha, 1 - \alpha)$ fragmentation operator of [17] is defined over the space $\mathcal{P}_\infty$. However, for further clarity we start with an explanation at the level of random distribution functions involving the representation in (4). Suppose that $G_{\alpha,1-\alpha}(y) := \sum_{k=1}^{\infty} Q_k \mathbb{I}_{\{U'_{k,1} \leq y\}}$, with $(Q_\ell) \sim \text{PD}(\alpha, 1 - \alpha)$ and, independent of this, $(U'_{\ell,1}) \stackrel{iid}{\sim} \text{Uniform}[0,1]$; hence, $G_{\alpha,1-\alpha} \stackrel{d}{=} F_{\alpha,1-\alpha}$. Suppose that $G_{\alpha,1-\alpha}$ is chosen independent of $F_{\alpha,\theta}$ in (4); then, it follows from [17] that

$$F_{\alpha,\theta+1}(y) \stackrel{d}{=} (1 - \tilde{P}_1) F_{\alpha,\theta+\alpha}(y) + \tilde{P}_1 G_{\alpha,1-\alpha}(y), \qquad (5)$$

and it is evident that the mass partition (Q_ℓ) shatters/fragments $\tilde{P}_1$ into a countably infinite number of pieces $(\tilde{P}_1(Q_\ell)) := (\tilde{P}_1 Q_\ell, \ell \geq 1) = (\tilde{P}_1 Q_1, \tilde{P}_1 Q_2, \ldots)$. It follows that, in this case, $\text{Rank}((P_\ell)_1, \tilde{P}_1(Q_\ell)) \sim \text{PD}(\alpha, \theta + 1)$, which is the featured case of the $\text{PD}(\alpha, 1 - \alpha)$ fragmentation described in [17]. Hence, for general $(P_\ell) = \text{Rank}(((P_\ell)_1, \tilde{P}_1)) \in \mathcal{P}_\infty$, a $\text{PD}(\alpha, 1 - \alpha)$ fragmentation of (P_ℓ) is defined as

$$\widehat{\text{Frag}}_{\alpha,1-\alpha}((P_\ell)) := \text{Rank}(((P_\ell)_1, \tilde{P}_1(Q_\ell))) \in \mathcal{P}_\infty,$$

where, independent of (P_ℓ), $(Q_\ell) \sim \text{PD}(\alpha, 1 - \alpha)$. Let $((Q_\ell^{(j)}); j \geq 1)$ denote an independent collection of $\text{PD}(\alpha, 1 - \alpha)$ mass partitions defining a sequence of independent fragmentation operators $(\widehat{\text{Frag}}_{\alpha,1-\alpha}^{(j)}(\cdot); j \geq 1)$. It follows from [17] that a version of the family $((P_{\ell,r}), Z_r; r \geq 0) \sim \text{MLMC}_{\text{frag}}(\alpha, \theta)$ may be constructed by the recursive fragmentation, for $r = 1, 2, \ldots$:

$$(P_{\ell,r}) = \widehat{\text{Frag}}_{\alpha,1-\alpha}^{(r)}((P_{\ell,r-1})) \qquad (6)$$

In particular, $(P_{\ell,r}) \sim \text{PD}(\alpha, \theta + r)$ when $(P_{\ell,0}) \sim \text{PD}(\alpha, \theta)$.

1.3. Remarks

We close this section with remarks related to some relevant work of Eugenio Regazzini and his students, arising in a Bayesian context. From [18], in regards to a species sampling context using $F_{\alpha,\theta}$ (see [21]), $N_{T_{\alpha,\theta}}(\lambda)$ interprets as the number of animals trapped and tagged up until time λ, and hence, $\Gamma_j / T_{\alpha,\theta}$ interprets as the time when the j-th animal is trapped for $j = 1, \ldots$. Ref. [18] indicates that this gives further interpretation to such types of quantities arising in [25,26]. Using a Chinese restaurant process metaphor, the animals may be replaced by customers arriving sequentially to a restaurant. More generically, $N_{T_{\alpha,\theta}}(\lambda)$ is the number of exchangeable samples drawn from $F_{\alpha,\theta}$ up until time λ. Furthermore, $F_{\alpha,n}(y) | N_{T_{\alpha,n}}(\lambda) = n$ for each $n = 0, 1, 2, \ldots$ is equivalent in distribution to $F_\alpha(y|\lambda) \stackrel{d}{=} \tau_\alpha(\lambda^\alpha y) / \tau_\alpha(\lambda^\alpha)$, which is now referred to in the Bayesian literature as a normalized generalized gamma process. While, according to [2], $F_\alpha(y|\lambda)$ appears in a relevant species sampling context in the 1965 thesis of McCloskey [27], and certainly elsewhere, the paper by Reggazzini, Lijoi, and Prünster [28] and subsequent works by Regazzini's students (see [29]) helped to popularize the usage of $F_\alpha(y|\lambda)$ in the modern literature on Bayesian non-parametrics. Our work presents a view of $F_\alpha(y|\lambda)$ subjected to the fragmentation operations in [17]. Although we do not consider specific Bayesian statistical applications in this work, we note that other types of fragmentation/coagulation of $PD(\alpha, \theta)$ models have been applied, for instance, in [30]. We anticipate the same will be true of the operations considered here.

2. Results

Hereafter, we shall focus on the case of $PD(\alpha, 0)$, as we will recover the general (α, θ) cases by applying gamma randomization as in ([4], Proposition 21) for $\theta \geq 0$ or ([19], Corollary 2.1) for $\theta > -\alpha$ and other results. See also ([6], Section 2.2.1). We first re-derive some relevant properties related to N_{T_α} that are easily verified by first conditioning on T_α and otherwise can be found in [18]. First, for fixed λ, and for $j = 0, 1, \ldots,$

$$\mathbb{P}(N_{T_\alpha}(\lambda) = j, T_\alpha \in ds) = \frac{\lambda^j}{j!} s^j e^{-\lambda s} f_\alpha(s) ds, \qquad (7)$$

and for $j = 1, 2, \ldots,$

$$\mathbb{P}\left(\frac{\Gamma_j}{T_\alpha} \in d\lambda, T_\alpha \in ds\right)/d\lambda = \frac{\lambda^{j-1}}{(j-1)!} s^j e^{-\lambda s} f_\alpha(s) ds. \qquad (8)$$

Note these simple results hold for any variable T with density f_T in place of T_α and f_α. It follows from (7) and (8) that $T_\alpha | N_{T_\alpha}(\lambda) = 0$ has the generalized gamma density $f_\alpha^{[0]}(t|\lambda) = e^{-\lambda t} e^{\lambda^\alpha} f_\alpha(t)$. Furthermore, for $j = 1, 2, \ldots$; $T_\alpha | N_{T_\alpha}(\lambda) = j$ has the same distribution as $T_\alpha | \Gamma_j / T_\alpha = \lambda$ with density $f_\alpha^{[j]}(t|\lambda)$. Since it is assumed that $(\Gamma_\ell; \ell \geq 1)$ is independent of (P_ℓ), it follows that for $(P_\ell) \sim PD(\alpha, 0)$, the conditional distribution of $(P_\ell)|T_\alpha = t, N_{T_\alpha}(\lambda) = n$ is $PD(\alpha|t)$, and hence, $(P_\ell)|N_{T_\alpha}(\lambda) = n$ has distribution $\mathbb{P}_\alpha^{[n]}(\lambda)$ for $n = 0, 1, \ldots,$ as mentioned previously.

Remark 2. *For the next results, which are extensions to $((P_{\ell,r}), Z_r; r \geq 0) \sim MLMC_{frag}(\alpha, 0)$, conditioned on $N_{T_\alpha}(\lambda) = n$, we note, as in [19], that the densities $f_\alpha^{[n]}(t|\lambda)$ are well-defined for any real number ϱ in place of $[n]$, with density $f_\alpha^{[\varrho]}(t|\lambda)$, provided that $\lambda > 0$, and for $\lambda = 0$ only in the case where $\varrho = -\theta < \alpha$, which corresponds to $f_{\alpha, \theta}(t)$. Ref. ([19], Corollary 2.1) shows that distributions for ϱ can be expressed as randomized (over λ) distributions for any $n > \varrho$.*

For clarity, with respect to $((P_{\ell,r}), Z_r; r \geq 0) \sim MLMC_{frag}(\alpha, 0)$, $B_j = Z_{j-1}/Z_j$ are independent Beta$(\frac{\alpha+j-1}{\alpha}, \frac{1-\alpha}{\alpha})$ variables for $j = 1, 2, \ldots,$ and $(B_1, \ldots, B_r)$ is independent of $Z_r = T_{\alpha,r}^{-\alpha}$ and $(P_{\ell,r})$ for each $r = 1, 2, \ldots$.

Proposition 1. *Consider $((P_{\ell,r}), Z_r; r \geq 0) \sim MLMC_{frag}(\alpha, 0)$, formed by the fragmentation operations in (6), when $(P_{\ell,0}) \sim PD(\alpha, 0)$. Denote the conditional distribution of $((P_{\ell,r}), Z_r; r \geq 0)|N_{T_\alpha}(\lambda) = n$ as $MLMC_{frag}^{[n]}(\alpha|\lambda)$ and its corresponding component values as $((P_{\ell,r}(\lambda)), Z_r(\lambda); r \geq 0)$. Then, the distribution has the following properties.*

(i) $(P_{\ell,0})|N_{T_\alpha}(\lambda) = n$ is equivalent in distribution to $(P_{\ell,0}(\lambda)) \sim \mathbb{P}_\alpha^{[n]}(\lambda) = \int_0^\infty PD(\alpha|t) f_\alpha^{[n]}(t|\lambda) dt$.

(ii) $(P_{\ell,r})|N_{T_\alpha}(\lambda) = n, \prod_{i=1}^r B_i = \mathbf{b}_r$ has distribution $\mathbb{P}_\alpha^{[n-r]}(\lambda \mathbf{b}_r^{-\frac{1}{\alpha}})$, for $r = 1, 2, \ldots$.

(iii) $(P_{\ell,r})|N_{T_\alpha}(\lambda) = n, \prod_{i=1}^r B_i = \mathbf{b}_r$ has the same distribution as $(P_{\ell,r})|N_{T_{\alpha,r}}(\lambda \mathbf{b}_r^{-\frac{1}{\alpha}}) = n$.

Proof. Statement (i) has already been established. For (ii) and equivalently (iii), we use $T_\alpha = T_{\alpha,r} \times \prod_{i=1}^r B_i^{-\frac{1}{\alpha}}$, to obtain $N_{T_\alpha}(\lambda) = N_{T_{\alpha,r}}(\lambda \prod_{i=1}^r B_i^{-\frac{1}{\alpha}})$. Use (7) and (8) with $T_{\alpha,r}$, with density $f_{\alpha,r}(t)$, in place of T_α, to conclude that $T_{\alpha,r}|N_{T_{\alpha,r}}(\lambda \mathbf{b}_r^{-\frac{1}{\alpha}}), \prod_{i=1}^r B_i = \mathbf{b}_r$ has density $f_\alpha^{[n-r]}(t|\lambda \mathbf{b}_r^{-\frac{1}{\alpha}})$. Then, apply $(P_{\ell,r})|T_{\alpha,r} = t, N_{T_\alpha}(\lambda) = n, \prod_{i=1}^r B_i = \mathbf{b}_r$ is $\mathrm{PD}(\alpha|t)$ for $(P_{\ell,r}) \sim \mathrm{PD}(\alpha, r)$. □

3. Results for $n = 0, 1$

We will now focus on results for $(B_1, \ldots, B_r, T_{\alpha,r})$, given $N_{T_\alpha}(\lambda) = n$, in the cases where $n = 0, 1$, and $((P_{\ell,r}), Z_r; r \geq 0) \sim \mathrm{MLMC}_{\mathrm{frag}}(\alpha, 0)$. This is equivalent to providing more explicit distributional results than Proposition 1 for the generalized gamma and its size-biased case, where $(P_{\ell,0}(\lambda)) \sim \mathbb{P}_\alpha^{[n]}(\lambda)$, for $n = 0, 1$, subjected to the fragmentation operations in (6). We first highlight a class of random variables that will play an important role in our descriptions.

Throughout, we define $\gamma_\theta \sim \mathrm{Gamma}(\theta, 1)$ for $\theta \geq 0$, with $\gamma_0 := 0$. Let $(\mathbf{e}^{(\ell)})$ and $(\gamma_{\frac{1-\alpha}{\alpha}}^{(\ell)})$ denote, respectively, iid collections of exponential(1) and $\mathrm{Gamma}(\frac{1-\alpha}{\alpha}, 1)$ random variables that are mutually independent. Use this to form iid sums $\gamma_{\frac{1}{\alpha}}^{(k)} := \mathbf{e}^{(k)} + \gamma_{\frac{1-\alpha}{\alpha}}^{(k)} \sim \mathrm{Gamma}(\frac{1}{\alpha}, 1)$, and construct increasing sums $\Gamma_{\alpha,k} := \sum_{j=1}^k \gamma_{\frac{1}{\alpha}}^{(j)} \sim \mathrm{Gamma}(\frac{k}{\alpha}, 1)$ for $k = 1, 2, \ldots$.

Lemma 1. *For $k = 1, 2, \ldots$, set $Y_k(\lambda) = (\Gamma_{\alpha,k-1} + \lambda^\alpha)/(\Gamma_{\alpha,k} + \lambda^\alpha)$, with $\Gamma_{\alpha,0} = 0$, and hence $Y_1(\lambda) = \lambda^\alpha/(\Gamma_{\alpha,1} + \lambda^\alpha)$. Then, for any $r = 1, 2, \ldots$ and $\lambda > 0$, the joint density of $(Y_1(\lambda), \ldots, Y_r(\lambda))$ can be expressed as*

$$\vartheta_{\alpha,r}^{[0]}(y_1, \ldots, y_r|\lambda) = \frac{\lambda^r}{[\Gamma(\frac{1}{\alpha})]^r} e^{-\lambda^\alpha/(\prod_{j=1}^r y_j)} e^{\lambda^\alpha} \prod_{l=1}^r y_l^{-\frac{(r-l+1)}{\alpha}-1}(1-y_l)^{\frac{1}{\alpha}-1}. \qquad (9)$$

Furthermore, $\lambda^\alpha / \prod_{j=1}^r Y_j(\lambda) = \Gamma_{\alpha,r} + \lambda^\alpha$.

3.1. Results for $(P_{\ell,0}(\lambda)) \sim \mathbb{P}_\alpha^{[0]}(\lambda)$, the Generalized Gamma Case

Let $(\beta_{(\frac{1-\alpha}{\alpha},1)}^{(k)})$ denote a collection of iid $\mathrm{Beta}(\frac{1-\alpha}{\alpha}, 1)$ variables, and independent of this, let $(\tau_\alpha^{(r)}(y))$ denote, for each fixed $y \geq 0$, a collection of iid variables such that $\tau_\alpha^{(r)}(y) \stackrel{d}{=} \tau_\alpha(y)$. In addition, for each r, $(\beta_{(\frac{1-\alpha}{\alpha},1)}^{(1)}, \ldots, \beta_{(\frac{1-\alpha}{\alpha},1)}^{(r)}, \tau_\alpha^{(r)}(\lambda))$ is independent of $(Y_1(\lambda), \ldots, Y_r(\lambda))$.

Proposition 2. *Consider $((P_{\ell,r}), Z_r; r \geq 0) \sim \mathrm{MLMC}_{\mathrm{frag}}(\alpha, 0)$; then, for each r, the joint distribution of the random variables $(B_1, \ldots, B_r, T_{\alpha,r})|N_{T_\alpha}(\lambda) = 0$ is equivalent component-wise and jointly to the distribution of $(B_1^{[0]}(\lambda), \ldots, B_r^{[0]}(\lambda), T_{\alpha,r}^{[0]}(\lambda))$, where:*

(i) $B_k^{[0]}(\lambda) \stackrel{d}{=} 1 - \beta_{(\frac{1-\alpha}{\alpha},1)}^{(k)}[1 - Y_k(\lambda)]$, *with conditional density given $Y_k(\lambda) = y_k$,*

$$\frac{1-\alpha}{\alpha}(1-b_k)^{\frac{1-\alpha}{\alpha}-1}(1-y_k)^{1-\frac{1}{\alpha}} \mathbb{I}_{\{y_k \leq b_k \leq 1\}},$$

for $k = 1, 2, \ldots$.

(ii) *The conditional distribution of $T_{\alpha,r}|N_{T_\alpha}(\lambda) = 0$ is equivalent to that of*

$$T_{\alpha,r}^{[0]}(\lambda) \stackrel{d}{=} \frac{\tau_\alpha^{(r)}(\Gamma_{\alpha,r} + \lambda^\alpha)}{(\Gamma_{\alpha,r} + \lambda^\alpha)^{1/\alpha}}$$

where recall $\lambda^\alpha / \prod_{j=1}^r Y_j(\lambda) = \Gamma_{\alpha,r} + \lambda^\alpha$.

(iii) *The conditional density of $T_{\alpha,r}^{[0]}(\lambda)|\prod_{i=1}^r Y_i(\lambda) = \mathbf{y}_r$, is $f_\alpha^{[0]}(t|\lambda \mathbf{y}_r^{-\frac{1}{\alpha}})$.*

(iv) *Hence, $(P_{\ell,r})|N_{T_\alpha}(\lambda) = 0 \sim \mathbb{E}[\mathbb{P}_\alpha^{[0]}((\Gamma_{\alpha,r} + \lambda^\alpha)^{1/\alpha})]$.*

(v) $(B_1^{[0]}(\lambda), \ldots, B_r^{[0]}(\lambda), T_{\alpha,r}^{[0]}(\lambda))|Y_1(\lambda), \ldots, Y_r(\lambda)$ *are independent.*

Corollary 1. Suppose that $(P_{\ell,0}(\lambda)) \overset{d}{=} (P_\ell^{[0]}(\lambda)) \sim \mathbb{P}_\alpha^{[0]}(\lambda) = \int_0^\infty \mathrm{PD}(\alpha|t) e^{-\lambda t} e^{\lambda^\alpha} f_\alpha(t) dt$, then for $r = 1, 2, \ldots$,

$$(P_{\ell,r}(\lambda)) = \widehat{\mathrm{Frag}}_{\alpha,1-\alpha}^{(r)}((P_{\ell,r-1}(\lambda))) \overset{d}{=} (P_\ell^{[0]}((\Gamma_{\alpha,r} + \lambda^\alpha)^{1/\alpha})) \quad (10)$$

where $\Gamma_{\alpha,r} = \sum_{j=1}^r \gamma_{\frac{1}{\alpha}}^{(j)} \sim \mathrm{Gamma}(\frac{r}{\alpha})$

Proof. This follows from statement (iv) of Proposition 2. □

The corollary shows that the fragmentation operations in (6) lead to a nested family of (mixed) normalized generalized gamma distributed mass partitions, with λ^α replaced by the random quantities $\lambda^\alpha / \prod_{j=1}^r Y_j(\lambda) = \Gamma_{\alpha,r} + \lambda^\alpha$. In other words, $(P_{\ell,r})|N_{T_{\alpha,0}}(\lambda) = 0$ equates in distribution to the ranked masses of the random distribution function, for $v \in [0,1]$:

$$F_\alpha(v|(\Gamma_{\alpha,r} + \lambda^\alpha)^{1/\alpha}) \overset{d}{=} \frac{\tau_\alpha([\Gamma_{\alpha,r} + \lambda^\alpha]v)}{\tau_\alpha(\Gamma_{\alpha,r} + \lambda^\alpha)}.$$

Now, in order to recover $\mathrm{MLMC}_{\mathrm{frag}}(\alpha, \theta)$ for $\theta \geq 0$, when $(P_{\ell,0}(\lambda)) \sim \mathbb{P}_\alpha^{[0]}(\lambda)$, set, for $\theta \geq 0$, $\tilde{G}_{\alpha,\theta} \overset{d}{=} G_{\frac{\theta}{\alpha}}^{\frac{1}{\alpha}} \overset{d}{=} \frac{\gamma_\theta}{T_{\alpha,\theta}}$, where $G_{\frac{\theta}{\alpha}} \sim \mathrm{Gamma}(\frac{\theta}{\alpha}, 1)$. When $(P_{\ell,0}(\lambda)) \overset{d}{=} (P_\ell^{[0]}(\lambda)) \sim \mathbb{P}_\alpha^{[0]}(\lambda)$, as in Corollary 1, it follows from ([4], Proposition 21) that $(P_{\ell,0}(\tilde{G}_{\alpha,\theta})) \sim \mathrm{PD}(\alpha, \theta)$. Hence $((P_{\ell,r}(\tilde{G}_{\alpha,\theta})), Z_r(\tilde{G}_{\alpha,\theta}); r \geq 0) \sim \mathrm{MLMC}_{\mathrm{frag}}(\alpha, \theta)$. It follows from Proposition 2 that, $B_k^{[0]}(\tilde{G}_{\alpha,\theta}) \overset{ind}{\sim} \mathrm{Beta}(\frac{\theta+\alpha+k-1}{\alpha}, \frac{1-\alpha}{\alpha})$ for $k = 1, 2, \ldots$. Notably, $(Y_1(\tilde{G}_{\alpha,\theta}), \ldots, Y_r(\tilde{G}_{\alpha,\theta}))$ are independent variables, such that $1 - Y_r(\tilde{G}_{\alpha,\theta}) \sim \mathrm{Beta}(\frac{1}{\alpha}, \frac{\theta+r-1}{\alpha})$ for $r = 1, 2, \ldots$. When $\theta = 0$, or equivalently $\lambda = 0$, $Y_1(0) = 0$, and $1 - Y_r(0) \sim \mathrm{Beta}(\frac{1}{\alpha}, \frac{r-1}{\alpha})$ for $r = 2, \ldots$.

3.2. Results for $(P_{\ell,0}(\lambda)) \sim \mathbb{P}_\alpha^{[1]}(\lambda)$

Proposition 3. Consider $((P_{\ell,r}), Z_r; r \geq 0)|N_{T_\alpha}(\lambda) = 1 \sim \mathrm{MLMC}_{\mathrm{frag}}^{[1]}(\alpha|\lambda)$; then, for each r, the joint distribution of the random variables $(B_1, \ldots, B_r, T_{\alpha,r})|N_{T_\alpha}(\lambda) = 1$ is equivalent component-wise and jointly to the distribution of $(B_1^{[1]}(\lambda), \ldots, B_r^{[1]}(\lambda), T_{\alpha,r}^{[1]}(\lambda))$, where:

(i) $B_1^{[1]}(\lambda) \overset{d}{=} \lambda^\alpha / (\gamma_{\frac{1-\alpha}{\alpha}} + \lambda^\alpha)$, where $\gamma_{\frac{1-\alpha}{\alpha}} \sim \mathrm{Gamma}(\frac{1-\alpha}{\alpha}, 1)$.

(ii) $B_k^{[1]}(\lambda) \overset{d}{=} B_{k-1}^{[0]}((\gamma_{\frac{1-\alpha}{\alpha}} + \lambda^\alpha)^{1/\alpha})$ for $k = 2, 3, \ldots$, component-wise and jointly.

(iii) $T_{\alpha,r}^{[1]}(\lambda)$ is equivalent in distribution to $T_{\alpha,r}|N_{T_\alpha}(\lambda) = 1$ and equivalent in distribution to

$$T_{\alpha,r-1}^{[0]}((\gamma_{\frac{1-\alpha}{\alpha}} + \lambda^\alpha)^{1/\alpha}) \overset{d}{=} \frac{\tau_\alpha^{(r-1)}(\Gamma_{\alpha,r-1} + \gamma_{\frac{1-\alpha}{\alpha}} + \lambda^\alpha)}{(\Gamma_{\alpha,r-1} + \gamma_{\frac{1-\alpha}{\alpha}} + \lambda^\alpha)^{1/\alpha}},$$

$r = 1, 2, \ldots$.

Corollary 2. The distributions of the components of $((P_{\ell,r}(\lambda)), Z_r(\lambda); r \geq 0) \sim \mathrm{MLMC}_{\mathrm{frag}}^{[1]}(\alpha|\lambda)$, where $(P_{\ell,0}(\lambda)) \overset{d}{=} (P_\ell^{[1]}(\lambda)) \sim \mathbb{P}_\alpha^{[1]}(\lambda)$, for $\lambda > 0$, satisfies for $r = 1, 2, \ldots$,

$$(P_{\ell,r}(\lambda)) = \widehat{\mathrm{Frag}}_{\alpha,1-\alpha}^{(r)}((P_{\ell,r-1}(\lambda))) \overset{d}{=} (P_\ell^{[1]}((\Gamma_{\alpha,r} + \lambda^\alpha)^{1/\alpha})), \quad (11)$$

where $(P_\ell^{[1]}((\mathbf{e}_1 + \Gamma_{\alpha,r-1} + \gamma_{\frac{1-\alpha}{\alpha}} + \lambda^\alpha)^{1/\alpha})) \overset{d}{=} (P_\ell^{[0]}((\Gamma_{\alpha,r-1} + \gamma_{\frac{1-\alpha}{\alpha}} + \lambda^\alpha)^{1/\alpha}))$ for $\mathbf{e}_1 \sim$ exponential(1) independent of the other variables. In this case, $\Gamma_{\alpha,r} \overset{d}{=} \mathbf{e}_1 + \Gamma_{\alpha,r-1} + \gamma_{\frac{1-\alpha}{\alpha}}$.

Proof. $(P_{\ell,r})|N_{T_\alpha}(\lambda) = 1$, has the same distribution as $(P_{\ell,r}(\lambda))$ in (11), and (iii) of Proposition 3 shows that they are equivalent in distribution to $(P_\ell^{[0]}((\Gamma_{\alpha,r-1} + \gamma_{\frac{1-\alpha}{\alpha}} + \lambda^\alpha)^{1/\alpha}))$. From ([19], Corollary 2.1, Proposition 3.2), there is the equivalence $(P_\ell^{[1]}((\mathbf{e}_1 + \lambda^\alpha)^{1/\alpha})) \stackrel{d}{=} (P_\ell^{[0]}(\lambda))$ for any $\lambda \geq 0$, yields (11). □

Now, in order to recover $\mathrm{MLMC}_{\mathrm{frag}}(\alpha, \theta)$ for $\theta > -\alpha$, when $(P_{\ell,0}(\lambda))) \sim \mathbb{P}_\alpha^{[1]}(\lambda)$, use $\hat{G}_{\alpha,\theta} \stackrel{d}{=} G_{\frac{\theta+\alpha}{\alpha}}^{\frac{1}{\alpha}} \stackrel{d}{=} \frac{\gamma_{1+\theta}}{T_{\alpha,\theta}}$, where $G_{\frac{\theta+\alpha}{\alpha}} \sim \mathrm{Gamma}(\frac{\theta+\alpha}{\alpha}, 1)$, and, $((P_{\ell,r}(\lambda)), Z_r(\lambda); r \geq 0) \sim \mathrm{MLMC}_{\mathrm{frag}}^{[1]}(\alpha|\lambda)$. It follows from ([19], Corollary 2.1) that $((P_{\ell,r}(\hat{G}_{\alpha,\theta})), Z_r(\hat{G}_{\alpha,\theta}); r \geq 0) \sim \mathrm{MLMC}_{\mathrm{frag}}(\alpha, \theta)$, for $\theta > -\alpha$.

3.3. Proofs of Propositions 2 and 3

Although the joint conditional density of $(B_1, \ldots, B_r, T_{\alpha,r})|N_{T_\alpha}(\lambda) = 0$ in the $\mathrm{MLMC}(\alpha, 0)$ setting can be easily obtained from ([6], p. 324), with $h(t) = e^{-\lambda t}e^{\lambda^\alpha}$, for clarity, we derive it here. Since $\mathbb{P}(N_{T_\alpha}(\lambda) = 0|T_{\alpha,r} = s, \prod_{i=1}^r B_i = \mathbf{b}_r) = e^{-\lambda s/\mathbf{b}_r^{1/\alpha}}$, and $\mathbb{P}(N_{T_\alpha}(\lambda) = 0) = e^{-\lambda^\alpha}$, it follows that the desired conditional density of $(B_1, \ldots, B_r, T_{\alpha,r})|N_{T_\alpha}(\lambda) = 0$, can be expressed as,

$$\frac{\alpha^r}{[\Gamma(\frac{1-\alpha}{\alpha})]^r}\prod_{i=1}^r b_i^{\frac{\alpha+i-1}{\alpha}-1}(1-b_i)^{\frac{1-\alpha}{\alpha}-1} \times s^{-r}f_\alpha(s)e^{-\lambda s/\mathbf{b}_r^{1/\alpha}}e^{\lambda^\alpha}. \qquad (12)$$

Now, a joint density of $(B_1^{[0]}(\lambda), \ldots, B_r^{[0]}(\lambda), T_{\alpha,r}^{[0]}(\lambda), Y_1(\lambda), \ldots, Y_r(\lambda))$ follows from the descriptions in Proposition 2 and Lemma 3.1 and can be expressed, for $0 \leq y_k \leq b_k \leq 1, k = 1, \ldots, r$, as

$$e^{\lambda^\alpha}f_\alpha(s)\frac{\lambda^r}{[\Gamma(\frac{1-\alpha}{\alpha})]^r}\prod_{k=1}^r (1-b_k)^{\frac{1-\alpha}{\alpha}-1} \times e^{-\lambda s/\mathbf{y}_r^{1/\alpha}}\prod_{l=1}^r y_l^{-\frac{(r-l+1)}{\alpha}-1}, \qquad (13)$$

for $\mathbf{y}_r = \prod_{i=1}^r y_i$. Proposition 2 is verified by showing that integrating over $(y_1, \ldots, y_r)$ in (13) leads to (12). This is equivalent to showing that

$$\int_0^{b_1}\cdots\int_0^{b_r} e^{-\lambda s/\mathbf{y}_r^{1/\alpha}}\prod_{l=1}^r y_l^{-\frac{(r-l+1)}{\alpha}-1}dy_r\cdots dy_1 = \alpha^r\lambda^{-r}s^{-r}e^{-\lambda s/\mathbf{b}_r^{1/\alpha}}\prod_{i=1}^r b_i^{\frac{i-1}{\alpha}}.$$

which follows by elementary calculations involving the change of variable $v_i = y_i^{-1/\alpha}$, for $i = 1, \ldots, r$ and exponential integrals. Now, to establish Proposition 3, first note that since $\mathbb{P}(N_{T_\alpha}(\lambda) = 1|T_{\alpha,1} = s, B_1 = b_1) = \lambda s b_1^{-\frac{1}{\alpha}}e^{-\lambda s/b_1^{\frac{1}{\alpha}}}$, and $\mathbb{P}(N_{T_\alpha}(\lambda) = 1) = \alpha\lambda^\alpha e^{-\lambda^\alpha}$, the joint density of $B_1, T_{\alpha,1}|N_{T_\alpha}(\lambda) = 1$ can be expressed as

$$\frac{\lambda^{1-\alpha}}{\Gamma(\frac{1-\alpha}{\alpha})}b_1^{-\frac{1}{\alpha}}(1-b_1)^{\frac{1-\alpha}{\alpha}-1} \times e^{-\lambda s/b_1^{1/\alpha}}e^{\lambda^\alpha}f_\alpha(s). \qquad (14)$$

Hence, the conditional density of $B_1|N_{T_\alpha}(\lambda) = 1$ can be expressed as,

$$\frac{\lambda^{1-\alpha}}{\Gamma(\frac{1-\alpha}{\alpha})}b_1^{-\frac{1}{\alpha}}(1-b_1)^{\frac{1-\alpha}{\alpha}-1} \times e^{-\lambda^\alpha/b_1}e^{\lambda^\alpha}. \qquad (15)$$

which corresponds to $B_1^{[1]}(\lambda) \stackrel{d}{=} \lambda^\alpha/(\gamma_{\frac{1-\alpha}{\alpha}} + \lambda^\alpha)$, verifying statement (i) of Proposition 3. Refs. (14) and (15) show that $T_{\alpha,1}|N_{T_\alpha}(\lambda) = 1, B_1 = b_1$ is $f_\alpha^{[0]}(s|\lambda b_1^{-\frac{1}{\alpha}})$, which leads to $(P_{\ell,1})|N_{T_\alpha}(\lambda) = 1, B_1 = b_1$ having distribution $\mathbb{P}_\alpha^{[0]}(\lambda b_1^{-\frac{1}{\alpha}})$. This agrees with statement (ii) of Proposition 1, with $n = r = 1$. Using $\lambda^{\alpha}/B_1(\lambda) \stackrel{d}{=} \gamma_{\frac{1-\alpha}{\alpha}} + \lambda^{\alpha}$ and applying Proposition 2 starting with $(P_{\ell,1})|N_{T_{\alpha}}(\lambda) = 1, B_1 = b_1$ subject to (6) concludes the proof of Proposition 3.

Funding: This research was supported in part by grants RGC-GRF 16301521, 16300217 and 601712 217 of the Research Grants Council (RGC) of the Hong Kong SAR. This research also received funding 218 from the European Research Council (ERC) under the European Union's Horizon 2020 research 219 and innovation programme under grant agreement No. 817257.

Institutional Review Board Statement: Not applicable.

Informed Consent Statement: Not applicable.

Acknowledgments: This article is dedicated to Eugenio Regazzini on the occasion of his 75th birthday.

Conflicts of Interest: The author declares no conflict of interest.

References

1. Perman, M.; Pitman, J.; Yor, M. Size-biased sampling of Poisson point processes and excursions. *Probab. Theory Relat. Fields.* **1992**, *92*, 21–39. [CrossRef]
2. Pitman, J. Poisson-Kingman partitions. In *Science and Statistics: A Festschrift for Terry Speed*; Goldstein, D.R., Ed.; Institute of Mathematical Statistics: Hayward, CA, USA, 2003; pp. 1–34.
3. Pitman, J. Combinatorial Stochastic Processes. In *Lectures from the 32nd Summer School on Probability Theory Held in Saint-Flour, July 7–24. 2002. With a Foreword by Jean Picard. Lecture Notes in Mathematics, 1875*; Springer: Berlin/Heidelberg, Germany, 2006.
4. Pitman, J.; Yor, M. The two-parameter Poisson-Dirichlet distribution derived from a stable subordinator. *Ann. Probab.* **1997**, *25*, 855–900. [CrossRef]
5. Rembart, F.; Winkel, M. A binary embedding of the stable line-breaking construction. *arXiv* **2016**, arXiv:1611.02333.
6. Ho, M.-W.; James, L.F.; Lau, J.W. Gibbs Partitions, Riemann-Liouville Fractional Operators, Mittag-Leffler Functions, and Fragmentations derived from stable subordinators. *J. Appl. Prob.* **2021**, *58*, 314–334. [CrossRef]
7. Goldschmidt, C.; Haas, B. A line-breaking construction of the stable trees. *Electron. J. Probab.* **2015**, *20*, 1–24. [CrossRef]
8. Haas, B.; Miermont, G.; Pitman, J.; Winkel, M. Continuum tree asymptotics of discrete fragmentations and applications to phylogenetic models. *Ann. Probab.* **2008**, *36*, 1790–1837. [CrossRef]
9. James, L.F. Stick-breaking PG(α, ζ)-Generalized Gamma Processes. Unpublished manuscript. *arXiv* **2013**, arXiv:1308.6570.
10. Aldous, D. The continuum random tree. I. *Ann. Probab.* **1991**, *19*, 1–28. [CrossRef]
11. Aldous, D. The continuum random tree III. *Ann. Probab.* **1993**, *21*, 248–289. [CrossRef]
12. Móri, T.F. The maximum degree of the Barabási-Albert random tree. *Combin. Probab. Comput.* **2005**, *14*, 339–348. [CrossRef]
13. Peköz, E.; Röllin, A.; Ross, N. Generalized gamma approximation with rates for urns, walks and trees. *Ann. Probab.* **2016**, *44*, 1776–1816. [CrossRef]
14. Peköz, E.; Röllin, A.; Ross, N. Joint degree distributions of preferential attachment random graphs. *Adv. Appl. Probab.* **2017**, *49*, 368–387. [CrossRef]
15. van der Hofstad, R. *Random Graphs and Complex Networks*; Cambridge University Press, New York, NY, USA, 2016; Volume I.
16. Bertoin, J. *Random Fragmentation and Coagulation Processes*; Cambridge University Press: Cambridge, UK, 2006.
17. Dong, R.; Goldschmidt, C.; Martin, J. Coagulation-fragmentation duality, Poisson-Dirichlet distributions and random recursive trees. *Ann. Appl. Probab.* **2006**, *16*, 1733–1750. [CrossRef]
18. Pitman, J. Mixed Poisson and negative binomial models for clustering and species sampling. Unpublished manuscript. 2017.
19. James, L.F. Stick-breaking Pitman-Yor processes given the species sampling size. *arXiv* **2019**, arXiv:1908.07186.
20. Favaro, S.; James, L.F. A note on nonparametric inference for species variety with Gibbs-type priors. *Electron. J. Statist.* **2015**, *9*, 2884–2902. [CrossRef]
21. Pitman, J. *Some Developments of the Blackwell-MacQueen urn Scheme*; Statistics, Probability and Game Theory, IMS Lecture Notes Monogr. Ser. 30; Institute of Mathematical Statistics: Hayward, CA, USA, 1996; pp. 245–267.
22. Ishwaran, H.; James, L.F. Gibbs sampling methods for stick-breaking priors. *J. Am. Statist. Assoc.* **2001**, *96*, 161–173. [CrossRef]
23. Ferguson, T.S. A Bayesian analysis of some nonparametric problems. *Ann. Statist.* **1973**, *1*, 209–230. [CrossRef]
24. James, L.F. Lamperti type laws. *Ann. Appl. Probab.* **2010**, *20*, 1303–1340. [CrossRef]
25. James, L.F.; Lijoi, A.; Prünster, I. Posterior analysis for normalized random measures with independent increments. *Scand. J. Stat.* **2009**, *36*, 76–97. [CrossRef]
26. Zhou, M.; Favaro, S.; Walker, S.G. Frequency of Frequencies Distributions and Size-Dependent Exchangeable Random Partitions. *J. Am. Statist. Assoc.* **2017**, *112*, 1623–1635. [CrossRef]
27. McCloskey, J.W. A Model for the Distribution of Individuals by Species in an Environment. Ph.D. Thesis, Michigan State University, East Lansing, MI, USA, 1965.
28. Regazzini, E.; Lijoi, A.; Prünster, I. Distributional results for means of normalized random measures with independent increments. *Ann. Statist.* **2003**, *31*, 560–585. [CrossRef]

29. Lijoi, A.; Prünster, I. Models Beyond the Dirichlet Process. In *Bayesian Nonparametrics*; Hjort, N.L., Holmes, C., Müller, P., Walker, S., Eds.; Cambridge University Press: Cambridge, UK, 2010; pp. 80–136.
30. Wood, F.; Gasthaus, J.; Archambeau, C.; James, L.F.; Teh, Y.W. The Sequence Memoizer. *Commun. ACM* **2011**, *54*, 91–98. [CrossRef]

Article

Asymptotic Efficiency of Point Estimators in Bayesian Predictive Inference

Emanuele Dolera [1,2]

1 Department of Mathematics, University of Pavia, Via Adolfo Ferrata 5, 27100 Pavia, Italy; emanuele.dolera@unipv.it
2 Collegio Carlo Alberto, Piazza V. Arbarello 8, 10134 Torino, Italy

Abstract: The point estimation problems that emerge in Bayesian predictive inference are concerned with random quantities which depend on both observable and non-observable variables. Intuition suggests splitting such problems into two phases, the former relying on estimation of the random parameter of the model, the latter concerning estimation of the original quantity from the distinguished element of the statistical model obtained by plug-in of the estimated parameter in the place of the random parameter. This paper discusses both phases within a decision theoretic framework. As a main result, a non-standard loss function on the space of parameters, given in terms of a Wasserstein distance, is proposed to carry out the first phase. Finally, the asymptotic efficiency of the entire procedure is discussed.

Keywords: asymptotic efficiency; bayesian predictive inference; compatibility equations; decision theory; de Finetti's representation theorem; exchangeability; Wasserstein distance

MSC: 62A01; 62C10; 62C12; 60F17

1. Introduction

This paper carries on a project—conceived by Eugenio Regazzini some years ago, and partially developed in collaboration with Donato M. Cifarelli—which aims at proving why and how some classical, frequentist algorithms from the theory of point estimation can be justified, under some regularity assumptions, within the Bayesian framework. See [1–4]. This project was inspired, in turn, by the works and the thoughts of Bruno de Finetti about the foundation of statistical inference, substantially based on the following principles.

1. De Finetti's vision of statistics is grounded on the irrefutable fact that the Bayesian standpoint—intended as the use of basic tools of probability theory and, especially, of conditional distributions—becomes a necessity for those who intend statistical inference as the utilization of observed data to update their original beliefs about other quantities of interest, not yet observed. See [5,6].
2. Rigorous notions of point estimation and optimality of an estimator can be achieved only within a decision-theoretic framework (see, e.g., [7]), at least if we admit all estimators into competition and disregard distinguished restrictions such as unbiasedness or equivariance. In turn, decision theory proves to be genuinely Bayesian, thanks to a well-known result by Abraham Wald. See [8] [Chapter 4].
3. At least from a mathematical stance, the existence of the prior distribution can be drawn from various representation theorems which, by pertaining to the more basic act of modeling incoming information, stand before the problem of point estimation. The most luminous example is the celebrated de Finetti representation theorem for exchangeable observations. See [6,9] and, for a predictive approach [10,11].

Indeed, these principles do not force the assessment of a specific prior distribution, but just lead the statistician to take cognizance that some prior has, in any case, to exist.

This fact agrees with de Finetti's indication to keep the concepts of "Bayesian standpoint" and "Bayesian techniques" as distinguished. See also [12].

Despite their robust logical coherence, orthodox Bayesian solutions to inferential problems suffer two main drawbacks on the practical, operational side, which may limit their use. On the one hand, it is rarely the case that a prior distribution is fully specified due to a lack of prior information, this phenomenon even being amplified by the choice of complex statistical models (e.g., of nonparametric type). On the other hand, the numerical tractability of the Bayesian solutions often proves to be a serious hurdle, especially in the presence of large datasets. For example, it suffices to mention those algorithms from Bayesian nonparametrics that involve tools from combinatorics (like permutations or set/integer partitions) having exponential algorithmic complexity. See, e.g., [13]. Finally, the implicit nature of the notion of *Bayesian estimator*, although conceptually useful, makes it hard to employ in practical problems, especially in combination with non-quadratic loss functions, even if noteworthy progress has been achieved from the numerical side in the last decade. All these issues still pervade modern statistical literature while, historically, they have paved the way firstly to the "Fisherian revolution" and then to more recent techniques such as empirical Bayes and objective Bayes methods. The ultimate result has been a proliferation of many *ad hoc* algorithms, often of limited conceptual value, that provide focused and operational solutions to very specific problems.

Aware of this trend, Eugenio Regazzini conceived his project with the aims of: reframing the algorithms of modern statistics—especially those obtained by frequentist techniques—within the Bayesian theory as summarized in points 1–3 above, showing whether they can be re-interpreted as good approximations of Bayesian algorithms. The rationale is that orthodox Bayesian theory could be open to accept even non-Bayesian solutions (hence, suboptimal ones if seen "through the glass of the prior") as long as such solutions prove to be more operational than the Bayesian ones and, above all, *asymptotically almost efficient*, in the Bayesian sense. This concept means that, for a fixed prior, the Bayesian risk function evaluated at the non-Bayesian estimator is approximately equal to the overall minimum of such risk function (achieved when evaluated at the Bayesian estimator), the error of approximation going to zero as the sample size increases. Of course, these goals can be carried out after providing quantitative estimates for the risk function, as done, for example, in some decision-theoretic work on the empirical Bayes approach to inference. See, e.g., the seminal work [14]. Indeed, Regazzini's project has much in common with the empirical Bayes theory, although the former strictly remains on the "orthodox Bayesian main way" whilst the latter mixes Bayesian and frequentist techniques. As to more practical results, an archetype of Regazzini's line of reasoning can be found in a previous statement from [15] [Section 5] which proves that the maximum likelihood estimator (MLE)—obtained in the classical context of n i.i.d. observations, driven by a regular parametric model—has the same Bayesian efficiency (coinciding with the *mean square error*, in this case) as the Bayesian estimator up to $O(1/n)$-terms, provided that the prior is smooth enough. Another example can be found in [16] where the authors, while dealing with species sampling problems, rediscover the so-called Good–Turing estimator for the probability of finding a new species (which is obtained via empirical Bayes arguments) within the Bayesian nonparametric setting described in [17]. Other examples are contained in [2,4]. In any case, Regazzini's project is not only a matter of "rigorously justifying" a given algorithm, but rather of logically conceiving an estimation problem from the beginning to the end by quantifying coherent degrees of approximation in terms of the Bayesian risk or, more generally, in terms of *speed of shrinkage of the posterior distribution* with respect to distances on the space of probability measures, these goals being proved *uniformly with respect to an entire class of priors*. Hence, this plan of action is conceptually antipodal to that of (nowadays called) "Bayesian consistency", i.e., to justify a Bayesian algorithm from the point of view of classical statistics.

1.1. Main Contributions and General Strategy

In this paper, we pursue Regazzini's project by considering some *predictive problems* where the quantity $U_{n,m}$ to be estimated depends explicitly on new (hitherto unobserved) variables $X_{n+1}, \ldots, X_{n+m}$, possibly besides the original sample variables $X_1, \ldots, X_n$ and an unobservable parameter T. Thus, $U_{n,m} = u_{n,m}(X_{n+1}, \ldots, X_{n+m}; X_1, \ldots, X_n; T)$. For simplicity, we confine ourselves to the simplest case in which both $(X_1, \ldots, X_n)$ and $(X_{n+1}, \ldots, X_{n+m})$ are segments of a whole sequence $\{X_i\}_{i \geq 1}$ of exchangeable $\mathbb{X}$-valued random variables, while T is a random parameter that makes the X_i's conditionally i.i.d. with a common distribution depending on T, in accordance with de Finetti's representation theorem. From the statistical point of view, the exchangeability assumption just reveals a supposed homogeneity between the observable quantities while, from a mathematical point of view, it simply states that the joint distribution of any k-subset of the X_i's depends only on k and not on the specific k-subset, for any $k \in \mathbb{N}$. Thus, we are setting our estimation problem within an orthodox Bayesian framework where, independently of the fact that we are able or not to precisely assess the prior distribution, such a prior has to exist for mere mathematical reasons. This solid theoretical background provides all the elements to logically formulate the original predictive estimation problem as the following decision-theoretic question: find

$$\hat{U}_{n,m} = \operatorname{Argmin}_Z \mathbb{E}[\mathcal{L}_{\mathbb{U}}(U_{n,m}, Z)], \qquad (1)$$

where: $\mathcal{L}_{\mathbb{U}}$ is a suitable loss function on the space $\mathbb{U}$ in which $U_{n,m}$ takes its values; Z runs over the space of all $\mathbb{U}$-valued, $\sigma(X_1, \ldots, X_n)$-measurable random variables; the expectation is taken with respect to the joint distribution of $(X_1, \ldots, X_{n+m})$ and T. It is remarkable that the same estimation problem would have been meaningless in classical (Fisherian) statistics, which can solely consider the estimation of (a function of) the parameter, and not of random quantities. Now, the solution displayed in (1) depends of course on the prior and it is the optimal one when seen, in terms of the Bayesian risk, "with the glass of that prior". However, the above-mentioned difficulties about the assessment of a specific prior can diminish the practical (but not the conceptual) value of this solution, in the sense that it could prove to be non-operational in the case of a lack of prior information. Sometimes, when the prior is known up to further unknown parameters, another estimation problem is needed.

Our research is then focused on formalizing a general strategy aimed at producing, under regularity conditions, alternative estimators $U^*_{n,m}$ which prove to be asymptotically nearly optimal (as specified above), uniformly with respect to any prior in some class. More precisely, for any fixed prior in that class, we aim at proving the validity of the asymptotic expansions (as $n \to +\infty$),

$$\mathbb{E}[\mathcal{L}_{\mathbb{U}}(U_{n,m}, \hat{U}_{n,m})] = \hat{R}_{0,m} + \frac{1}{n}\hat{R}_{1,m} + o\left(\frac{1}{n}\right) \qquad (2)$$

$$\mathbb{E}[\mathcal{L}_{\mathbb{U}}(U_{n,m}, U^*_{n,m})] = R^*_{0,m} + \frac{1}{n}R^*_{1,m} + o\left(\frac{1}{n}\right), \qquad (3)$$

along with $\hat{R}_{i,m} = R^*_{i,m}$ for $i = 0, 1$, where $\hat{U}_{n,m}$ is the same as in (1). This is exactly the content of Theorem 5.1 and Corollary 5.1 in [15], which deal with the case where: $U_{n,m} = T$ (estimation of the parameter of the model), so that $\mathbb{U}$ coincides with the parameter space $\Theta \subseteq \mathbb{R}$; $\mathcal{L}_{\mathbb{U}}$ is the quadratic loss function, so that the risk function coincides with the mean square error; $\hat{U}_{n,m} = \mathbb{E}[T \mid X_1, \ldots, X_n]$ is the Bayesian estimator with respect to $\mathcal{L}_{\mathbb{U}}$; $U^*_{n,m}$ coincides with the MLE; $\hat{R}_{0,m} = R^*_{0,m} = 0$ and $\hat{R}_{1,m} = R^*_{1,m} = \int_\Theta [\mathrm{I}(\theta)]^{-1} \pi(\mathrm{d}\theta)$, I denoting the Fisher information of the model and π being any prior on Θ with positive and sufficiently smooth density (with respect to the Lebesgue measure). Moving to truly predictive problems, the main operational solutions come from the empirical Bayes theory, which shares Equation (1) with the approach we are going to present. However, the empirical Bayes theory very soon leaves the "Bayesian main way" by bringing some sort of

Law of Large Numbers into the game, in order to replace the unknown quantities (usually, the prior itself). Here, on the contrary, we pursue Regazzini's project by proposing a new method that remains on the Bayesian main way. It consists of the following six steps.

Step 1. Reformulate problem (1) into another (orthodox Bayesian) estimation problem about T, the random parameter of the model. Roughly speaking, start from the following de Finetti representation:

$$\mathbb{P}[X_1 \in A_1, \ldots, X_k \in A_k \mid T = \theta] = \mu^{\otimes k}(A_1 \times \ldots \times A_k \mid \theta) := \prod_{i=1}^{k} \mu(A_i \mid \theta), \quad (4)$$

valid for all $k \in \mathbb{N}$, Borel sets $A_1, \ldots, A_k$, $\theta \in \Theta$, and some probability kernel $\mu(\cdot \mid \cdot)$, which coincides with the statistical model for the single observation. Then, consider the following estimation problem: find

$$\hat{T}_{n,m} = \text{Argmin}_W \mathbb{E}\left[\mathcal{L}_{\Theta,(X_1,\ldots,X_n)}(T, W)\right], \quad (5)$$

where: $\mathcal{L}_{\Theta,(X_1,\ldots,X_n)}$ is a suitable loss function on Θ; W runs over the space of all Θ-valued, $\sigma(X_1, \ldots, X_n)$-measurable random variables; the expectation is taken with respect to the joint distribution of $(X_1, \ldots, X_n)$ and T. The explicit definition of $\mathcal{L}_{\Theta,(X_1,\ldots,X_n)}$ is given in terms of a Wasserstein distance, as follows:

$$\mathcal{L}_{\Theta,(x_1,\ldots,x_n)}(\theta, \tau) = \inf_{\Gamma} \int_{\mathbb{U}^2} \mathcal{L}_{\mathbb{U}}(u,v) \Gamma(dudv), \quad (6)$$

where Γ runs over the Fréchet class of all probability measures on $\mathbb{U}^2$ with marginals $\gamma_{\theta,(x_1,\ldots,x_n)}$ and $\gamma_{\tau,(x_1,\ldots,x_n)}$, respectively, and $\gamma_{\theta,(x_1,\ldots,x_n)}$ stands for the pull-back measure $\mu^{\otimes m}(\cdot \mid \theta) \circ u_{n,m}(\cdot; x_1, \ldots, x_n; \theta)^{-1}$ on $\mathbb{U}$.

Step 2. After getting the estimator $\hat{T}_{n,m}$ from (5), consider estimators $U^*_{n,m}$ that satisfy the following approximated version of problem (1): find

$$U^*_{n,m} = \text{Argmin}_Z \int_{\mathbb{X}^m} \mathcal{L}_{\mathbb{U}}\left(u_{n,m}(y_1, \ldots, y_m; X_1, \ldots, X_n; \hat{T}_{n,m}), Z\right) \mu^{\otimes m}(dy_1 \ldots dy_m \mid \hat{T}_{n,m}), \quad (7)$$

where Z runs over the space of all $\mathbb{U}$-valued, $\sigma(X_1, \ldots, X_n)$-measurable random variables.

Step 3. For the estimators $\hat{U}_{n,m}$ and $U^*_{n,m}$ that solve (1) and (7) respectively, prove that (2) and (3) hold along with $\hat{R}_{i,m} = R^*_{i,m}$ for $i = 0, 1$. This entails the asymptotic almost efficiency of $U^*_{n,m}$, which it is still a prior-dependent estimator. In any case, this step is crucial to show that the loss function $\mathcal{L}_{\Theta,(x_1,\ldots,x_n)}$ given in (6) is "Bayesianly well-conceived", that is, in harmony with the original aim displayed in (1).

Step 4. Identities (2) and (3) provides conditions on the statistical model $\mu(\cdot \mid \cdot)$ that possibly allows the existence of some *prior-free* estimator $\tilde{T}_{n,m}$ of T which turns out to be asymptotically almost efficient, with respect to the same risk function as that displayed on the right-hand side of (5). More precisely, this fact consists of proving the validity of the following identities (as $n \to +\infty$)

$$\mathbb{E}\left[\mathcal{L}_{\Theta,(X_1,\ldots,X_n)}(T, \hat{T}_{n,m})\right] = \hat{\rho}_{0,m} + \frac{1}{n}\hat{\rho}_{1,m} + o\left(\frac{1}{n}\right) \quad (8)$$

$$\mathbb{E}\left[\mathcal{L}_{\Theta,(X_1,\ldots,X_n)}(T, \tilde{T}_{n,m})\right] = \tilde{\rho}_{0,m} + \frac{1}{n}\tilde{\rho}_{1,m} + o\left(\frac{1}{n}\right), \quad (9)$$

along with $\hat{\rho}_{i,m} = \tilde{\rho}_{i,m}$ for $i = 0, 1$, where $\hat{T}_{n,m}$ is the same as in (5), for all prior distributions in a given class.

Step 5. After getting estimators $\tilde{T}_{n,m}$ as in **Step 4**, consider the *prior-free* estimators $\tilde{U}_{n,m}$ satisfying the analogous minimization problem as in (7), with $\hat{T}_{n,m}$ replaced by $\tilde{T}_{n,m}$.

Step 6. For any estimator $\tilde{U}_{n,m}$ found as in **Step 5**, prove the validity of the following identity (as $n \to +\infty$):

$$\mathbb{E}\left[\mathcal{L}_{\mathbb{U}}(U_{n,m}, \tilde{U}_{n,m})\right] = \tilde{R}_{0,m} + \frac{1}{n}\tilde{R}_{1,m} + o\left(\frac{1}{n}\right), \tag{10}$$

along with $\hat{R}_{i,m} = \tilde{R}_{i,m}$ for $i = 0, 1$, where the $\hat{R}_{i,m}$'s are the same as in (2), for all prior distributions in the same class as specified in **Step 4**. This last step shows why and how the frequentist (i.e., prior-free) estimator $\tilde{U}_{n,m}$ can be used, within the orthodox Bayesian framework, as a good approximation of the Bayesian estimator $\hat{U}_{n,m}$. This is particularly remarkable at least in two cases that do not exclude each other: when the estimator $\tilde{T}_{n,m}$ obtained from **Step 4** is much simpler and numerically manageable than $\hat{T}_{n,m}$; when prior information is sufficient to characterize only a class of priors, but not a specific element of it.

This plan of action obeys the following principles:

(A) The loss function $\mathcal{L}_{\Theta,(x_1,\ldots,x_n)}$ on Θ is harmoniously coordinated with the original choice of the loss function $\mathcal{L}_{\mathbb{U}}$ on $\mathbb{U}$. This principle is much aligned with de Finetti's thought (see [18]), since it remarks on the more concrete nature of the space $\mathbb{U}$ compared with the space Θ which is, in principle, only a set of labels. Hence, it is much more reasonable to firstly metrize the space $\mathbb{U}$ and then the space Θ accordingly (as in (6)), rather than directly metrize Θ—even without taking account of the original predictive aim.

(B) The Bayesian risk function associated with both $U^*_{n,m}$ and $\tilde{U}_{n,m}$ can be bounded from above by the sum of two quantities: the former taking account of the error in estimating T, the latter reflecting the fact that we are estimating both $U^*_{n,m}$ and $\tilde{U}_{n,m}$ from an "estimated distribution".

The former principle, whose formalization constitutes the main novelty of this work, is concerned with the geometrical structure of the space of the parameters Θ. This is what we call a *relativistic principle* in point estimation theory: the goal of estimating a random quantity that depends on the observations (possibly besides the parameter) yields a modification of the geometry of Θ, to be now thought of as a curved space according to a non-trivial geometry. Of course, this modified geometry entails a coordinated notion of mean square error, now referred to the Riemannian geodesic distance. The term relativistic principle just hints at the original main principle of General Relativity Theory according to which the presence of a massive body modifies the geometry of the physical surrounding space, by means of the well-known Einstein tensor equations. These equations formalize a sort of compatibility between the physical and the geometric structures of the space. Thus, the identities (43) and (44), as stated in Section 3 to properly characterize the (Riemannian) metric on Θ, we will call *compatibility equations*. Actually, the idea of metrizing the parameter space Θ in a non standard way is well-known since the pioneering paper [19] by Radhakrishna Rao, and has received so much attention in the statistical literature to give birth to a fertile branch called *Information Geometry*. See, e.g., [20]. In particular, the concepts of efficiency, unbiasedness, Cramér–Rao lower bounds, Rao–Blackwell and Lehmann–Scheffé theorems are by far best-understood in this non-standard (i.e., non-Euclidean) setting. See [21]. In any case, to the best of our knowledge, this is the first work which connects the use of a non-standard geometric setting on Θ with predictive estimation problems—even if some hints can be drawn from [22]. In our opinion, the lack of awareness about the aforesaid relativistic principle, combined with an abuse of the quadratic loss function on Θ, has produced a lot of actually sub-efficient algorithms, most of which focused on the estimation of certain probabilities, or of nonparametric objects. In these cases, the efficiency of the ensuing estimators is created artificially through a misuse of the quadratic loss, and it proves to be drastically downsized whenever these estimators are evaluated by means of other, more concrete loss functions which take account (as in (6)) of the natural geometry of the spaces of really observable quantities. To get an idea of this phenomenon, see the discussion about Robbins' estimators in Section 4.4 below.

1.2. Organization of the Paper

We conclude the introduction by summarizing the main results of the paper, which are threefold. The first block of results, including Theorem 1, Proposition 1 and Lemma 1 in Section 2.2, concerns some refinement of de Finetti's Law of Large Numbers for the log-likelihood process. The second block of theoretical results, developed in Section 3, contains:

(i)   Proposition 2, which shows how to bound from above the Bayesian risk of any estimator of $U_{n,m}$ by using the Wasserstein distance;
(ii) Proposition 3, which explains how to use the Laplace method of the approximation of integrals to get asymptotic expansions of the Bayesian risk functions;
(iii) the formulation of the compatibility Equations (43) and (44);
(iv) the proof of the "asymptotic almost efficiency" of the estimator $U_{n,m}^*$ obtained in **Step 2**, via verification of identities (2) and (3);
(v)   the successful completion of **Step 6**, that is, the proof of the "asymptotic almost efficiency" of estimators $\tilde{U}_{n,m}$ obtained in **Step 5**, via verification of identity (10).

The last block of results, contained in Section 4, consists of explicit verifications of the compatibility equations for some simple statistical models (Sections 4.1–4.3), and also the adaptation of our plan of action to the same Poisson-mixture model used by Herbert Robbins in [23] to illustrate his empirical Bayes approach to predictive inference (Section 4.4). Finally, all the proofs of the theoretical results are deferred to Section 5, while some conclusions and future developments are hinted at in Section 6.

2. Technical Preliminaries

We begin by rigorously fixing the mathematical setting, split into two subsections. The former will contain a very general framework which will serve to give a precise meaning to the questions presented in the Introduction and to state in full generality one of the main results, that is, Proposition 2 in Section 3. In fact, this statement will include some inequalities that, by carrying out the goal described in point (B) of the Introduction will constitute the starting point for all the results presented in Section 3. The second subsection will deal with a simplification of the original setting—essentially based on additional regularity conditions for the spaces $\mathbb{U}$ and Θ and for the statistical model $\mu(\cdot|\cdot)$—aimed at introducing the novel compatibility equations without too many technicalities.

2.1. The General Framework

Let $(\mathbb{X}, \mathscr{X})$ and $(\Theta, \mathscr{T})$ be standard Borel spaces called *sample space* (for any single observation) and *parameter space*, respectively. Consider a sequence $\{X_i\}_{i\geq 1}$ of $\mathbb{X}$-valued random variables (r.v.'s, from now on) along with another Θ-valued r.v. T, all the X_i's and T being defined on a suitable probability space $(\Omega, \mathscr{F}, \mathbb{P})$. Assume that (4) holds for all $k \in \mathbb{N}$, $A_1, \ldots, A_k \in \mathscr{X}$ and $\theta \in \Theta$ with some given probability kernel $\mu(\cdot|\cdot) : \mathscr{X} \times \Theta \to [0,1]$, called *statistical model* (for any single observation). The validity of (4) entails that the X_i's are *exchangeable* and that

$$\mathbb{P}[X_1 \in A_1, \ldots, X_k \in A_k] = \int_\Theta \mu^{\otimes k}(A_1 \times \ldots \times A_k \mid \theta)\pi(d\theta) =: \alpha_k(A_1 \times \ldots \times A_k) \quad (11)$$

holds for all $k \in \mathbb{N}$ and $A_1, \ldots, A_k \in \mathscr{X}$ with some given probability measure (p.m.) π on $(\Theta, \mathscr{T})$ called *prior distribution*. Identity (11) uniquely characterizes the p.m. α_k on $(\mathbb{X}^k, \mathscr{X}^k)$ for any $k \in \mathbb{N}$, this p.m. being called *law of k-observations*, where $\mathbb{X}^k$ ($\mathscr{X}^k$, respectively) denotes the k-fold cartesian product (σ-algebra product, respectively) of k copies of $\mathbb{X}$ ($\mathscr{X}$, respectively). Moreover, let

$$\pi_k(B \mid x_1, \ldots, x_k) := \mathbb{P}[T \in B \mid X_1 = x_1, \ldots, X_k = x_k]$$
$$\beta_k(A \mid x_1, \ldots, x_k) := \mathbb{P}[X_{k+1} \in A \mid X_1 = x_1, \ldots, X_k = x_k]$$

be two probability kernels, with $\pi_k(\cdot|\cdot): \mathscr{T} \times \mathbb{X}^k \to [0,1]$ and $\beta_k(\cdot|\cdot): \mathscr{X} \times \mathbb{X}^k \to [0,1]$, defined as respective solutions of the following disintegration problems

$$\mathbb{P}[X_1 \in A_1, \ldots, X_k \in A_k, T \in B] = \int_{A_1 \times \ldots \times A_k} \pi_k(B \mid x_1, \ldots, x_k) \alpha_k(dx_1 \ldots dx_k)$$

$$\mathbb{P}[X_1 \in A_1, \ldots, X_k \in A_k, X_{k+1} \in A] = \int_{A_1 \times \ldots \times A_k} \beta_k(A \mid x_1, \ldots, x_k) \alpha_k(dx_1 \ldots dx_k)$$

for any $k \in \mathbb{N}$, $A_1, \ldots, A_k, A \in \mathscr{X}$ and $B \in \mathscr{T}$. The probability kernels $\pi_k(\cdot|\cdot)$ and $\beta_k(\cdot|\cdot)$ are called *posterior distribution* and *predictive distribution*, respectively.

Let $(\mathbb{U}, d_\mathbb{U})$ be a Polish metric space and, for fixed $n, m \in \mathbb{N}$, let $u_{n,m}: \mathbb{X}^m \times \mathbb{X}^n \times \Theta \to \mathbb{U}$ be a measurable map. Let $U_{n,m} := u_{n,m}(X_{n+1}, \ldots, X_{n+m}; X_1, \ldots, X_n; T)$ be the random quantity to be estimated with respect to the loss function $\mathcal{L}_\mathbb{U}(u,v) := d_\mathbb{U}^2(u,v)$. Now, recall the notion of *barycenter* (also known as Fréchet mean) of a given p.m.. Let $(\mathbb{S}, d_\mathbb{S})$ be a Polish metric space, endowed with its Borel σ-algebra $\mathscr{B}(\mathbb{S})$. Given a p.m. μ on $(\mathbb{S}, \mathscr{B}(\mathbb{S}))$, define

$$\mathrm{Bary}_\mathbb{S}[\mu; d_\mathbb{S}] := \mathrm{Argmin}_{y \in \mathbb{S}} \int_\mathbb{S} d_\mathbb{S}^2(x,y) \mu(dx)$$

provided that μ has finite second moment ($\mu \in \mathcal{P}_2(\mathbb{S}, d_\mathbb{S})$, in symbols) and that at least one minimum point exists. See [24–26] for results on existence, uniqueness and some characterizations of barycenters. Then, put

$$\rho_{n,m}(C \mid x_1, \ldots, x_n) := \mathbb{P}[U_{n,m} \in C \mid X_1 = x_1, \ldots, X_n = x_n],$$

meaning that $\rho_{n,m}(\cdot|\cdot): \mathscr{B}(\mathbb{U}) \times \mathbb{X}^n \to [0,1]$ is a probability kernel that solves the disintegration problem

$$\mathbb{P}[X_1 \in A_1, \ldots, X_n \in A_n, U_{n,m} \in C] = \int_{A_1 \times \ldots \times A_n} \rho_{n,m}(C \mid x_1, \ldots, x_n) \alpha_k(dx_1 \ldots dx_k)$$

for any $A_1, \ldots, A_n \in \mathscr{X}$ and $C \in \mathscr{B}(\mathbb{U})$. If $\mathbb{E}[d_\mathbb{U}^2(U_{n,m}, u_0)] < +\infty$ for some $u_0 \in \mathbb{U}$ and $\mathrm{Bary}_\mathbb{U}[\rho_{n,m}(\cdot \mid x_1, \ldots, x_n); d_\mathbb{U}]$ exists uniquely for α_n-almost all $(x_1, \ldots, x_n)$, then

$$\hat{U}_{n,m} = \mathrm{Bary}_\mathbb{U}[\rho_{n,m}(\cdot \mid X_1, \ldots, X_n); d_\mathbb{U}] \tag{12}$$

solves the minimization problem (1). To give an analogous formalization to the minimization problem (7), define

$$\gamma_{\theta,(x_1,\ldots,x_n)}(C) := \mu^{\otimes m}\Big(\{(y_1,\ldots,y_m) \in \mathbb{X}^m \mid u_{n,m}(y_1,\ldots,y_m; x_1,\ldots,x_n;\theta) \in C\} \Big| \theta\Big)$$

for any $\theta \in \Theta$, $(x_1, \ldots, x_n) \in \mathbb{X}^n$ and $C \in \mathscr{B}(\mathbb{U})$. Again, if $\gamma_{\theta,(x_1,\ldots,x_n)} \in \mathcal{P}_2(\mathbb{U}, d_\mathbb{U})$ and $\mathrm{Bary}_\mathbb{U}[\gamma_{\theta,(x_1,\ldots,x_n)}(\cdot); d_\mathbb{U}]$ exists uniquely for any $\theta \in \Theta$ and α_n-almost all $(x_1, \ldots, x_n)$, then

$$U_{n,m}^* = \mathrm{Bary}_\mathbb{U}[\gamma_{\hat{T}_{n,m},(X_1,\ldots,X_n)}; d_\mathbb{U}] \tag{13}$$

solves the minimization problem (7). By the way, notice that a combination of de Finetti's representation theorem with basic properties of conditional distributions entails that

$$\rho_{n,m}(C \mid x_1, \ldots, x_n) = \int_\Theta \gamma_{\theta,(x_1,\ldots,x_n)}(C) \pi_n(d\theta \mid x_1, \ldots, x_n) \tag{14}$$

for α_n-almost all $(x_1, \ldots, x_n)$. It remains to formalize the minimization problem (5). If $\gamma_{\theta,(x_1,\ldots,x_n)}, \gamma_{\tau,(x_1,\ldots,x_n)} \in \mathcal{P}_2(\mathbb{U}, d_\mathbb{U})$, then the loss function in (6) satisfies

$$\mathcal{L}_{\Theta,(x_1,\ldots,x_n)}(\theta, \tau) = \mathcal{W}_\mathbb{U}^2\Big(\gamma_{\theta,(x_1,\ldots,x_n)}; \gamma_{\tau,(x_1,\ldots,x_n)}\Big),$$

where $\mathcal{W}_\mathbb{U}$ denotes the 2-Wasserstein distance on $\mathcal{P}_2(\mathbb{U}, d_\mathbb{U})$. See [27] [Chapters 6–7] for more information on the Wasserstein distance. Therefore, if $\pi_n(\cdot \mid x_1, \ldots, x_n) \in \mathcal{P}_2(\Theta, \mathcal{L}^{1/2}_{\Theta,(x_1,\ldots,x_n)})$ and $\mathrm{Bary}_\Theta[\pi_n(\cdot \mid x_1, \ldots, x_n); \mathcal{L}^{1/2}_{\Theta,(x_1,\ldots,x_n)}]$ exists uniquely for α_n-almost all $(x_1, \ldots, x_n)$, then

$$\hat{T}_{n,m} = \mathrm{Bary}_\Theta\left[\pi_n(\cdot \mid X_1, \ldots, X_n); \mathcal{L}^{1/2}_{\Theta,(X_1,\ldots,X_n)}\right] \tag{15}$$

solves the minimization problem (5).

To conclude, it remains to formalize the definition of various Bayesian risk functions, that will appear in the formulation of the main results. For any estimator $U^\dagger_{n,m} = u^\dagger_{n,m}(X_1, \ldots, X_n)$ of $U_{n,m}$, obtained with a measurable $u^\dagger_{n,m} : \mathbb{X}^n \to \mathbb{U}$, put

$$\mathfrak{R}_\mathbb{U}[U^\dagger_{n,m}] := \mathbb{E}\left[\mathcal{L}_\mathbb{U}(U_{n,m}, U^\dagger_{n,m})\right]$$

$$= \int_\Theta \int_{\mathbb{X}^{n+m}} \mathcal{L}_\mathbb{U}\left(u_{n,m}(\mathbf{y}; \mathbf{x}; \theta), u^\dagger_{n,m}(\mathbf{x})\right) \mu^{\otimes n+m}(\mathbf{dydx} \mid \theta) \pi(d\theta)$$

$$= \int_{\mathbb{X}^n} \int_\Theta \int_{\mathbb{X}^m} \mathcal{L}_\mathbb{U}\left(u_{n,m}(\mathbf{y}; \mathbf{x}; \theta), u^\dagger_{n,m}(\mathbf{x})\right) \mu^{\otimes m}(\mathbf{dy} \mid \theta) \pi_n(d\theta \mid \mathbf{x}) \alpha_n(\mathbf{dx}) \tag{16}$$

provided that the integrals are finite. Here and throughout, the bold symbols $\mathbf{x}, \mathbf{y}$ are just short-hands to denote the vectors $(x_1, \ldots, x_n)$ and $(y_1, \ldots, y_m)$, respectively. Analogously, for any estimator $T^\dagger_{n,m} = t^\dagger_{n,m}(X_1, \ldots, X_n)$ of T, obtained with a measurable $t^\dagger_{n,m} : \mathbb{X}^n \to \Theta$, put

$$\mathfrak{R}_\Theta[T^\dagger_{n,m}] := \mathbb{E}\left[\mathcal{L}_{\Theta,(X_1,\ldots,X_n)}(T, T^\dagger_{n,m})\right]$$

$$= \int_\Theta \int_{\mathbb{X}^n} \mathcal{L}_{\Theta,\mathbf{x}}\left(\theta, t^\dagger_{n,m}(\mathbf{x})\right) \mu^{\otimes n}(\mathbf{dx} \mid \theta) \pi(d\theta)$$

$$= \int_{\mathbb{X}^n} \int_\Theta \mathcal{L}_{\Theta,\mathbf{x}}\left(\theta, t^\dagger_{n,m}(\mathbf{x})\right) \pi_n(d\theta \mid \mathbf{x}) \alpha_n(\mathbf{dx}) \tag{17}$$

provided that the integrals are finite.

2.2. The Simplified Framework

Start by assuming that $\mathbb{U} = \mathbb{R}$ and $\mathcal{L}_\mathbb{U}(u, v) = |u - v|^2$. Then, restrict the attention to those predictive problems in which the quantity to be estimated depends only on the new observations $X_{n+1}, \ldots, X_{n+m}$ and on the random parameter T, but not on the observable variables $X_1, \ldots, X_n$. This restriction is actually non-conceptual, and it is made only to diminish the mathematical complexity of the ensuing asymptotic expansions (valid as $n \to +\infty$), having this way fewer sources of dependence from the variable n. Thus, the quantity to be estimated has the form $u_m(X_{n+1}, \ldots, X_{n+m}; T)$ for some measurable $u_m : \mathbb{X}^m \times \Theta \to \mathbb{R}$. From now on, it will be assumed that

$$\mathbb{E}\left[(u_m(X_{n+1}, \ldots, X_{n+m}; T))^2\right] < +\infty. \tag{18}$$

Whence, for the Bayesian estimator $\hat{U}_{n,m}$ in (12) existence and uniqueness are well-known: its explicit form is given by $\hat{U}_{n,m} = \hat{u}_{n,m}(X_1, \ldots, X_n)$ with

$$\hat{u}_{n,m}(x_1, \ldots, x_n) = \mathbb{E}[u_m(X_{n+1}, \ldots, X_{n+m}; T) \mid X_1 = x_1, \ldots, X_n = x_n]$$

$$= \int_\Theta \int_{\mathbb{X}^m} u_m(y_1, \ldots, y_m; \theta) \mu^{\otimes m}(dy_1 \ldots dy_m \mid \theta) \pi_n(d\theta \mid x_1, \ldots, x_n),$$

which is finite for α_n-almost all $(x_1, \ldots, x_n)$. The risk function $\mathfrak{R}_\mathbb{U}$ evaluated at $\hat{U}_{n,m}$ achieves its overall minimum value and, from (16), it takes the form:

$$\mathfrak{R}_\mathbb{U}[\hat{U}_{n,m}] = \int_{\mathbb{X}^n} \left\{ \int_\Theta \mathsf{v}(\theta) \pi_n(\mathrm{d}\theta \mid \mathbf{x}) + \int_\Theta [\mathsf{m}(\theta)]^2 \pi_n(\mathrm{d}\theta \mid \mathbf{x}) \right. \\ \left. - \left(\int_\Theta \mathsf{m}(\theta) \pi_n(\mathrm{d}\theta \mid \mathbf{x}) \right)^2 \right\} \alpha_n(\mathrm{d}\mathbf{x}), \qquad (19)$$

with

$$\mathsf{m}(\theta) := \int_{\mathbb{X}^m} u_m(y_1, \ldots, y_m; \theta) \mu^{\otimes m}(\mathrm{d}y_1 \ldots \mathrm{d}y_m \mid \theta)$$

$$\mathsf{v}(\theta) := \int_{\mathbb{X}^m} [u_m(y_1, \ldots, y_m; \theta) - \mathsf{m}(\theta)]^2 \mu^{\otimes m}(\mathrm{d}y_1 \ldots \mathrm{d}y_m \mid \theta)$$

thanks to the well-known "Law of Total Variance". See, e.g., [28] [Problem 34.10(b)]. As to the issue of estimating T, the first remarkable simplification induced by the above assumptions is that the p.m. $\gamma_{\theta,(x_1,\ldots,x_n)}$ is independent of $(x_1, \ldots, x_n)$. Whence,

$$\Delta(\theta, \tau) := [\mathcal{L}_{\Theta,(x_1,\ldots,x_n)}(\theta, \tau)]^{1/2} = \mathcal{W}_\mathbb{U}\left(\gamma_{\theta,(x_1,\ldots,x_n)}; \gamma_{\tau,(x_1,\ldots,x_n)}\right), \qquad (20)$$

is, in turn, independent of $(x_1, \ldots, x_n)$ and *defines a distance* on Θ provided that

$$\gamma_{\theta,(x_1,\ldots,x_n)} = \gamma_{\tau,(x_1,\ldots,x_n)}$$

entails $\theta = \tau$. Thus, for any estimator $T_{n,m}^\dagger = t_{n,m}^\dagger(X_1, \ldots, X_n)$ of T, obtained with a measurable $t_{n,m}^\dagger : \mathbb{X}^n \to \Theta$, (17) becomes

$$\mathfrak{R}_\Theta[T_{n,m}^\dagger] = \int_{\mathbb{X}^n} \int_\Theta [\Delta(\theta, t_{n,m}^\dagger(\mathbf{x}))]^2 \pi_n(\mathrm{d}\theta \mid \mathbf{x}) \alpha_n(\mathrm{d}\mathbf{x}). \qquad (21)$$

The last simplifications concern the basic object of the inference, i.e., the statistical model $\mu(\cdot|\cdot)$ and the prior π. First, assume that $\Theta = (a,b) \subseteq \mathbb{R}$ and that π has a density p (with respect to the Lebesgue measure). Even if this one-dimensionality assumption can seem a drastic simplification, it is again of a non-conceptual nature, and it is made to diminish the mathematical complexity of the ensuing statements. In fact, one of the goals of this work is to provide a Riamannian-like characterization of the metric space (Θ, Δ), and this is particularly simple in such a one-dimensional setting. The following arguments should be quite easily reproduced at least in a finite-dimensional setting (i.e., when $\Theta \subseteq \mathbb{R}^d$) by using basic tools of Riemannian geometry, such as local expansions of the geodesic distance. See, e.g., [29] [Chapter 5]. As to the statistical model $\mu(\cdot|\cdot)$, consider the following:

Assumption 1. *$\mu(\cdot|\cdot)$ is dominated by some σ-finite measure χ on $(\mathbb{X}, \mathscr{X})$ with a (distinguished version of the) density $f(\cdot|\theta)$ that satisfies:*

(i) $f(x|\theta) > 0$ for all $x \in \mathbb{X}$ and $\theta \in \Theta$;
(ii) *for any fixed $x \in \mathbb{X}$, $\theta \mapsto f(x|\theta)$ belongs to $C^4(\Theta)$;*
(iii) *there exists a separable Hilbert space $\mathcal{H}$ for which $\log f(x|\cdot) \in \mathcal{H}$ for all $x \in \mathbb{X}$, and such that, for any open Θ' whose closure is compact in Θ ($\overline{\Theta'} \Subset \Theta$, in symbols), the restriction operators $\mathcal{R}_{\Theta'} : h \mapsto h_{|\overline{\Theta'}}$ are continuous from $\mathcal{H}$ to $C^0(\overline{\Theta'})$;*
(iv) $\int_\mathbb{X} |\log f(x|\theta)|^2 \mu(\mathrm{d}x|\theta) < +\infty$ *for π-a.e. θ, and the Kullback-Leibler divergence*

$$\mathsf{K}(t \| \theta) := \int_\mathbb{X} \left(\frac{\log f(x|t)}{\log f(x|\theta)} \right) \mu(\mathrm{d}x|t) \qquad (22)$$

is well-defined.

A canonical choice for the Hilbert space $\mathcal{H}$ is in the form of a *weighted Sobolev space* $H^r(\Theta; \pi)$ for some $r \geq 1$. See, e.g., [30,31] for definition and further properties of weighted Sobolev spaces, such as embedding theorems. By the way, it is worth remarking that such assumptions are made to easily state the following results. It is plausible they could be relaxed in future works.

In this regularity setting, introduce the sequence $\{H_n\}_{n \geq 1}$, where $H_n : \Omega \to \mathcal{H}$ represents the (normalized) *log-likelihood process*, that is

$$H_n := \frac{1}{n} \ell_n(\cdot; X_1, \ldots, X_n) := \frac{1}{n} \sum_{i=1}^n \log f(X_i | \cdot) = \int_\mathbb{X} \log f(\xi | \cdot) \mathfrak{e}_n^{(X_1, \ldots, X_n)}(\mathrm{d}\xi) \quad (23)$$

the symbol $\mathfrak{e}_n^{(X_1, \ldots, X_n)}$ standing for the *empirical measure* based on $(X_1, \ldots, X_n)$, i.e.,

$$\mathfrak{e}_n^{(X_1, \ldots, X_n)} := \frac{1}{n} \sum_{i=1}^n \delta_{X_i}.$$

For completeness, any notation like $\ell_n(\cdot; X_1, \ldots, X_n)$ is just a short-hand to denote the entire function $\theta \mapsto \ell_n(\theta; X_1, \ldots, X_n)$. First of all, observe that H_n is a *sufficient statistics* in both classical and Bayesian sense. See [11]. Then, a version of de Finetti's Law of Large Numbers (see [9,32]) for the log-likelihood process can be stated as follows:

Theorem 1. *Under Assumption 1, define the following $\mathcal{H}$-valued r.v.*

$$H := \int_\mathbb{X} \log f(z | \cdot) \mu(\mathrm{d}z | T) = -\mathsf{K}(T \| \cdot) + \int_\mathbb{X} \log f(z | T) \mu(\mathrm{d}z | T)$$

along with $\nu_n(D) := \mathbb{P}[H_n \in D]$ and $\nu(D) := \mathbb{P}[H \in D]$, for any $D \in \mathscr{B}(\mathcal{H})$. Then, it holds that

$$H_n \xrightarrow{L^2} H \quad (24)$$

which, in turn, yields that $\nu_n \Rightarrow \nu$, where $\Rightarrow$ denotes weak convergence of p.m.'s on $(\mathcal{H}, \mathscr{B}(\mathcal{H}))$.

Then, to carry out the objectives mentioned in the Introduction, a quantitative refinement of the thesis $\nu_n \Rightarrow \nu$ is needed, as stated in the following proposition.

Proposition 1. *Let $C_b^2(\mathcal{H})$ denote the space of bounded, C^2 functionals on $\mathcal{H}$. Besides Assumption 1, suppose there exists a function $\Gamma(\cdot; \mu, \pi) : \mathcal{H} \to \mathbb{R}$ such that*

$$\frac{1}{2} \mathbb{E}\left[\mathrm{Hess}[\Psi]_H \otimes \mathrm{Cov}_T[\log f(X_i | \cdot)]\right] = \mathbb{E}[\Psi(H) \Gamma(H; \mu, \pi)] \quad (25)$$

holds for all functional $\Psi \in C_b^2(\mathcal{H})$, where $\mathrm{Hess}[\Psi]_h$ denotes the Hessian of Ψ at $h \in \mathcal{H}$, $\otimes$ is the tensor product between quadratic forms (operators) and $\mathrm{Cov}_t[\log f(X_i | \cdot)]$ stands for the covariance operator of the $\mathcal{H}$-valued r.v.'s $\log f(X_i | \cdot)$ with respect to the p.m. $\mu(\cdot | t)$. Then,

$$\int_\mathcal{H} \Psi(h) \nu_n(\mathrm{d}h) = \int_\mathcal{H} \Psi(h) \nu(\mathrm{d}h) + \frac{1}{n} \int_\mathcal{H} \Psi(h) \Gamma(h; \mu, \pi) \nu(\mathrm{d}h) + o\left(\frac{1}{n}\right) \quad (26)$$

holds as $n \to +\infty$ for all continuous $\Psi : \mathcal{H} \to \mathbb{R}$ for which the above integrals are convergent.

For further information on second-order differentiability in Hilbert/Banach spaces, see [33,34]. By the way, the above identity (26) is a quantitative strengthening of de Finetti's theorem similar to the identities stated in Theorem 1.1 of [8] [Chapter 6], valid in a finite-dimensional setting. Later on, we will resort to *uniform* versions of (26), meaning that the $o(\frac{1}{n})$-term is uniformly bounded with respect to h. However, such a kind of results—much more in the spirit of the Central Limit Theorem—are very difficult to prove and, to the best of the author's knowledge, there are no known results in infinite-dimension. Examples in finite-dimensional settings are given in [35,36], which prove Berry–Esseen like inequalities in the very specific context of Bernoulli r.v.s. See also [37]. Anyway, since one merit of [35] is to show how to use the classical Central Limit Theorem to prove an expansion as in (26), one could hope to follow that very same line of reasoning by resorting to some version of the central limit theorem for Banach spaces, such as that stated in [38]. Research on this is ongoing.

Now, to make the above Proposition 1 a bit more concrete, it is worth noticing the case in which $f(\cdot|\theta)$ is in exponential form. In fact, in this case, the identity (26) can be rewritten in a simpler form, condensed in the following statement.

Lemma 1. *Besides Assumption 1, suppose that* $f(x \mid \theta) = \exp\{\theta S(x) - M(\theta)\}$, *with some measurable* $S : \mathbb{X} \to \mathbb{R}$ *and* $M(\theta) := \log\left(\int_{\mathbb{X}} e^{\theta S(x)} \chi(dx)\right) \in \mathbb{R}$ *for all* $\theta \in \Theta$. *Then,* (26) *holds with*

$$\nu(D) := \mathbb{P}\big[(\theta \mapsto \theta M'(T) - M(\theta)) \in D\big]$$

and

$$\Gamma\big((\theta \mapsto \theta M'(t) - M(\theta)); \mu, \pi\big) = \frac{M''(t)}{p(t)} \frac{d^2}{dy^2} \big[M''(V(y))p(V(y))V'(y)\big]\big|_{y=M'(t)},$$

where $V(M'(t)) = t$ *for any* $t \in \Theta$.

To conclude this subsection, consider the expressions (19)–(21) and notice that they depend explicitly on the posterior distribution $\pi_n(\cdot \mid x_1, \ldots, x_n)$. Now, thanks to Assumption 1, the mapping $t \mapsto \delta_t$ can be seen as defined on Θ and taking values in the dual space $\mathcal{H}^*$, with Riesz representative $\mathfrak{h}_t \in \mathcal{H}$. More formally, for any $h \in \mathcal{H}$ and $t \in \Theta$, it holds that $h(t) = {}_\mathcal{H}\langle h, \delta_t \rangle_{\mathcal{H}^*} = \langle h, \mathfrak{h}_t \rangle$, where $\langle \cdot, \cdot \rangle$ stands for the scalar product on $\mathcal{H}$ while ${}_\mathcal{H}\langle \cdot, \cdot \rangle_{\mathcal{H}^*}$ denotes the pairing between $\mathcal{H}$ and $\mathcal{H}^*$. In this notation, the posterior distribution can be rewritten in exponential form as:

$$\pi_n(B \mid X_1, \ldots, X_n) = \frac{\int_B \exp\{n\langle H_n, \mathfrak{h}_\theta\rangle\}\pi(d\theta)}{\int_\Theta \exp\{n\langle H_n, \mathfrak{h}_\theta\rangle\}\pi(d\theta)} = \pi_n^*(B \mid H_n) \qquad (27)$$

for any $B \in \mathscr{T}$, the probability kernel $\pi_n^*(\cdot \mid \cdot) : \mathscr{T} \times \mathcal{H} \to [0, 1]$ being defined by

$$\pi_n^*(B \mid h) := \frac{\int_B \exp\{n\langle h, \mathfrak{h}_\theta\rangle\}\pi(d\theta)}{\int_\Theta \exp\{n\langle h, \mathfrak{h}_\theta\rangle\}\pi(d\theta)} . \qquad (28)$$

This is particularly interesting because it shows that the posterior distribution can always be thought of, in the presence of a dominated statistical model characterized by strictly positive, smooth densities, as an element of an exponential family, even if the original statistical model $\mu(\cdot \mid \cdot)$ is not in exponential form. By utilizing the kernel π_n^* in combination with the p.m. ν_n, the following re-writings of (19)–(21) are valid:

$$\mathfrak{R}_\mathbb{U}[\hat{U}_{n,m}] = \int_\mathcal{H} \bigg\{ \int_\Theta \mathsf{v}(\theta)\pi_n^*(d\theta \mid h) + \int_\Theta [\mathsf{m}(\theta)]^2 \pi_n^*(d\theta \mid h) \\ - \bigg(\int_\Theta \mathsf{m}(\theta)\pi_n^*(d\theta \mid h)\bigg)^2 \bigg\} \nu_n(dh) \qquad (29)$$

$$\mathfrak{R}_\Theta[T_{n,m}^\dagger] = \int_\mathcal{H} \int_\Theta [\Delta(\theta, \mathfrak{T}_{n,m}^\dagger(h))]^2 \pi_n^*(d\theta \mid h)\nu_n(dh), \qquad (30)$$

where the mapping $\mathfrak{T}_{n,m}^\dagger$ is such that $\mathfrak{T}_{n,m}^\dagger(H_n) = t_{n,m}^\dagger(X_1, \ldots, X_n)$ holds $\mathbb{P}$-a.s.

3. Main Results

The first result establishes a relationship between the Bayesian risk functions $\mathfrak{R}_\mathbb{U}$ and $\mathfrak{R}_\Theta$ defined in (16) and (17), respectively. Due to the central role of this relationship, it will be formulated within the general framework described in Section 2.1.

Proposition 2. *Consider any estimator $U_{n,m}^\dagger = u_{n,m}^\dagger(X_1, \ldots, X_n)$ of $U_{n,m}$ and any estimator $T_{n,m}^\dagger = t_{n,m}^\dagger(X_1, \ldots, X_n)$ of T such that $\mathbb{E}[d_\mathbb{U}^2(U_{n,m}^\dagger, u_0)] < +\infty$ holds for some $u_0 \in \mathbb{U}$ along with $\mathbb{E}\left[\mathcal{L}_{\Theta,(X_1,\ldots,X_n)}(T_{n,m}^\dagger, t_0)\right] < +\infty$ for some $t_0 \in \Theta$. Then, it holds*

$$\mathfrak{R}_\mathbb{U}[U_{n,m}^\dagger] \leq \mathfrak{R}_\Theta[T_{n,m}^\dagger] + \mathbb{E}\left[\int_\mathbb{U} d_\mathbb{U}^2(U_{n,m}^\dagger, u)\gamma_{T_{n,m}^\dagger,(X_1,\ldots,X_n)}(du)\right]$$
$$+ 2\mathbb{E}\left[\mathcal{L}_{\Theta,(X_1,\ldots,X_n)}^{1/2}(T, T_{n,m}^\dagger)\left(\int_\mathbb{U} d_\mathbb{U}^2(U_{n,m}^\dagger, u)\gamma_{T_{n,m}^\dagger,(X_1,\ldots,X_n)}(du)\right)^{1/2}\right]. \quad (31)$$

In particular, if the Bayesian risk function $\mathfrak{R}_\Theta$ is optimized by choosing $T_{n,m}^\dagger = \hat{T}_{n,m}$, where $\hat{T}_{n,m}$ is as in (15), and $U_{n,m}^\dagger$ is chosen equal to $U_{n,m}^*$, where $U_{n,m}^*$ is as in (13), then (31) becomes

$$\mathfrak{R}_\mathbb{U}[U_{n,m}^*] \leq \mathfrak{R}_\Theta[\hat{T}_{n,m}] + \mathbb{E}\left[\int_\mathbb{U} d_\mathbb{U}^2(U_{n,m}^*, u)\gamma_{\hat{T}_{n,m},(X_1,\ldots,X_n)}(du)\right]$$
$$+ 2\mathbb{E}\left[\mathcal{L}_{\Theta,(X_1,\ldots,X_n)}^{1/2}(T, \hat{T}_{n,m})\left(\int_\mathbb{U} d_\mathbb{U}^2(U_{n,m}^*, u)\gamma_{\hat{T}_{n,m},(X_1,\ldots,X_n)}(du)\right)^{1/2}\right]$$
$$= \inf_{T_{n,m}^\dagger} \mathfrak{R}_\Theta[T_{n,m}^\dagger] + \mathbb{E}\left[\inf_{U_{n,m}^\dagger} \int_\mathbb{U} d_\mathbb{U}^2(U_{n,m}^\dagger, u)\gamma_{\hat{T}_{n,m},(X_1,\ldots,X_n)}(du)\right]$$
$$+ 2\mathbb{E}\left[\mathcal{L}_{\Theta,(X_1,\ldots,X_n)}^{1/2}(T, \hat{T}_{n,m})\left(\int_\mathbb{U} d_\mathbb{U}^2(U_{n,m}^*, u)\gamma_{\hat{T}_{n,m},(X_1,\ldots,X_n)}(du)\right)^{1/2}\right]. \quad (32)$$

As an immediate remark, notice that the last member of (32) is obtained by first optimizing the risk $\mathfrak{R}_\Theta$ with respect to the choice of $T_{n,m}^\dagger$ and then, after getting $\hat{T}_{n,m}$, the term $\mathbb{E}\left[\int_\mathbb{U} d_\mathbb{U}^2(U_{n,m}^\dagger, u)\gamma_{\hat{T}_{n,m},(X_1,\ldots,X_n)}(du)\right]$ is optimized with respect to the choice of $U_{n,m}^\dagger$. Of course, it can be argued about the convenience of this procedure—and it is actually due—even if, in most problems, it seems that the strategy proposed in Proposition 2 proves indeed to be the simplest and the most feasible one, above all if computational issues are taken into account. In fact, the absolute best theoretical strategy—consisting of optimizing the right-hand side of (31) jointly with respect to the choice of $(U_{n,m}^\dagger, T_{n,m}^\dagger)$—turns out to be very often too complex and onerous to carry out. Therefore, it seems reasonable to quantify, at least approximately, how far the strategy of Proposition 2 is from absolute optimality, in terms of efficiency. Finally, the additional term

$$2\mathbb{E}\left[\mathcal{L}_{\Theta,(X_1,\ldots,X_n)}^{1/2}(T, \hat{T}_{n,m})\left(\int_\mathbb{U} d_\mathbb{U}^2(U_{n,m}^*, u)\gamma_{\hat{T}_{n,m},(X_1,\ldots,X_n)}(du)\right)^{1/2}\right] \quad (33)$$

will be reconsidered in next statement, within the simplified setting of Section 2.2. Indeed, by arguing asymptotically, it will be shown that it is essentially negligible, proving in this way a sort of "Pythagorean inequality".

Henceforth, to make the above remark effective, we will formulate the subsequent results within the simplified setting introduced in Section 2.2. Indeed, **Steps 1–3** mentioned in the Introduction are worthy of being reconsidered in light of Proposition 2. On the one hand, **Steps 1** and **2** boil down to checking the existence and uniqueness of the barycenters appearing in (15) and (13), for instance by using the results contained in [24–26]. On the other hand, **Step 3** hinges on the validity of (2) and (3), which are somewhat related to inequality (32). More precisely, (2) will be proved directly by resorting to identity (29), while (3) will be obtained by estimating the right-hand side of (32). Here is a precise statement.

Proposition 3. *Besides Assumptions 1 and (18), suppose that $p > 0$ and $p \in C^1(\Theta)$, $m, v \in C^2(\Theta)$, $\Delta^2 \in C^2(\Theta^2)$, and κ_t is any element of $\mathcal{H} \cap C^3(\Theta)$ with a unique minimum point at $t \in \Theta$. Then, it holds*

$$\int_\Theta \{v(\theta) + [m(\theta)]^2\} \pi_n^*(d\theta \mid -\kappa_t) - \left(\int_\Theta m(\theta) \pi_n^*(d\theta \mid -\kappa_t) \right)^2$$
$$= v(t) + \frac{1}{n\kappa_t''(t)} \left\{ [m'(t)]^2 + \frac{1}{2} v''(t) + v'(t) \left[\frac{p'(t)}{p(t)} - \frac{1}{2} \frac{\kappa_t'''(t)}{\kappa_t''(t)} \right] \right\} + o\left(\frac{1}{n}\right) \quad (34)$$

$$\int_\Theta \Delta^2(\theta, \tau) \pi_n^*(d\theta \mid -\kappa_t)$$
$$= \Delta^2(t, \tau) + \frac{1}{n\kappa_t''(t)} \left\{ \frac{1}{2} \frac{\partial^2}{\partial \theta^2} \Delta^2(\theta, \tau)_{|\theta=t} + \frac{\partial}{\partial \theta} \Delta^2(\theta, \tau)_{|\theta=t} \left[\frac{p'(t)}{p(t)} - \frac{1}{2} \frac{\kappa_t'''(t)}{\kappa_t''(t)} \right] \right\} + o\left(\frac{1}{n}\right) \quad (35)$$

as $n \to +\infty$, for any $\tau \in \Theta$.

Here, it is worth noticing that the asymptotic expansions derived in the above proposition are obtained by means of the Laplace method, as first proposed in [39]. See also [40] [Chapter 20]. At this stage, we face the problem of optimizing the left-hand side of (35) with respect to τ. Since the explicit expression of $\Delta^2(t, \tau)$ will be hardly known in closed form, a reasonable strategy considers, for fixed $t \in \Theta$, the optimization of the right-hand side of (35) with respect to τ, disregarding the remainder term $o(1/n)$. If $\Delta^2 \in C^3(\Theta^2)$, this attempt leads to considering the equation

$$\frac{\partial}{\partial \tau} \left[\Delta^2(t, \tau) + \frac{1}{n\kappa_t''(t)} \left\{ \frac{1}{2} \frac{\partial^2}{\partial \theta^2} \Delta^2(\theta, \tau)_{|\theta=t} + \frac{\partial}{\partial \theta} \Delta^2(\theta, \tau)_{|\theta=t} \left[\frac{p'(t)}{p(t)} - \frac{1}{2} \frac{\kappa_t'''(t)}{\kappa_t''(t)} \right] \right\} \right] = 0 \quad (36)$$

and, since

$$\frac{\partial}{\partial \tau} \Delta^2(t, \tau)_{|\tau=t} = 0, \quad (37)$$

we have that any solution of (36) is of the form $\hat{\tau}_n = t + \epsilon_n$, with some ϵ_n that goes to zero as $n \to +\infty$. For completeness, the validity of (37) could be obtained by using the explicit expression of the Wasserstein distance due to Dall'Aglio. See [41].

If $\Delta^2 \in C^4(\Theta^2)$, we can plug the expression of $\hat{\tau}_n$ into (35), and expand further the right-hand side. Exploiting that

$$\Delta^2(t, \hat{\tau}_n) = \frac{1}{2} \frac{\partial^2}{\partial \tau^2} \Delta^2(t, \tau)_{|\tau=t} \cdot \epsilon_n^2 + o(\epsilon_n^2)$$

$$\frac{\partial}{\partial t} \Delta^2(t, \hat{\tau}_n) = \frac{\partial}{\partial \tau} \left[\frac{\partial}{\partial t} \Delta^2(t, \tau) \right]_{|\tau=t} \cdot \epsilon_n + \frac{1}{2} \frac{\partial^2}{\partial \tau^2} \left[\frac{\partial}{\partial t} \Delta^2(t, \tau) \right]_{|\tau=t} \cdot \epsilon_n^2 + o(\epsilon_n^2)$$

$$\frac{\partial^2}{\partial t^2} \Delta^2(t, \hat{\tau}_n) = \frac{\partial^2}{\partial t^2} \Delta^2(t, \tau)_{|\tau=t} + \frac{\partial}{\partial \tau} \left[\frac{\partial^2}{\partial t^2} \Delta^2(t, \tau) \right]_{|\tau=t} \cdot \epsilon_n$$
$$+ \frac{1}{2} \frac{\partial^2}{\partial \tau^2} \left[\frac{\partial^2}{\partial t^2} \Delta^2(t, \tau) \right]_{|\tau=t} \cdot \epsilon_n^2 + o(\epsilon_n^2),$$

we get

$$\int_\Theta \Delta^2(\theta, \hat{\tau}_n) \pi_n^*(d\theta \mid -\kappa_t)$$

$$= \frac{1}{2} \frac{\partial^2}{\partial \tau^2} \Delta^2(t,\tau)\Big|_{\tau=t} \cdot \epsilon_n^2 + \frac{1}{n\kappa_t''(t)} \left\{ \frac{1}{2} \frac{\partial^2}{\partial t^2} \Delta^2(t,\tau)\Big|_{\tau=t} + \frac{1}{2} \frac{\partial}{\partial \tau} \left[\frac{\partial^2}{\partial t^2} \Delta^2(t,\tau) \right]\Big|_{\tau=t} \cdot \epsilon_n \right.$$

$$+ \frac{1}{4} \frac{\partial^2}{\partial \tau^2} \left[\frac{\partial^2}{\partial t^2} \Delta^2(t,\tau) \right]\Big|_{\tau=t} \cdot \epsilon_n^2 + \left[\frac{p'(t)}{p(t)} - \frac{1}{2} \frac{\kappa_t'''(t)}{\kappa_t''(t)} \right] \cdot \frac{\partial}{\partial \tau} \left[\frac{\partial}{\partial t} \Delta^2(t,\tau) \right]\Big|_{\tau=t} \cdot \epsilon_n$$

$$+ \frac{1}{2} \left[\frac{p'(t)}{p(t)} - \frac{1}{2} \frac{\kappa_t'''(t)}{\kappa_t''(t)} \right] \cdot \frac{\partial^2}{\partial \tau^2} \left[\frac{\partial}{\partial t} \Delta^2(t,\tau) \right]\Big|_{\tau=t} \cdot \epsilon_n^2 \bigg\} + o(\epsilon_n^2) + o\left(\frac{1}{n}\right). \quad (38)$$

The right-hand side of this expression has the form

$$a \cdot \epsilon_n^2 + \frac{1}{n} \left[A \cdot \epsilon_n^2 + B \cdot \epsilon_n + C \right] + o(\epsilon_n^2) + o\left(\frac{1}{n}\right),$$

so that the choice

$$\epsilon_n = -\frac{B}{2na}\left(1 + \frac{A}{na}\right) + o\left(\frac{1}{n^2}\right) = -\frac{B}{2na} + o\left(\frac{1}{n}\right)$$

optimizes its expression. Whence,

$$\hat{\tau}_n = t - \frac{1}{n\kappa_t''(t)} \left(\frac{\partial^2}{\partial \tau^2} \Delta^2(t,\tau)\Big|_{\tau=t} \right)^{-1} \cdot \left\{ \frac{1}{2} \frac{\partial}{\partial \tau} \left[\frac{\partial^2}{\partial t^2} \Delta^2(t,\tau) \right]\Big|_{\tau=t} \right.$$

$$+ \left[\frac{p'(t)}{p(t)} - \frac{1}{2} \frac{\kappa_t'''(t)}{\kappa_t''(t)} \right] \cdot \frac{\partial}{\partial \tau} \left[\frac{\partial}{\partial t} \Delta^2(t,\tau) \right]\Big|_{\tau=t} \bigg\} + o\left(\frac{1}{n}\right) \quad (39)$$

and consequently

$$\int_\Theta \Delta^2(\theta, \hat{\tau}_n) \pi_n^*(d\theta \mid -\kappa_t) = \frac{1}{2n\kappa_t''(t)} \frac{\partial^2}{\partial t^2} \Delta^2(t,\tau)\Big|_{\tau=t} + o\left(\frac{1}{n}\right). \quad (40)$$

A first consequence of these computations is that the (Bayesian) estimator $\hat{T}_{n,m}$ in (15) has the same form as (39) with t and κ_t replaced by the MLE, denoted by $\hat{\theta}_n$, and $-H_n$, respectively. Of course, this fact has some relevance only in the case that $\hat{\theta}_n$ exists and is unique. Moreover, coming back to (32), it is worth noticing that

$$\inf_{U_{n,m}^\dagger} \int_\mathbb{U} |U_{n,m}^\dagger - u|^2 \gamma_{\hat{\tau}_n}(du) = v(\hat{\tau}_n) = v(t) + v'(t)\epsilon_n + o\left(\frac{1}{n}\right), \quad (41)$$

where we have dropped the dependence on $(X_1, \ldots, X_n)$ in the expression of $\gamma_{\hat{\tau}_n}$, in agreement with the simplified setting of Section 2.2 we are following. The last preliminary remark is about the additional term (33) that appears in the last member of (32). In fact, exploiting from the beginning that $\mathbb{U} = \mathbb{R}$ and $\mathcal{L}_\mathbb{U}(u,v) = |u-v|^2$, we find that it reduces to

$$2\mathbb{E}\bigg[\int_\Theta \int_{\mathbb{X}^m} \Big(u_m(\mathbf{y},\theta) - u_m(\mathbf{y}, \hat{T}_{n,m})\Big) \Big(u_m(\mathbf{y}, \hat{T}_{n,m}) - m(\hat{T}_{n,m})\Big) \times$$

$$\times \mu^{\otimes m}(d\mathbf{y} \mid \theta) \pi_n(d\theta \mid X_1, \ldots, X_n) \bigg] \quad (42)$$

by which we notice that it also involves "covariance terms". The way is now paved to state the following

Theorem 2. *Besides Assumptions 1 and (18), suppose that* $m, v \in C^2(\Theta)$ *and* $\Delta^2 \in C^4(\Theta^2)$. *Then, the identities*

$$\frac{\partial^2}{\partial \tau^2}\Delta^2(t,\tau)\Big|_{\tau=t} = -\frac{\partial}{\partial \tau}\left[\frac{\partial}{\partial t}\Delta^2(t,\tau)\right]_{\tau=t} \tag{43}$$

$$\frac{1}{2}v''(t) + [m'(t)]^2 = \frac{1}{2}\frac{\partial^2}{\partial t^2}\Delta^2(t,\tau)\Big|_{\tau=t}$$
$$- \frac{1}{2}\left(\frac{\partial^2}{\partial \tau^2}\Delta^2(t,\tau)\Big|_{\tau=t}\right)^{-1}\frac{\partial}{\partial \tau}\left[\frac{\partial^2}{\partial t^2}\Delta^2(t,\tau)\right]_{\tau=t}v'(t) \tag{44}$$

entail that

$$\int_\Theta \{v(\theta) + [m(\theta)]^2\}\pi_n^*(d\theta \mid -\kappa_t) - \left(\int_\Theta m(\theta)\pi_n^*(d\theta \mid -\kappa_t)\right)^2$$
$$= \int_\Theta \Delta^2(\theta, \hat{\tau}_n)\pi_n^*(d\theta \mid -\kappa_t) + v(t) + v'(t)\epsilon_n + o\left(\frac{1}{n}\right) \tag{45}$$

for any $t \in \Theta$, *any* κ_t *in* $\mathcal{H} \cap C^3(\Theta)$ *with a unique minimum point at* $t \in \Theta$, *and any* $p > 0$ *with* $p \in C^1(\Theta)$, *provided that the term in (42) is of* $o(\frac{1}{n})$-*type. Thus, if either*

(A1) (26) holds uniformly with respect to some class $\mathfrak{F}$ *of continuous functionals* $\Psi : \mathcal{H} \to \mathbb{R}$, *in the sense that*

$$\sup_{\Psi \in \mathfrak{F}}\left|\int_\mathcal{H} \Psi(h)\nu_n(dh) = \int_\mathcal{H} \Psi(h)\nu(dh) + \frac{1}{n}\int_\mathcal{H} \Psi(h)\Gamma(h;\mu,\pi)\nu(dh)\right| = o\left(\frac{1}{n}\right)$$

(A2) both the functionals $h \mapsto \int_\Theta\{v(\theta) + [m(\theta)]^2\}\pi_n^*(d\theta \mid h) - (\int_\Theta m(\theta)\pi_n^*(d\theta \mid h))^2$ *and* $h \mapsto \inf_{\mathfrak{T}_{n,m}^\dagger} \int_\Theta [\Delta(\theta, \mathfrak{T}_{n,m}^\dagger(h))]^2 \pi_n^*(d\theta \mid h)$ *belong to* $\mathfrak{F}$, *for all* $n \in \mathbb{N}$

or

(B1) (34) and (40) hold uniformly for all κ_t *belonging to a given subset* $\mathcal{D}$ *of* $\mathcal{H}$
(B2) $\nu_n(\mathcal{D}) = 1$ *for all* $n \in \mathbb{N}$

then (2)–(3) are in force with

$$\hat{R}_{0,m} = R_{0,m}^* = \int_\Theta v(t)\pi(dt) \tag{46}$$

$$\hat{R}_{1,m} = R_{1,m}^* = \int_\Theta \frac{1}{\overline{\kappa}_t''(t)}\left\{[m'(t)]^2 + \frac{1}{2}v''(t) + v'(t)\left[\frac{p'(t)}{p(t)} - \frac{1}{2}\frac{\overline{\kappa}_t'''(t)}{\overline{\kappa}_t''(t)}\right]\right\}\pi(dt)$$
$$+ \int_\Theta v(t)\Gamma(\overline{\kappa}_t;\mu,\pi)\pi(dt), \tag{47}$$

where $\overline{\kappa}_t(\theta) := K(t \parallel \theta)$, *for any* $p > 0$ *with* $p \in C^1(\Theta)$.

As announced in the Introduction, here we have minted the term *compatibility equations* to refer to identities (43) and (44). They actually constitute two "compatibility conditions" that involve only the statistical model, without any mention to the prior. The dependence on the quantity to be estimated is indeed hidden in the expression of Δ^2. More deeply, these equations can be viewed as a check on the compatibility between the original estimation problem (1) and the fact that we have metrized the space of the parameters Θ as in (20). Actually, they could have a more general value if interpreted as relations aimed at *characterizing* Δ^2, rather than imposing that this distance is given in terms of the Wasserstein distance as in (20). However, for a distance that is characterized differently from (20), an analogous of inequality (32) should be checked in terms of this new distance on Θ. As to the concrete check of the compatibility equations, we notice that the former identity (43) is generally valid as a consequence of the representation formula or the Wasserstein distance due to Dall'Aglio (see [41]), as long as the exchange between derivatives and integrals is allowed. For the other identity (44), we have instead collected in Section 4 some examples of simple statistical models for which its verification proves to be quite simple. Finally, the issue of extending these equations in a higher dimension, including the infinite dimension, is deferred to Section 6.

Apropos of the other assumptions, the verification that the term in (42) is of $o(\frac{1}{n})$-type is generally straightforward. For instance, such a term is even equal to zero if u_m is independent of θ. As to the two groups of assumptions which are needed to prove (46) and (47), the latter block, formed by (B1) and (B2), is certainly easier to check. However, (B1) and (B2) can prove to be rather strong since they require the existence of the MLE for any $n \in \mathbb{N}$. On the other hand, checking (A1) and (A2) is generally harder since it constitutes a strong reinforcement of de Finetti's Law of Large Number for the log-likelihood process, similar in its conception to those stated in [35,36]. Moreover, the check of (A2) is more or less equivalent to prove a uniform regularity of the mapping $h \mapsto \pi_n^*(d\theta \mid h)$, as a map from $\mathcal{H}$ into the space of p.m.'s on $(\Theta, \mathcal{T})$ metrized with a Wasserstein distance. This theory is presented and developed in [42,43]. In any case, these lines of research deserve further investigations, to be deferred to a forthcoming paper.

Finally, we consider **Steps 4–6** mentioned in the Introduction, in light of the previous results. In fact, the compatibility Equations (43) and (44) suggest two new compatibility conditions, which are necessary to get (10) along with $\hat{R}_{i,m} = \tilde{R}_{i,m}$ for $i = 0, 1$. A formal statement reads as follows.

Theorem 3. *Besides Assumptions 1 and (18), suppose that* m, v $\in C^2(\Theta)$, $\Delta^2 \in C^4(\Theta^2)$. *Assume also that either (A1) and (A2) or (B1) and (B2) of Theorem 2 are in force. Then, any solution $\hat{\tau}_n$ of the following equations:*

$$v(\hat{\theta}_n) = v(\hat{\tau}_n) + \Delta^2(\hat{\tau}_n, \hat{\theta}_n) + o\left(\frac{1}{n}\right) \tag{48}$$

$$v'(\hat{\theta}_n) = \frac{\partial}{\partial t}\Delta^2(\hat{\tau}_n, t)\Big|_{t=\hat{\theta}_n} + o\left(\frac{1}{n}\right) \tag{49}$$

$$\frac{1}{2}v''(\hat{\theta}_n) + [m'(\hat{\theta}_n)]^2 = \frac{1}{2}\frac{\partial^2}{\partial t^2}\Delta^2(\hat{\tau}_n, t)\Big|_{t=\hat{\theta}_n} + o\left(\frac{1}{n}\right), \tag{50}$$

where $\hat{\theta}_n$ stands for the MLE, yields a prior-free estimator $\hat{T}_{n,m}$ and, through **Step 5**, *another prior-free estimator $\tilde{U}_{n,m}$ that satisfies (10) along with $\hat{R}_{i,m} = \tilde{R}_{i,m}$ for $i = 0, 1$, where $\hat{R}_{0,m}$ and $\hat{R}_{1,m}$ are as in (46) and (47), respectively, provided that the term in (42) is of $o(\frac{1}{n})$-type.*

The derivation of new prior free-estimators via this procedure represents a novel line of research that we would like to pursue in forthcoming works.

4. Applications and Examples

This section is split into four subsections, and has two main purposes. In fact, Sections 4.1–4.3 just contain explicit examples of very simple statistical models for which the compatibility equations are satisfied. These models are the one-dimensional Gaussian, the exponential and the Pareto model. Section 4.4 has a different nature, since it is devoted to a more concrete application of our approach to the original Poisson-mixture setting used by Herbert Robbins to introduce his own approach to empirical Bayes theory. Finally, Section 4.5 carries on the discussion initiated in Section 4.4 by showing a concrete application relative to one year of claims data for an automobile insurance company.

4.1. The Gaussian Model

Here, we have $\mathbb{X} = \Theta = \mathbb{R}$ and

$$\mu(A \mid \theta) = \int_A \frac{1}{\sqrt{2\pi\sigma^2}} \exp\{-\frac{1}{2\sigma^2}(x-\theta)^2\} dx \qquad (A \in \mathscr{B}(\mathbb{R}))$$

for some known $\sigma^2 > 0$. For simplicity, we put $m = 1$ and $u_1(y, \theta) = y$, which is tantamount to saying that the original predictive aim was focused on the estimation of X_{n+1}. In this setting, it is very straightforward to check that $m(\theta) = \theta$ and $v(\theta) = \sigma^2$. Moreover, in view of well-know computations on the Wasserstein distance (see [44,45]), it is also straightforward to check that $\Delta^2(\theta, \tau) = |\theta - \tau|^2$. Therefore, (43) becomes $2 = 2$, while (44) reduces to $1 = 1$. Finally, it is also possible to check the validity of (48)–(50) with the simplest choice $\hat{\tau}_n = \hat{\theta}_n$.

The case of constant mean and unknown variance will not be dealt with here because its treatment is substantially included in the following subsection. Apropos of the multidimensional variant of this model, very important in many statistical applications, we just mention the interesting paper [46] which paves the way, mathematically speaking, to write down the multidimensional analogous of the compatibility equations in a full Riemannian context.

4.2. The Exponential Model

Here, we have $\mathbb{X} = \Theta = (0, \infty)$ and

$$\mu(A \mid \theta) = \int_A \theta e^{-\theta x} dx \quad (A \in \mathcal{B}(0, +\infty)).$$

Again, for simplicity, we put $m = 1$ and $u_1(y, \theta) = y$, which is tantamount to saying that the original predictive aim was focused on the estimation of X_{n+1}. In this setting, it is very straightforward to check that $m(\theta) = 1/\theta$ and $v(\theta) = 1/\theta^2$. Moreover, by resorting to Dall'Aglio representation of the Wasserstein distance (see [41]), it is also straightforward to check that $\Delta^2(\theta, \tau) = 2|1/\theta - 1/\tau|^2$. Although very simple, this is a very interesting example of non-Euclidean distance on $\Theta = (0, \infty)$. As to the validity of the compatibility equations, we easily see that (43) yields $4/t^4 = 4/t^4$, while (44) becomes:

$$\frac{3}{t^4} + \left(\frac{1}{t^2}\right)^2 = \frac{1}{2} \cdot \frac{4}{t^4} - \frac{1}{2}\left(\frac{8}{t^5}\right) \cdot \left(\frac{4}{t^4}\right)^{-1} \cdot \left(-\frac{2}{t^3}\right).$$

4.3. The Pareto Model

Here, we have $\mathbb{X} = \Theta = (0, \infty)$ and

$$\mu(A \mid \theta) = \int_{A \cap (\theta, +\infty)} \frac{\alpha \theta^\alpha}{x^{\alpha+1}} dx \quad (A \in \mathcal{B}(0, +\infty))$$

for some known $\alpha > 2$. Again, for simplicity, we put $m = 1$ and $u_1(y, \theta) = y$, which is tantamount to saying that the original predictive aim was focused on the estimation of X_{n+1}. In this setting, it is very straightforward to check that $m(\theta) = \frac{\alpha}{\alpha-1}\theta$ and $v(\theta) = \frac{\alpha}{(\alpha-2)(\alpha-1)^2}\theta^2$. Moreover, by resorting to the Dall'Aglio representation of the Wasserstein distance (see [41]), it is also straightforward to check that $\Delta^2(\theta, \tau) = \frac{\alpha}{\alpha-2}|\theta - \tau|^2$. Of course, this is not a regular model since the support of $\mu(\cdot|\theta)$ varies with θ. Anyway, it is interesting to notice that the compatibility equations are still also valid in this case. Therefore, the analysis of such non-regular models should motivate further investigations about their intrinsic value.

4.4. Robbins Approach to Empirical BAYES

In his seminal paper [23], Herbert Robbins introduced the following model to present his own approach to empirical Bayes theory. The problem that he considers is inspired by car insurance data analysis, and it is only slightly different from a "standard" predictive problem. We start by putting $\mathbb{X} = \mathbb{N}_0^2$ and $\mathbb{U} = \mathbb{N}_0$, and considering exchangeable random variables X_i's with $X_i = (\xi_i, \eta_i)$. The practical meaning is that ξ_i represents the number of accidents experienced by the i-th customer in the past year, while η_i represents the number of accidents that the same i-th customer will experience in the current year. Then, Robbins (in his own notation) attaches to each customer a random parameter, say $\lambda_i > 0$ to the i-th customer, which represents the rate of a Poisson distribution for that customer. Moreover, he considers the λ_i's as i.i.d. and, conditionally on the λ_i's, the X_i's become independent, and in addition ξ_i and η_i become i.i.d. with distribution $\text{Poi}(\lambda_i)$, for all $i \in \mathbb{N}$. Robbins calls G the common distribution of the λ_i's and interpret it as a "prior distribution". However, if we strictly follow the Bayesian main way, we should call this distribution θ to avoid confusion, and just realize that we have, this way, defined the statistical model, that is

$$\mu(\{(k,h)\} \mid \theta) = \int_0^{+\infty} \frac{e^{-z}z^k}{k!} \frac{e^{-z}z^h}{h!} \theta(\mathrm{d}z) \qquad ((k,h) \in \mathbb{N}_0^2). \tag{51}$$

Thus, the actual prior (Bayesianly speaking) is some p.m. π on the space of all p.m.'s on $((0, +\infty), \mathscr{B}(0, +\infty))$, while the random parameter T considered in the present paper is some *random probability measure*. Here, the objective—actually very practical and intuitively logic—is to estimate η_1 on the basis of the sample $(\xi_1, \ldots, \xi_n)$. Thus, our $U_{n,m}$ coincides with η_1 and the loss function is just, as usual, the quadratic loss. Throughout his paper, Robbins works under the conditioning to $T = \theta$ (that his under a fixed prior, in his own terminology). Hence, his "theoretical estimator" reads

$$\mathbb{E}_\theta[\eta_1 \mid (\xi_1, \ldots, \xi_n)] = \mathbb{E}_\theta[\eta_1 \mid \xi_1] = (\xi_1 + 1) \frac{p_\theta(\xi_1 + 1)}{p_\theta(\xi_1)}, \tag{52}$$

where $p_\theta(k) := \mu(\{k\} \times \mathbb{N}_0 \mid \theta)$. To get rid of the unobservable θ, Robbins exploits that $\theta = \mathbb{E}_\theta[\xi_1] = \sum_{k=0}^{+\infty} k p_\theta(k)$ to bring the Strong Law of Large Numbers into the game. Indeed, since

$$\hat{p}(k) := \frac{1}{n} \sum_{i=1}^n \mathbb{1}\{\xi_i = k\} \stackrel{\mathbb{P}_\theta - a.s.}{\longrightarrow} p_\theta(k)$$

holds for any θ, then it could be worth considering the (prior-free) estimator:

$$\tilde{U}_{n,m} = (\xi_1 + 1) \frac{\hat{p}(\xi_1 + 1)}{\hat{p}(\xi_1)}. \tag{53}$$

At this stage, if we want to maintain the Bayesian main way, we should make three basic considerations. First, given the statistical model (51), independently of the estimation problem, the assumption of exchangeability of the X_i's entails the existence of some prior distribution π, by de Finetti's representation theorem. Second, given the quadratic loss function on $\mathbb{U}$, the best (i.e., the most efficient) estimator is given by:

$$\hat{U}_{n,m} := \mathbb{E}[\eta_1 \mid (\xi_1, \ldots, \xi_n)],$$

where the expectation $\mathbb{E}$ depends of course on the prior π. Third, if we consider the above estimator as useless, because of an effective ignorance about the prior π, we are justified to consider the above $\tilde{U}_{n,m}$ as a possible approximation of $\hat{U}_{n,m}$, in the sense expressed by the joint validity of (2) and (10), with $\hat{R}_{i,m} = \tilde{R}_{i,m}$ for $i = 0, 1$, uniformly with respect to a whole (possible very large) class of priors π. Unfortunately, it is not the case. Or rather, we could actually achieve this goal, in the presence of distinguished choices of π. Therefore, if there is ignorance on π, we can only consider the Robbins estimator as efficient "at zero-level", and not also "at $O(\frac{1}{n})$-level". If we follow the approach presented in this paper, the natural choice for an estimator is given by:

$$U_{n,m}^* = (\xi_1 + 1) \frac{\int_0^{+\infty} \left(\frac{e^{-z}z^{\xi_1+1}}{(\xi_1+1)!} \right) \hat{T}_{n,m}(\mathrm{d}z)}{\int_0^{+\infty} \left(\frac{e^{-z}z^{\xi_1}}{\xi_1!} \right) \hat{T}_{n,m}(\mathrm{d}z)}, \tag{54}$$

where the estimator $\hat{T}_{n,m}$ belongs to the effective space of the parameters Θ, that is the space of all p.m.'s on $((0, +\infty), \mathscr{B}(0, +\infty))$, and is identified as:

$$\hat{T}_{n,m} = \text{Argmin}_\tau \int_\Theta \mathcal{W}_2^2\left(\mu_{\xi_1,\theta}, \mu_{\xi_1,\tau}\right) \pi_n(\mathrm{d}\theta \mid \xi_1, \ldots, \xi_n), \tag{55}$$

with

$$\mu_{k,\theta}(A) := \frac{\int_A \left(\frac{e^{-z}z^k}{k!}\right) \theta(\mathrm{d}z)}{\int_0^{+\infty} \left(\frac{e^{-z}z^k}{k!}\right) \theta(\mathrm{d}z)} \quad (A \in \mathscr{B}(0, +\infty)).$$

The proof of the fact that our estimator is more efficient than Robbins estimator—at least asymptotically and uniformly with respect to a whole class of priors—will be given in a forthcoming paper. Indeed, such a proof will constitute only a first step towards a complete vindication of our approach. The crowing achievement of the project would be represented by the production of some prior-free approximation of $\hat{T}_{n,m}$ that could lead, through (54), to an efficient estimator $\tilde{U}_{n,m}$ up to the "$O(\frac{1}{n})$-level". Research on this is ongoing.

4.5. An Example of Real Data Analysis

This subsection represents a continuation of the analysis of Robbins' approach to empirical Bayes theory, hinting at some concrete applications. We display below a Table 1 from [47] which is relative to one year of claims data for a European automobile insurance company. The original source of the data is the actuarial work [48].

Table 1. Table reporting, in the second line, the exact counts of claimed accidents. Third and fourth lines display estimated numbers of accidents.

Claims	0	1	2	3	4	5	6	7
Counts	7840	1317	239	42	14	4	4	1
Robbins estimator	0.168	0.363	0.527	1.33	1.43	6.00	1.25	0
Gamma MLE	0.164	0.398	0.633	0.87	1.10	1.34	1.57	0

Here, a population of 9461 automobile insurance policy holders is considered. Out of these, 7840 made no claims during the year; 1317 made a single claim; 239 made two claims each and so forth, continuing to the one person who made seven claims. The insurance company is concerned about the claims each policy holder will make in the next year. The third and the fourth lines provide estimations of such numbers by following the original Robbins method (based on (53)) and another compound model discussed in Section 6.1 of [47], respectively. In particular, the Robbins estimator predicts that the 7840 policy holders that made no claims during the year will contribute to an amount of $7840 \times 0.168 \approx 1317$ accidents, and so on. Analogously, the compound model predicts that the same 7840 policy holders will contribute to an amount of $7840 \times 0.164 \approx 1286$ accidents, and so on. Moreover, it is worth noticing that the original Robbins estimator suffers the lack of certain regularity properties, such as monotonicity, so that various smoothed versions of it have been provided by other authors. See [49]. See also [50] [Chapter 5] for a comprehensive treatment.

Here, we seize the opportunity to give the reader a taste of our approach, as explained in Section 4.4. A detailed treatment would prove, in any case, too complex to be thoroughly developed in this paper, due to the significant amount of numerical techniques which are necessary to carry out our strategy. Indeed, the big issue is concerned with the implementation of the infinite-dimensional minimization problem (55), which is still under investigation. However, we can simplify the treatment by restricting the attention on prior distributions π that put the total unitary mass, for example, on the set $\mathcal{E}$ of exponential distributions, so that $\theta(dz) = \beta e^{-\beta z} dz$ for $z > 0$ and some hyper-parameter $\beta > 0$. Thus, given some hyper-prior ζ on the hyper-parameter β, we can easily see that (55) boils down to a simple, one-dimensional minimization problem. Its solution $\hat{T}_{n,m}$ is provided by the distribution

$$\hat{\beta}_n e^{-\hat{\beta}_n z} dz$$

with $\hat{\beta}_n$ coinciding with the harmonic mean of the posterior distribution of the hyper-parameter β. On the basis of the theory developed in the paper, this solution will prove asymptotically nearly optimal uniformly with respect to the (narrow) class of prior distributions that put the total unitary mass on $\mathcal{E}$. Whence, the estimator $U_{n,m}^*$ in (54) assumes the form

$$U_{n,m}^* = \frac{\zeta_1 + 1}{\hat{\beta}_n + 1}.$$

This last estimator is, of course, not prior-free, because $\hat{\beta}_n$ depends on the prior ζ. However, to get a quick result, we can approximate $\hat{\beta}_n$ by means of the Laplace methods again yielding

$$\frac{\zeta_1 + 1}{\hat{\beta}_n + 1} \approx (\zeta_1 + 1) \frac{S_n}{S_n + n} := \tilde{U}_{n,m},$$

where S_n represents the total amount of accidents. Since $n = 9461$ and $S_n = 2028$ in the dataset under consideration, we provide the following new Table 2,

Table 2. Table reporting, in the second line, the exact counts of claimed accidents. Third line displays estimated numbers of accidents.

Claims	0	1	2	3	4	5	6	7
Counts	7840	1317	239	42	14	4	4	1
Estimator $\tilde{U}_{n,m}$	0.176	0.353	0.53	0.706	0.882	1.06	1.23	1.41

which is indeed comparable with the previous one. To give an idea, the Robbins estimator predicts 2019 total accidents for the next year, while the estimator $\tilde{U}_{n,m}$ above predicts 2033 total accidents for the next year.

In any case, a thorough analysis of this specific example deserves more attention, and will be developed in a forthcoming new paper.

5. Proofs

Gathered here are the proofs of the results stated in the main body of the paper.

5.1. Theorem 1

First, by following the same line of reasoning as in [32], conclude that the sequence $\{H_n\}_{n \geq 1}$ is a Cauchy sequence in $L^2(\Omega; \mathcal{H}) := \{W : \Omega \to \mathcal{H} \mid \mathbb{E}[\|W\|_{\mathcal{H}}^2] < +\infty\}$. Thus, by completeness, there exists a random element H^* in $L^2(\Omega; \mathcal{H})$ such that $H_n \xrightarrow{L^2} H^*$. Now, exploit the continuous embedding $\mathcal{H} \subset C^0(\overline{\Theta})$. By de Finetti's Strong Law of Large Numbers (see [9]), $H_n(\theta)$ converges $\mathbb{P}$-a.s. to $-K(T \| \theta) + \int_{\mathbb{X}} \log f(z|T) \mu(dz \mid T) = H(\theta)$, for any fixed $\theta \in \Theta$. Since $H \in \mathcal{H}$ by Assumption 1, then $H = H^*$ as elements of $\mathcal{H}$. At this stage the conclusion that $\nu_n \Rightarrow \nu$ follows by the standard implication that L^2-convergence implies convergence in distribution, which is still true for random elements taking values in a separable Hilbert space. See [51].

5.2. Proposition 1

Start by considering a functional Ψ in $C_b^2(\mathcal{H})$. Notice that

$$\int_{\mathcal{H}} \Psi(h) \nu_n(dh) = \mathbb{E}[\Psi(H_n)]$$

and then expand the term $\Psi(H_n - H + H)$ by the Taylor formula (see [33,34]) to get

$$\Psi(H_n) = \Psi(H) + \langle \nabla \Psi(H), H_n - H \rangle + \frac{1}{2} \langle \text{Hess}[\Psi]_H (H_n - H), H_n - H \rangle + o(\|H_n - H\|^2).$$

Observe that H is $\sigma(T)$-measurable while, by de Finetti's representation theorem, the distribution of $H_n - H$, given T, coincides with the distribution of a sum of n i.i.d. random elements. Whence, the tower property of the conditional expectation entails:

$$\mathbb{E}[\Psi(H_n)] = \mathbb{E}[\Psi(H)] + \frac{1}{2n} \mathbb{E}[\text{Hess}[\Psi]_H \otimes \text{Cov}_T[\log f(X_i|\cdot)]] + o\left(\frac{1}{n}\right)$$

since $\mathbb{E}[H_n - H \mid T] = 0$ and, then,

$$\mathbb{E}[\langle \nabla \Psi(H), H_n - H \rangle \mid T] = \langle \nabla \Psi(H), \mathbb{E}[H_n - H \mid T] \rangle = 0$$

the expression $\mathbb{E}[H_n - H \mid T]$ being intended as a Bochner integral. Thus, the main identity (26) follows immediately from (25), for any $\Psi \in C_b^2(\mathcal{H})$. Once (26) is established for regular Ψ's, one can extend its validity to more general continuous Ψ's by standard approximation arguments.

5.3. Lemma 1

First, observe that:

$$-K(T \parallel \theta) + \int_{\mathbb{X}} \log f(z \mid T) \mu(dz \mid T) = \theta M'(T) - M(\theta).$$

Notice also that:

$$\int_{\mathcal{H}} \Psi(h) \nu_n(dh) = \mathbb{E}\left[\Psi\left(\theta \mapsto \frac{\theta}{n} \sum_{i=1}^n S(X_i) - M(\theta)\right)\right].$$

Then, repeat the same arguments as in the previous proof, getting

$$\int_{\mathcal{H}} \Psi(h) \nu_n(dh) = \int_{\mathcal{H}} \Psi(h) \nu(dh)$$
$$+ \frac{1}{2n} \int_{\Theta} \left[\frac{d^2}{dx^2} \Psi(\theta \mapsto x\theta - M(\theta))\right]_{x = M'(t)} M''(t) p(t) dt + o\left(\frac{1}{n}\right).$$

For standard exponential families, the function M' is ono-to-one, with inverse function V. Whence, by indicating the range of M' as $\text{Cod}(M')$,

$$\int_{\Theta} \left[\frac{d^2}{dx^2} \Psi(\theta \mapsto x\theta - M(\theta))\right]_{x=M'(t)} M''(t) p(t) dt$$
$$= \int_{\text{Cod}(M')} \left[\frac{d^2}{dx^2} \Psi(\theta \mapsto x\theta - M(\theta))\right] M''(V(x)) p(V(x)) V'(x) dx$$
$$= \int_{\text{Cod}(M')} \Psi(\theta \mapsto x\theta - M(\theta)) \left[\frac{d^2}{dx^2} [M''(V(x)) p(V(x)) V'(x)]\right] dx,$$

where, for the last identity, a double integration-by-parts has been used. Finally, changing the variable according to $x = M'(t)$ leads to the desired result.

5.4. Proposition 2

A disintegration argument shows that

$$\mathfrak{R}_\mathbb{U}[U_{n,m}^\dagger] = \int_{\mathbb{X}^n} \int_{\Theta \times \mathbb{X}^m} \mathcal{L}_\mathbb{U}\left(u_{n,m}(\mathbf{y};\mathbf{x};\theta), u_{n,m}^\dagger(\mathbf{x})\right) \times$$
$$\times \mathbb{P}[(X_{n+1},\ldots,X_{n+m}) \in d\mathbf{y}, T \in d\theta \mid (X_1,\ldots,X_n) = \mathbf{x}]\alpha_n(d\mathbf{x})$$
$$= \int_{\mathbb{X}^n} \int_{\Theta \times \mathbb{X}^m} \mathcal{L}_\mathbb{U}\left(u_{n,m}(\mathbf{y};\mathbf{x};\theta), u_{n,m}^\dagger(\mathbf{x})\right) \mu^{\otimes m}(d\mathbf{y} \mid \theta) \pi_n(d\theta \mid \mathbf{x}) \alpha_n(d\mathbf{x})$$
$$= \int_{\mathbb{X}^n} \int_\Theta \mathcal{W}_\mathbb{U}^2\left(\gamma_{\theta,\mathbf{x}}; \delta_{u_{n,m}^\dagger(\mathbf{x})}\right) \pi_n(d\theta \mid \mathbf{x}) \alpha_n(d\mathbf{x}).$$

Then, use the triangular inequality for the Wasserstein distance to obtain:

$$\mathcal{W}_\mathbb{U}\left(\gamma_{\theta,\mathbf{x}}; \delta_{u_{n,m}^\dagger(\mathbf{x})}\right) \leq \mathcal{W}_\mathbb{U}\left(\gamma_{\theta,\mathbf{x}}; \gamma_{\tau,\mathbf{x}}\right) + \mathcal{W}_\mathbb{U}\left(\gamma_{\tau,\mathbf{x}}; \delta_{u_{n,m}^\dagger(\mathbf{x})}\right)$$

for any $\tau \in \Theta$. Take the square of both side and observe that:

$$\mathcal{W}_\mathbb{U}\left(\gamma_{\tau,\mathbf{x}}; \delta_{u_{n,m}^\dagger(\mathbf{x})}\right) = \int_\mathbb{U} d_\mathbb{U}^2(u, u_{n,m}^\dagger(\mathbf{x}))\gamma_{\tau,\mathbf{x}}(du).$$

Now, (31) is proved by letting $\tau = T_{n,m}^\dagger$ after noticing that the latter summand in the above right-hand side is independent of θ.

Finally, (32) is obtained by first optimizing the risk $\mathfrak{R}_\Theta$ with respect to the choice of $T_{n,m}^\dagger$ and then, after getting $\hat{T}_{n,m}$, the term $\mathbb{E}\left[\int_\mathbb{U} d_\mathbb{U}^2(U_{n,m}^\dagger, u)\gamma_{\hat{T}_{n,m},(X_1,\ldots,X_n)}(du)\right]$ is optimized with respect to the choice of $U_{n,m}^\dagger$.

5.5. Proposition 3

Preliminarily, use Theorem 1 in [52] [Section II.1] to prove that:

$$\int_\Theta \varphi(\theta) e^{-\kappa_t(\theta)} d\theta = \frac{2\sqrt{\pi}}{\sqrt{n}} e^{-\kappa_t(t)} \left[c_0 + \frac{c_2}{2n} + o\left(\frac{1}{n}\right)\right]$$

holds for any $\varphi \in C^2(\Theta)$ such that $\varphi(t) \neq 0$, where

$$c_0 := \frac{b_0}{2a_0^{1/2}}$$

$$c_2 := \left\{\frac{b_2}{2} - \frac{3a_1 b_1}{a_0} + [5a_1^2 - 4a_0 a_2]\frac{3b_0}{16a_0^2}\right\} \times \frac{1}{a_0^{3/2}},$$

with $a_0 := \frac{1}{2}\kappa_t''(t)$, $a_1 := \frac{1}{3!}\kappa_t'''(t)$, $a_2 := \frac{1}{4!}\kappa_t''''(t)$, $b_0 = \varphi(t)$, $b_1 = \varphi'(t)$ and $b_2 = \frac{1}{2}\varphi''(t)$. Moreover, from that very same theorem, it holds that:

$$\int_\Theta \varphi(\theta) e^{-\kappa_t(\theta)} d\theta = \sqrt{\pi} e^{-\kappa_t(t)} \left[\frac{c_1}{n^{3/2}} + o\left(\frac{1}{n^{3/2}}\right)\right]$$

for any $\varphi \in C^2(\Theta)$ with a zero of order 1 at t, where

$$c_1 := \left[\frac{b_1^*}{2} - \frac{a_1 b_0^*}{2a_0}\right]\frac{1}{a_0}$$

with $b_0^* := \varphi'(t)$ and $b_1^* := \frac{1}{2}\varphi''(t)$. At this stage, application of this formulas gives:

$$\int_\Theta \mathsf{m}(\theta) \pi_n^*(d\theta \mid -\kappa_t) = \mathsf{m}(t) + \frac{1}{na_0}\left[\frac{1}{4}\mathsf{m}''(t) + \frac{1}{2}\mathsf{m}'(t)\frac{p'(t)}{p(t)} - \frac{3}{4}\frac{a_1}{a_0}\mathsf{m}'(t)\right] + o\left(\frac{1}{n}\right)$$

and

$$\int_\Theta m^2(\theta)\pi_n^*(d\theta \mid -\kappa_t) = m^2(t) + \frac{1}{na_0}\left[\frac{1}{2}(m'(t))^2 + \frac{1}{2}m''(t)m(t)\right.$$
$$\left. + m'(t)m(t)\frac{p'(t)}{p(t)} - \frac{3}{2}\frac{a_1}{a_0}m'(t)m(t)\right] + o\left(\frac{1}{n}\right).$$

Then, in addition,

$$\int_\Theta [v(\theta) - v(t)]\pi_n^*(d\theta \mid -\kappa_t) = \frac{1}{na_0}\left[\frac{1}{4}v''(t) + \frac{1}{2}v'(t)\frac{p'(t)}{p(t)} - \frac{3}{4}\frac{a_1}{a_0}v'(t)\right] + o\left(\frac{1}{n}\right)$$

and

$$\int_\Theta \Delta^2(\theta,\tau)\pi_n^*(d\theta \mid -\kappa_t) = \Delta^2(t,\tau)$$
$$+ \frac{1}{na_0}\left[\frac{1}{2}\frac{\partial}{\partial\theta}\Delta^2(\theta,\tau)_{|\theta=t}\frac{p'(t)}{p(t)}\right.$$
$$\left. + \frac{1}{2}\frac{\partial^2}{\partial\theta^2}\Delta^2(\theta,\tau)_{|\theta=t} - \frac{3}{4}\frac{a_1}{a_0}\frac{\partial}{\partial\theta}\Delta^2(\theta,\tau)_{|\theta=t}\right] + o\left(\frac{1}{n}\right)$$

completing the proof just by mere substitutions.

5.6. Theorem 2

The core of the proof hinges on the identity (45). Now, the asymptotic expansion of its left-hand side is provided by (34), while the analogous expansion for right-hand side follows from a combination of (40) with (41). It is now straightforward to notice that the validity of (43) and (44) entails (45). At this stage, the validity of (46) and (47) for $\hat{R}_{0,m}$ and $\hat{R}_{1,m}$ follows directly by substitution. As to the same identities for $R_{0,m}^*$ and $R_{1,m}^*$, the argument rests on the combination of (3) with (32), exploiting the fact that the additional term (33) is of $o(1/n)$-type. Thus, the asymptotic expansion of the left-hand side of (3) is given in terms of integrals with respect to ν of the sum of the two left-hand sides of (40) and (41), respectively. Resorting once again to (45), one gets the desired identities for $R_{0,m}^*$ and $R_{1,m}^*$ by substitution.

5.7. Theorem 3

The core is the proof of (10), with the same expressions (46) and (47) also for $\tilde{R}_{0,m}$ and $\tilde{R}_{1,m}$, respectively. As in the proof of Theorem 2, the left-hand side of (10) is analyzed by resorting to inequality (32), exploiting the fact that the ensuing additional term, similar to that in (33), is of $o(1/n)$-type. Now, the argument is very similar to that of the preceding proof, with the variant that now the expansion (35) is not optimized in τ, but it is just evaluated at $\tau = \hat{\tau}_n$. The conclusion reduces once again to a matter of substituting the expressions (48)–(50) into the two expansions (35) and (41).

6. Conclusions and Future Developments

This paper should be seen as a pioneering work in the field of predictive problems, whose main aim is to show how the practical construction of efficient estimators of random quantities (that depend on future and/or past observations) entails nonstandard metrizations of the parameter space Θ. This is the essence of the compatibility Equations (43) and (44). Of course, all the lines of research proposed in this paper deserve much more attention, in order to produce new results of wider validity.

The first issue deals with the extension of the compatibility equations to higher dimensions, including the infinite dimension. For finite dimensions, this is only a technical fact. Indeed, the question relies on extending the asymptotic expansion given in Proposition 3 from dimension 1 to dimension $d > 1$. This is done in [39] as far as the Bayesian setting, and in [53,54] for a general mathematical setting. See also [55] [Section 2.2]. For the infinite dimension, the mathematical literature is rather scant. Some interesting results on asymptotic expansions of Laplace type for separable Hilbert spaces with Gaussian measure are contained in [56]. Finally, the topic is still in its early stage as far as metric measure spaces (i.e., the full nonparametric setting) are concerned. See [57,58].

Another mathematical tool that proves to be critical to our study is the Wasserstein distance. As explained in specific monographs like [27,59], the Wasserstein distance has several connections with other fields of mathematical analysis, such as optimal transportation and the theory of PDEs. Actually, the achievement of some estimators within our theory (like the one in (55)) is tightly connected with some optimization issues in transport theory. In this respect, an interesting mathematical area to explore is represented by the theory of Wasserstein barycenters and the ensuing numerical algorithms. See [60]. Research on this is ongoing.

Then, all the extensions of de Finetti's Law of Large Numbers for the log-likelihood process, stated in Theorem 1, Proposition 1 and Lemma 1 in Section 2.2, are worth being reconsidered, independently of their use for the purposes of this paper. As to possible extensions, the first hint is concerned with the analysis of dominated, parametric non-regular models, as those considered in [61–63]. Here, in fact, we never used the properties of the MLE as the root of the gradient of the log-likelihood, so that the asymptotic results contained in the quoted works should be enough to extend our statements. Subsequently, it would be also very interesting to consider dominated models which are parametrized by infinite-dimensional objects, where typically the MLE does not exist. See, e.g., the recent book [64] for plenty of examples.

As to more statistical objectives, it would be interesting to further deepen the connection between our approach and some relevant achievements obtained within the empirical Bayes theory, such as those contained in [22,23,65–68]. See also the book [69] for plenty of applications. In particular, the discussion contained in Section 4.4 about the original Poisson-mixture setting considered by Herbert Robbins deserves more attention.

A very fertile area of the application of predictive inference is that of *species sampling problems*. The pioneering works on this topic can be identified with the works [66,67,70]. Nowadays, the Bayesian approach (especially of nonparametric type) has received much attention, and has produced noteworthy new results in this field. See [17,71–73] and also [55,74,75] for novel asymptotic results. Indeed, it would be interesting to investigate whether it is possible to derive, within the approach of this paper, both asymptotic results and new estimators, hopefully more competitive than the existing ones.

Another prolific field of application is that of *density estimation*, aimed at solving clustering and/or classification problems. See [76] for a Bayesian perspective. Here, there is an additional technical difficulty due to the fact that the parameter is an element of some infinite-dimensional manifold, so that the characterization of any metric on Θ will prove mathematically more complex.

A last mention is devoted to predictive problems with "compressed data". This kind of research comes directly from computer science, where the complexity of the observed data make the available sample essentially useless for statistical inference purposes. For this reason, many algorithms have been conceived to compress the information in order to make it useful in some sense. See, e.g., [77]. Here, the Bayesian approach is in its early stage (see [78]), and the results of this paper can provide a valuable contribution.

Funding: This research received funding from the European Research Council (ERC) under the European Union's Horizon 2020 research and innovation programme under grant agreement No 817257.

Institutional Review Board Statement: Not applicable.

Informed Consent Statement: Not applicable.

Data Availability Statement: Not applicable.

Acknowledgments: I wish to express my enormous gratitude and admiration to Eugenio Regazzini. He has represented for me a constant source of inspiration, transmitting enthusiasm and method for the development of my own research. This paper represents a small present for his 75-th birthday.

Conflicts of Interest: The author declares no conflict of interest.

References

1. Cifarelli, D.M.; Dolera, E.; Regazzini, E. Note on "Frequentist Approximations to Bayesian prevision of exchangeable random elements" [Int. J. Approx. Reason. 78 (2016) 138–152]. *Int. J. Approx. Reason.* **2017**, *86*, 26–27. [CrossRef]
2. Cifarelli, D.M.; Dolera, E.; Regazzini, E. frequentist approximations to Bayesian prevision of exchangeable random elements. *Int. J. Approx. Reason.* **2016**, *78*, 138–152. [CrossRef]
3. Dolera, E. On an asymptotic property of posterior distributions. *Boll. Dell'Unione Mat. Ital.* **2013**, *6*, 741–748. (In Italian)
4. Dolera, E.; Regazzini, E. Uniform rates of the Glivenko–Cantelli convergence and their use in approximating Bayesian inferences. *Bernoulli* **2019**, *25*, 2982–3015. [CrossRef]
5. de Finetti, B. Bayesianism: Its unifying role for both the foundations and applications of statistics. *Int. Stat. Rev.* **1974**, *42*, 117–130. [CrossRef]
6. de Finetti, B. La prévision: Ses lois logiques, ses sources subjectives. *Ann. L'Inst. Henri Poincaré* **1937**, *7*, 1–68.
7. Ferguson, T.S. *Mathematical Statistics: A Decision Theoretic Approach*; Academic Press: Cambridge, MA, USA, 1967.
8. Lehmann, E.L.; Casella, G. *Theory of Point Estimation*, 2nd ed.; Springer: Berlin/Heidelberg, Germany, 1998.
9. Aldous, D.J. *Exchangeability and Related Topics*; Ecole d'Eté de Probabilités de Saint-Flour XIII, Lecture Notes in Mathematics; Springer: Berlin/Heidelberg, Germany, 1985; pp. 1–198.
10. Berti, P.; Pratelli, L.; Rigo, P. Exchangeable sequences driven by an absolutely continuous random measure. *Ann. Probab.* **2013**, *78*, 138–152. [CrossRef]
11. Fortini, S.; Ladelli, L.; Regazzini, E. Exchangeability, predictive distributions and parametric models. *Sankhya* **2000**, *62*, 86–109.
12. Rubin, D.B. Bayesianly justifiable and relevant frequency calculations for the applied statisticians. *Ann. Stat.* **1984**, *12*, 1151–1172. [CrossRef]
13. Lijoi, A.; Prünster, I. Models beyond the Dirichlet process. In *Bayesian Nonparametrics*; Hjort, N.L., Holmes, C.C., Müller, P., Walker, S.G., Eds.; Cambridge University Press: Cambridge, UK, 2010; pp. 80–136.
14. Robbins, H. The empirical Bayes approach to statistical decision problems. *Ann. Math. Stat.* **1964**, *35*, 1–20. [CrossRef]
15. Ghosh, J.K.; Sinha, B.K.; Joshi, S.N. Expansions for posterior probability and integrated Bayes risk. In *Statistical Decision Theory and Related Topics III*; Gupta, S., Berger, J., Eds.; Academic Press: Cambridge, MA, USA, 1982; pp. 403–456.
16. Favaro, S.; Nipoti, B.; Teh, Y.W. Rediscovery of Good-Turing estimators via Bayesian nonparametrics. *Biometrics* **2016**, *72*, 136–45. [CrossRef] [PubMed]
17. Lijoi, A.; Mena, R.H.; Prünster, I. Bayesian Nonparametric Estimation of the Probability of Discovering New Species. *Biometrika* **2009**, *94*, 769–786. [CrossRef]
18. de Finetti, B. Probabilità di una teoria e probabilità dei fatti. In *Studi di Probabilità, Statistica e Ricerca Operativa in onore di Giuseppe Pompilj*; Oderisi: Gubbio, Italy, 1971; pp. 86–101. (In Italian)
19. Rao, R.C. Information and the accuracy attainable in the estimation of statistical parameters. *Bull. Calcutta Math. Soc.* **1945**, *37*, 81–91.
20. Amari, S.-I. *Information Geometry and Its Applications*; Applied Mathematical Sciences; Springer: Berlin/Heidelberg, Germany, 2016; Volume 194.
21. Oller, J.M.; Corcuera, J.M. Intrinsic analysis of statistical estimation. *Ann. Stat.* **1995**, *23*, 1562–1581. [CrossRef]
22. Zhang, C.-H. Estimation of sums of random variables: Example and information bounds. *Ann. Stat.* **2005**, *33*, 2022–2041. [CrossRef]
23. Robbins, H. An empirical Bayes approach to statistics. In *Proceedings of the Third Berkeley Symposium on Mathematical Statistics and Probability*; Statistical Laboratory of the University of California: Davis Davis, CA, USA, 1956; Volume I, pp. 157–163.
24. Berezin, S.; Miftakhov, A. On barycenters of probability measures. *Bull. Pol. Acad. Sci. Math.* **2020**, *68*, 11–20. [CrossRef]
25. Karcher, H. Riemannian center of mass and mollifier smoothing. *Commun. Pure Appl. Math.* **1977**, *30*, 509–541. [CrossRef]
26. Kim, Y.-H.; Pass, B. Nonpositive curvature, the variance functional, and the Wasserstein barycenter. *Proc. Am. Math. Soc.* **2000**, *148*, 1745–1756. [CrossRef]
27. Ambrosio, L.; Gigli, N.; Savaré, G. *Gradient Flows in Metric Spaces and in the Space of Probability Measures*, 2nd ed.; Birkhäuser: Basel, Switzerland, 2008.
28. Billingsley, P. *Probability and Measure*, 3rd ed.; John Wiley & Sons: Hoboken, NJ, USA, 1995.
29. do Carmo, M.P. *Riemannian Geomerty*; Birkhäuser: Basel, Switzerland, 2013.
30. Heinonen, J.; Kilpeläinen, T.; Martio, O. *Nonlinear Potential Theory of Degenerate Elliptic Equations*; Oxford Science Publications: Oxford, UK, 2008.
31. Kufner, A. *Weighted Sobolev Spaces*; John Wiley & Sons: Hoboken, NJ, USA, 1985.
32. de Finetti, B. La legge dei grandi numeri nel caso dei numeri aleatori equivalenti. *Rend. Della R. Accad. Naz. Lincei* **1933**, *18*, 203–207. (In Italian)

33. Bauschke, H.H.; Combettes, P.L. *Convex Analysis and Monotone Operator Theory in Hilbert Spaces*, 2nd ed.; Springer: Berlin/Heidelberg, Germany, 2017.
34. Borwein, J.M.; Noll, D. Second order differentiability of convex functions in Banach spaces. *Trans. Am. Math. Soc.* **1994**, *132*, 43–81. [CrossRef]
35. Dolera, E.; Favaro, S. Rates of convergence in de Finetti's representation theorem, and Hausdorff moment problem. *Bernoulli* **2020**, *26*, 1294–1322. [CrossRef]
36. Mijoule, G.; Peccati, G.; Swan, Y. On the rate of convergence in de Finetti's representation theorem. *Lat. Am. J. Probab. Math. Stat.* **2016**, *13*, 1–23. [CrossRef]
37. Dolera, E. Estimates of the approximation of weighted sums of conditionally independent random variables by the normal law. *J. Inequal. Appl.* **2013**, *2013*, 320. [CrossRef]
38. Götze, F. On the rate of convergence in the central limit theorem in Banach Spaces. *Ann. Probab.* **1986**, *14*, 922–942. [CrossRef]
39. Tierney, L.; Kadane, J.B. Accurate approximations for posterior moments and marginal densities. *J. Am. Stat. Assoc.* **1986**, *81*, 82–86. [CrossRef]
40. DasGupta, A. *Asymptotic Theory of Statistics and Probability*; Springer: Berlin/Heidelberg, Germany, 2008.
41. Dall'Aglio, G. Sugli estremi dei momenti delle funzioni di ripartizione doppia. *Ann. Della Sc. Norm. Super. Pisa Cl. Sci.* **1956**, *10*, 35–74. (In Italian)
42. Dolera, E.; Mainini, E. On Uniform Continuity of Posterior Distributions. *Stat. Probab. Lett.* **2020**, *157*, 108627. [CrossRef]
43. Dolera, E.; Mainini, E. Lipschitz continuity of probability kernels in the optimal transport framework. *arXiv* **2020**, arXiv:2010.08380.
44. Dowson, D.C.; Landau, B.V. The Fréchet distance between multivariate normal distributions. *J. Multivar. Anal.* **1982**, *12*, 450–455. [CrossRef]
45. Olkin, I.; Pukelsheim, F. The distance between two random vectors with given dispersion matrices. *Linear Algebra Its Appl.* **1982**, *48*, 257–263. [CrossRef]
46. Malagó, L.; Montrucchio, L.; Pistone, G. Wasserstein Riemannian geometry of positive definite matrices. *Inf. Geom.* **2018**, *1*, 137–179. [CrossRef]
47. Efron, B.; Hastie, T. *Computer Age Statistical Inference. Algorithms, Evidence, and Data Science*; Cambridge University Press: Cambridge, UK, 2016.
48. Thyrion, P. Contribution à l'étude du bonus pour non sinistre en assurance automobile. *ASTIN Bull. J. IAA* **1960**, *1*, 142–162. (In French) [CrossRef]
49. van Houwelingen, J.C. Monotonizing empirical Bayes estimators for a class of discrete distributions with monotone likelihood ratio. *Stat. Neerl.* **1977**, *31*, 95–104. [CrossRef]
50. Carlin, B.P.; Louis, T.A. *Bayesian Methods for Data Analysis*, 3rd ed.; Chapman and Hall: Boca Raton, FL, USA, 2009.
51. Ledoux, M.; Talagr, M. *Probability in Banach Spaces*; Springer: Berlin/Heidelberg, Germany, 1991.
52. Wong, R. *Asymptotic Approximations of Integrals*; SIAM: Philadelphia, PA, USA, 2001.
53. McClure, J.P.; Wong, R. Error bounds for multidimensional Laplace approximation. *J. Approx. Theory* **1983**, *37*, 372–390. [CrossRef]
54. Olver, F.W.J. Error bounds for the Laplace approximation for definite integrals. *J. Approx. Theory* **1968**, *1*, 293–313. [CrossRef]
55. Dolera, E.; Favaro, S. A Berry–Esseen theorem for Pitman's α–diversity. *Ann. Appl. Probab.* **2020**, *30*, 847–869. [CrossRef]
56. Albeverio, S.; Steblovskaya, V. Asymptotics of infinite-dimensional integrals with respect to smooth measures. (I). *Infin. Dimens. Anal. Quantum Probab. Relat. Top.* **1999**, *2*, 529–556. [CrossRef]
57. Gigli, N. Second order analysis on $(\mathscr{P}_2(M), W_2)$. *Mem. Am. Math. Soc.* **2012**, *216*, xii+154.
58. Gigli, N.; Ohta, S.I. First variation formula in Wasserstein spaces over compact Alexandrov spaces. *Can. Math. Bull.* **2010**, *55*, 723–735. [CrossRef]
59. Villani, C. *Optimal Transport. Old and New*; Springer: Berlin/Heidelberg, Germany, 2009.
60. Cuturi, M.; Doucet, A. Fast Computation of Wasserstein Barycenters. In Proceedings of the 31st International Conference on Machine Learning, Beijing, China, 21–26 June 2014; Volume 32, pp. 685–693.
61. Smith, R.L. Maximum likelihood estimation in a class of nonregular cases. *Biometrika* **1985**, *72*, 67–90. [CrossRef]
62. Woodroofe, M. Maximum likelihood estimation of a translation parameter of a truncated distribution. *Ann. Math. Stat.* **1972**, *43*, 113–122. [CrossRef]
63. Woodroofe, M. Maximum likelihood estimation of a translation parameter of a truncated distribution (II). *Ann. Stat.* 1974, *2*, 474–488. [CrossRef]
64. Giné, E.; Nickl, R. *Mathematical Foundations of Infinite-Dimensional Statistical Models*; Cambridge Series in Statistical and Probabilistic Mathematics: Cambridge, UK, 2016.
65. Efron, B.; Thisted, R. Estimating the number of unseen species: How many words did Shakespeare know? *Biometrika* **1976**, *63*, 435–447. [CrossRef]
66. Good, I.J. The population frequencies of species and the estimation of population parameters. *Biometrika* **1953**, *40*, 237–264. [CrossRef]
67. Good, I.J.; Toulmin, G.H. The number of new species, and the increase in population coverage, when a sample is increased. *Biometrika* **1956**, *43*, 45–63. [CrossRef]
68. Orlitsky, A.; Suresh, A.T.; Wu, Y. Optimal prediction of the number of unseen species. *Proc. Natl. Acad. Sci. USA* **2016**, *113*, 13283–13288. [CrossRef]

69. Maritz, J.S.; Lwin, T. *Empirical Bayes Methods with Applications*; Chapman and Hall: Boca Raton, FL, USA, 1989
70. Fisher, R.A.; Corbet, A.S.; Williams, C.B. The relation between the number of species and the number of individuals in a random sample of an animal population. *J. Anim. Ecol.* **1943**, *12*, 42–58. [CrossRef]
71. Favaro, S.; Lijoi, A.; Mena, R.H.; Prünster, I. Bayesian nonparametric inference for species variety with a two parameter Poisson-Dirichlet process prior. *J. Roy. Statist. Soc. Ser. B* 2009, *71*, 993–1008. [CrossRef]
72. Favaro, S.; Lijoi, A.; Prünster, I. A new estimator of the discovery probability. *Biometrics* **2012**, *68*, 1188–1196. [CrossRef]
73. Arbel, J.; Favaro, S.; Nipoti, B.; Teh, Y.W. Bayesian nonparametric inference for discovery probabilities: Credible intervals and large sample asymptotic. *Stat. Sin.* **2017**, *27*, 839–858. [CrossRef]
74. Dolera, E.; Favaro, S. A compound Poisson perspective of Ewens–Pitman sampling model. *Mathematics* **2021**, *9*, 2820. [CrossRef]
75. Pitman, J. *Combinatorial Stochastic Processes*; Ecole d'Eté de Probabilités de Saint-Flour XXXII, Lecture Notes in Mathematics; Springer: Berlin/Heidelberg, Germany, 2006.
76. Sambasivan, R.; Das, S.; Sahu, S.K. A Bayesian perspective of statistical machine learning for big data. *Comput. Stat.* **2020**, *35*, 893–930. [CrossRef]
77. Cormode, G.; Yi, K. *Small Summaries for Big Data*; Cambridge University Press: Cambridge, UK, 2020
78. Dolera, E.; Favaro, S.; Peluchetti, S. Learning-augmented count-min sketches via Bayesian nonparametrics. *arXiv* **2021**, arXiv:2102.04462.

Article

Fisher, Bayes, and Predictive Inference †

Sandy Zabell

Department of Mathematics, Northwestern University, Evanston, IL 60208, USA; zabell@math.northwestern.edu
† Dedicato a Eugenio Regazzini, con ammirazione, affetto, e stima.

Abstract: We review historically the position of Sir R.A. Fisher towards Bayesian inference and, particularly, the classical Bayes–Laplace paradigm. We focus on his Fiducial Argument.

Keywords: Bayesian inference; Fisher fiducial argument; inverse probability; uniform distribution

MSC: 62F15

1. Introduction

R.A. Fisher was—famously—an outspoken critic of Bayesian inference. Yet, anyone who reads Fisher's later work cannot but be struck by Fisher's evident admiration for Bayes himself. This paper discusses this apparent paradox. There are really three threads to this story. One is to recognize that Fisher was a critic of "inverse probability" (in the narrow sense of the use of uniform priors to capture the notion of ignorance or lack of knowledge about a prior) and not "Bayesian inference" per se (if by the latter, we mean the use of prior distributions summarizing genuine knowledge). Indeed, Fisher appears to have introduced the use of the term "Bayesian"; see [1]. A second thread in our story is Fisher's discovery of fiducial inference in 1930 as an alternative to the Bayesian approach. Never widely accepted in the statistical community (unlike virtually all of Fisher's other contributions to statistics), its ultimately inference-based approach meant Fisher came to associate himself more with statisticians such as his fellow Cambridge don Harold Jeffreys (an "objective Bayesian") rather than his arch-enemy and rival, the frequentist Jerzy Neyman. The final thread is how all this played itself out in the 1950s. Sensing then that he was on the wrong side of history, during the last decade of his life, Fisher made a supreme effort to clarify and recast the logical basis of his approach to statistical inference, most notably in his 1956 book *Statistical Methods and Scientific Inference* [2–5]. This in turn led him to return to Bayes himself and his famous paper [6], for which Fisher had only praise.

2. Fisher on Inverse Probability in the 1920s

With two exceptions [7,8], Fisher's papers on statistics all date from 1920 on. By this point, he was already a dedicated opponent of the use of inverse probability, writing that Bayes' attempt ≪admittedly depended upon an arbitrary assumption, so that the whole method has been widely discredited≫ [9] (p. 4), and a year later was even more forceful, pointing with disdain to:

≪inverse probability, which like an impenetrable jungle arrests progress towards precision of statistical concepts.≫ [10] (p. 311)

However, it was not always so; this position represented a change in view, as Fisher later acknowledged:

≪I may myself say that I learned [inverse probability] at school as an integral part of the subject, and for some years saw no reason to question its validity.≫ [11] (p. 248)

Indeed, Fisher conceded that not only had he accepted the legitimacy of inverse probability, he had even employed it in his very first paper:

≪I must indeed plead guilty in my original statement of the Method of Maximum Likelihood [in 1912] to having based my argument upon the principle of inverse probability; in the same paper, it is true, I emphasized the fact that such inverse probabilities were relative only.≫ [12] (p. 326)

There has indeed been some discussion and debate about the extent to which Fisher's 1912 paper did employ "the principle of inverse probability" (see [13,14]), but for our purposes, the key point is Fisher's unambiguous statement that, initially, he "saw no reason to question its validity".

2.1. Fisher's Critique of the Uniform Prior

By inverse probability, Fisher really meant the classical assumption and use of uniform priors (rather than Bayesian inference in the more modern sense of the use of a prior summarizing initial information). Fisher had several grounds on which he criticized the use of such uniform priors; one of these was that uniform priors are not scale invariant. For example, if $0 \leq p \leq 1$ is the probability of a success in a sequence of Bernoulli trials, and:

$$\sin \theta = 2p - 1,$$

then the flat prior dp on the p scale corresponds to the prior:

$$\frac{\cos \theta}{2} d\theta$$

on the θ scale (see [10] (p. 325) and [4]) (pp. 16–17). Thus, this formulation of ignorance is scale dependent, even though, in a hypothetical state of "complete ignorance" (whatever that is), there should be no reason to prefer one scale over the other; instead of the parameter p, we "might, so far as cogent evidence is concerned, equally have taken any monotonic function of p" in its place.

Interestingly, Fisher was not the first to raise this objection; much earlier, the German physiologist Johannes von Kries (1853–1928), in his *Principien der Wahrscheinlichkeitsrechnung* (see [15] (p. 31) and also [16]), gave the scientifically more interesting example of

$$\sigma: \text{specific weight,} \qquad \omega = \frac{1}{\sigma}: \text{specific volume,}$$

noting that a uniform prior for one is not uniform for the other, and thereby concluding "Der Wahrscheinlichkeits-Ansatz ist also unbestimmt" (the probability assumption is thus indeterminate). Nor was this an isolated, special instance: other examples could easily be given, such as in the case of a pendulum, the length and the duration or frequency of oscillation.

2.2. Fisher vs. Pearson on the Correlation Coefficient

Although this objection (and other critiques given in von Kries and later reported by Keynes in his classic 1921 *Treatise on Probability*, see [17]) were certainly cogent, one might wonder at the vehemence with which the young Fisher pressed. However, here, there is a simple explanation, stemming from Fisher's conflicted relations with Karl Pearson (1857–1936), then the towering figure in English statistics, the only person at that point to hold a chair in statistics in England (at University College London).

In 1915, Fisher published his first great paper on the distribution of the sample correlation coefficient, in Pearson's journal *Biometrika*. In the last section of his paper (see [8] (pp. 520–521)), Fisher derived a point estimate for the theoretical correlation coefficient ρ using his earlier method of maximum likelihood (see [7]).

The MLE can be interpreted as the mode of the posterior distribution of a parameter starting from a uniform prior, although Fisher did not frame the issue this way in his 1912 paper. Two years later, however, a "Cooperative Study" in *Biometrika* with Pearson as a co-author (see [18]) interpreted Fisher as doing precisely this, criticizing his ostensible use of a uniform prior in this case as contrary to common experience, and going on to discuss several alternative priors (pp. 352–360). Angered, Fisher wrote in response:

≪[Pearson's] comments upon my methods imply such a serious misunderstanding of my meaning that a brief reply is necessary...

From this passage a reader, who did not refer my paper, which had appeared in the previous year, and to which the *Cooperative Study* was called an "Appendix", might imagine that I had used Boole's ironical phrase "equal distribution of ignorance", and that I had appealed to "Bayes' theorem". I must therefore state that I did neither.≫

These words did not appear in Pearson's journal *Biometrika*, for Pearson had refused to publish Fisher's response, and marked a clear and acrimonious break in relations between the two. It is easy to appreciate Fisher's bitterness given his relative youth and professional vulnerability at this early stage in his career. It is nevertheless surprising to see how much this still rankled twenty years after Pearson's death, when Fisher wrote in his book *Statistical Methods and Scientific Inference* ([2] (pp. 2–3), cited below as SMSI, page references to the third, 1973 edition unless otherwise noted):

≪Pearson's energy was unbounded. In the course of his long life he gained the devoted service of a number of able assistants, some of whom he did not treat particularly well. He was prolific in magnificent, or grandiose, schemes capable of realization perhaps by an army of industrious robots responsive to a magic wand.

The terrible weakness of his mathematical and scientific work flowed from his incapacity in self-criticism, and his unwillingness to admit the possibility that he had anything to learn from others, even in biology, of which he knew very little. His mathematics, consequently, though always vigorous, were usually clumsy, and often misleading. In controversy, to which he was much addicted, he constantly showed himself to be without a sense of justice.≫

Another revealing episode was an ill-advised invitation to Fisher to write an entry on Pearson for the *Dictionary of National Biography* shortly after Pearson's death in 1936; see [19]. After several drafts in which the editors were unable to persuade Fisher to tone down his evident contempt for Pearson, the project was abandoned by mutual agreement.

3. The Fiducial Argument

Fisher's harsh criticism of inverse probability was ultimately grounded in principle, even if it was in part sharpened by his animus towards Pearson. However, it is difficult to replace something by nothing, and as long as a credible alternative to inverse methods was missing, inverse probability could be expected to survive. As the Harvard mathematician Julian Lowell Coolidge noted, ≪defective as it is, Bayes' formula is the only thing we have to answer certain important questions which do arise in the calculus of probability... we use Bayes' formula with a sigh, as the only thing available under the circumstances≫ (see [20] p. 100). However, in 1930, Fisher believed he had in fact found such an alternative.

3.1. The Original Fiducial Argument

In 1930, Fisher wrote a short paper titled "Inverse probability" [21], which was the origin of his "fiducial argument", ultimately leading to more than three decades of controversy. After revisiting his oft-stated attack on the inverse probability approach, Fisher wrote:

≪There are, however, certain cases in which statements in terms of probability can be made with respect to the parameters of the population.≫

How did Fisher succeed in this, breaking the "Bayesian omelet without breaking the Bayesian eggs"? From a modern perspective, what Fisher discovered was a general method of constructing confidence intervals in the case of a continuous one-parameter family of probability distributions. Curiously and remarkably, there is (almost) nothing controversial in this paper. Indeed, according to Charles Stein (who was in no way a Fisherian), there was only a single objectionable passage in the entire paper (personal communication from Charles Stein to SZ, c. 1980; this would have been the reference on p. 534 to "the fiducial distribution of a parameter θ for a given statistic T").

Here is a brief *précise* of Fisher's discovery recast in modern language. Let:

- T be a sample statistic having a continuous distribution;
- θ be a one-dimensional parameter for the distribution of T;
- $t_p(\theta)$ be the p-th quantile for T under θ:

$$P_\theta[T < t_p(\theta)] = p.$$

If t_p is strictly increasing in θ, then t_p^{-1} is also strictly increasing, and

$$P_\theta[t_p^{-1}(T) < \theta] = p.$$

Therefore, $(t_p^{-1}(T), \infty)$ is a (lower) $100p\%$ confidence interval for θ.

Perhaps Fisher's clearest statement of what he had in mind here was a letter from Fisher to Tukey, dated 27 April 1955 ([22] pp. 220–222):

≪The probability integral of the exact frequency distribution, in finite samples, of an exhaustive statistic is used to form a continuum of probability-statements, of the form

$$Pr\{T < T_P(\theta)\} = P,$$

and using the monotonic property of the functions T_p for all P, this is transformed to the equivalent

$$Pr\{\theta_{1-P}(T) < \theta\} = P,$$

a complete set of probability statements about θ, in terms known for a given sample value T.≫

3.2. Example

The following simple example should clarify the basic conceptual issues. Suppose T is an exponentially distributed random variable with parameter $\theta > 0$ and cumulative distribution function:

$$F(t, \theta) := P_\theta(T \leq t) = 1 - e^{-\theta t} \quad (0 < t < \infty).$$

If we let $F(t, \theta) = p$, then the relation $p = 1 - e^{-\theta t}$ implicitly defines functions

$$t_p(\theta) := -\frac{1}{\theta} \log(1-p), \qquad \theta_p(t) := -\frac{1}{t} \log(1-p).$$

It is immediate that t_p and θ_p are strictly decreasing functions mapping $(0, \infty)$ to itself and inverses of each other:

$$\theta_p(t_p(\theta)) = \theta, \qquad t_p(\theta_p(\theta)) = \theta.$$

In modern parlance, $t_p(\theta)$ is the *critical value* for a $100(1-p)\%$ test of significance for the parameter θ in which one rejects θ whenever $T > t_p(\theta)$. Since $T > t_p(\theta) \iff \theta_p(T) < \theta$, the random interval $(0, \theta_p(T))$ consists of precisely those θ *consistent* with the observed value of T in the sense that it contains those θ not rejected by the test of significance. Put another way, this random interval contains the true value of θ $100p\%$ of the time; in Neyman's terminology, it is a $100p\%$ confidence interval for θ conveniently summarizing the corresponding continuum of tests of significance.

For any θ, $F(t, \theta)$ is the cumulative distribution function of the random variable T under θ, but equally, given $1 - e^{-\theta t}$ is symmetric in t and θ, for every t, the function

$F(t, \theta)$ is a mathematical distribution function (in the sense that it is a left-continuous increasing function such that $\lim_{\theta \to 0} F(t, \theta) = 0$ and $\lim_{\theta \to \infty} F(t, \theta) = 1$). In the statistical model initially described, however ($\{P_\theta, 0 < \theta < \infty\}$), it is not immediately evident what random variable it is the distribution of. Although we are given a family of distributions P_θ on $\mathbb{R}^+$, the sample space of possible values of T, we do not have a corresponding family of distributions P_t on $\Theta := (0, \infty)$, the space of possible values of θ, let alone a joint distribution on $\mathbb{R}^+ \times \Theta$.

Fisher recognized this, for he was at pains to explain (in the context of his more complex example involving the sample correlation coefficient) the intimate relationship between the interpretation of a fiducial percentile and the original sampling distribution of the underlying statistic, and that the former only has a meaning when expressed in terms of the latter. Using the concrete example of four pairs of observations drawn from a bivariate normal and using a table that had been computed in this case, Fisher wrote:

≪From the table we can read off the 95 per cent. r for any given ρ, or equally the fiducial 5 per cent. ρ for any given r. Thus if a value $r = 0.99$ were obtained from the sample, we should have a fiducial 5 per cent. ρ equal to about 0.765. The value of ρ can then only be less than 0.765 in the event that r has exceeded its 95 per cent. point, an event which is known to occur just once in 20 trials. In this sense ρ has a probability of just 1 in 20 of being less than 0.765.≫ [21] (p. 534)

Can such an approach be extended to the case of several parameters? Fisher tried to do this after Jerzy Neyman advanced his own multi-parameter theory of confidence intervals ([23]), but the resulting difficulties led to an important shift in Fisher's views, including almost immediately the abandonment of the sampling interpretation of fiducial probabilities; see generally [24]. This led after Fisher's break with Neyman to a surprising detente with the objective Bayesian Sir Harold Jeffreys. On one occasion, Fisher told Jeffreys he agreed with Jeffreys' approach more than the current school of Neyman, and Jeffreys very emphatically replied, ≪Yes, we are closer in our approach≫, as related by S. K. Runcorn, who knew Fisher as a Fellow at Fisher's College Gonville and Caius; see [25] (p. 441). This change is evident in Fisher's last book, *Statistical Methods and Scientific Inference* (1956).

3.3. A Short Period of Peaceful Coexistence

Another interesting aspect of Fisher's views in this early period was his willingness to accept that Bayesian and fiducial probabilities can co-exist.

≪The fiducial frequency distribution will in general be different numerically from the inverse probability distribution obtained from any particular hypothesis as to a priori probability. Since such an hypothesis may be true, it is obvious that the two distributions must differ not only numerically, but in their logical meaning.

There is ... no contradiction between the two statements.≫

Fisher later changed his mind on this, arguing at the end of his life that both statements had the same logical meaning (referring to probabilities of the same nature), the fiducial distribution corresponding to the case of the complete lack of information, the Bayesian prior representing a state of very specific information. This is discussed below in Section 5.

4. Bayes and Predictive Inference

Fisher's 1956 book *Statistical Methods and Scientific Inference* (SMSI) was an attempt to explain and justify his statistical methods and views. It is not in a class with his other, prewar books: *Statistical Methods for Research Workers* (1925) gave a coherent framework for statistical theory and methods that dominated the field for more than half a century; *The Genetical Theory of Natural Selection* (1930) established Fisher as one of the three leading population geneticists of the first half of the 20th Century; *The Design of Experiments* (1935) created an entirely new field of statistics. Nevertheless, SMSI is filled with interesting ideas and examples that reward a close reading. Here, we focus primarily on what it has to say about the two (interrelated) topics of Bayes and predictive inference. Page references are to the third (posthumous) 1973 edition. See the note at the beginning of the bibliography.

4.1. Bayes' "Billiard Ball"

Fisher (SMSI, p. 132) distinguished between what Bayes actually wrote in his *Essay*, and the later, uncritical use of uniform priors by other writers:

≪In mathematical teaching the mistake is often made of overlooking the fact that Bayes obtained his probabilities *a priori* by an appropriate *experiment*, and that he specifically rejected the alternative of introducing them axiomatically on the ground that this "might not perhaps be looked at by all as reasonable"; moreover, he did not wish to "take into his mathematical reasoning anything that might admit dispute". This passage (and additional text) was added in the 2nd edition, pp. 127–128, and still further further material was added to this section (5.26, "Observations of two kinds") in the 3rd edition as well.≫

What was Bayes' experimental alternative to the axiomatic assumption of a prior? Bayes envisaged the following procedure:

- Pick a point uniformly on a rectangular table.
- Project this point onto the horizontal axis of the table.
- Posit the position of this point O on the axis to be uniformly distributed.
- Record whether or not subsequent points X selected on the axis in this way lie to the left of O.

This hypothetical process generates a sequence of independent Bernoulli p-trials:

$$X_1, X_2, \ldots, X_n, \ldots \quad \text{having} \quad p \sim \text{Unif}[0,1].$$

Proceeding in this way has some immediate, but simple mathematical consequences: if $S_n = X_1 + \cdots + X_n$, $q = 1 - p$ and $n = a + b$, then:

1. $P(S_n = a) = \dfrac{1}{n+1}$, $0 \leq a \leq n$ (the discrete uniform prior);

2. The posterior of p given $S_n = a$: $\dfrac{(a+b+1)!}{a!b!} p^a q^b \, dp$;

3. $P(X_{n+1} = 1 | S_n = a) = \dfrac{a+1}{n+2}$ (Laplace's "rule of succession");

4. $P(c, d | a, b) = \dfrac{(a+b+1)!}{a!b!} \cdot \dfrac{(a+c)!(b+d)!}{(a+b+c+d+1)!} \cdot \dfrac{(c+d)!}{c!d!}$.

Fisher had no problem with this approach because Bayes was assuming the uniform prior on p was the result of a physical mechanism, rather than the claim that it captured the idea of a supposed state of ignorance. The point is that if the distribution of p is uniform, then $P(S_n = a)$ is also uniform, and Bayes argued it was this that captured the idea of the absence of knowledge concerning the probability of the event; see [26,27] and [28] (pp. 743–753) for further discussion. Bayes [6] (p. 393) refers to ≪an event concerning the probability of which we absolutely know nothing antecedently to any trials made concerning it≫, making it clear the ignorance in question is about the probability of the event, not the event itself. Similarly, Fisher (SMSI, p. 58) refers to the "absence of knowledge *a priori* of the distribution of θ" as the precondition for the applicability of the fiducial argument.

4.1.1. Karl Pearson Enters the Fray

However, to virtually all readers of Bayes, the subtlety of this passage from the uniformity of the distribution of the continuous parameter p to that of the discrete parameter a went unappreciated, and thus, there remained the nagging problem of justifying the assumption of uniformity. Against this background, in 1920 Karl Pearson, in his paper "The fundamental problem of practical statistics" in *Biometrika*, thought he had at last discovered a solution to this conundrum ([29] p. 4):

> ≪It has occurred to me that possibly the bull itself is a chimera, and there may be no need whatever to master it. In short, is it not possible that any continuous distribution of a priori chances would lead us equally well to the Bayes-Laplace result? If this be so, then the main line of attack of its critics fails.≫

Then, after a page of dense mathematical argumentation, Pearson triumphantly concluded:

> ≪Thus it would appear that the fundamental formula of Laplace ...in no way depends on the equal distribution of ignorance. It is sufficient to assume any continuous distribution–which may vary from one type of a priori probability problem to a second.≫

This bold claim did not go unchallenged, and soon after, both Edgeworth [30] (pp. 82–83) and the mathematician William Burnside [31] wrote to challenge Pearson's bold claim. In fact, Pearson's assertion is obviously absurd: one need only use any mathematically convenient prior (such as the Dirichlet) to compute the resulting posterior predictive distribution and see it is *not* uniform. Therefore, what did Pearson actually discover? Suppose:

- X is the position of the initial point O;
- $F(x) = P(X \leq x)$ is the CDF of X;
- F is continuous, *but not necessarily uniform*.

Then, as is well known, the distribution of the probability integral is uniform: $p = F(X) \sim \text{Unif}[0, 1]$. That is, the distribution of the chance p is still uniform even though the location of X (the mechanism for generating it) need not be.

Karl Pearson eventually understood this crucial distinction [32,33], but only with some help:

> ≪I owe to Miss Ethel Newbold this insight into the exact relation between the two hypotheses.≫ [33] (p. 192)

4.1.2. Fisher's Version

Fisher discusses this example in SMSI (not mentioning Pearson), playfully invoking radioactive decay in place of Bayes' "billiard ball". Letting $p = e^{-\xi\theta}$ be the probability (for some fixed $\theta > 0$) that an atomic particle does not decay up to time ξ, Fisher [2] (p. 132) considered the case where ξ is a multiple c of an earlier observation x, $\xi = cx$ (so that when $c = 1$, we are back in Bayes' original setting):

> ≪In many continental countries this distinction, which Bayes made perfectly clear, has been overlooked, and the axiomatic approach which he rejected has actually been taught as Bayes' method. The example of this Section exhibits Bayes' own method, replacing the billiard table by a radioactive source, as an apparatus more suitable for the 20th century.≫

4.2. The Rule of Succession

Fisher, as a harsh critic of inverse probability, would presumably not feel kindly towards one of its most famous applications, Laplace's rule of succession, but somewhat surprisingly, he effectively endorses it. In [2] (Chapter III, Section 5), Fisher contrasts three posterior means of p:

$$\frac{a}{N} + \frac{b-a}{2N^2} - \frac{3(b-a)}{2N^3} + \ldots \quad \text{(the fiducial approximation),}$$

$$\frac{a+1}{N+2} = \frac{a}{N} + \frac{b-a}{N^2} - \frac{2(b-a)}{2N^3} + \ldots \quad \text{(arising from the flat prior),}$$

$$\frac{a+\frac{1}{2}}{N+1} = \frac{a}{N} + \frac{b-a}{2N^2} - \frac{(b-a)}{2N^3} + \ldots \quad \text{(arising from the Jeffreys prior} \frac{1}{\pi\sqrt{pq}}dp\text{).}$$

All of these agree to first order, and so, as Fisher notes, "to this extent a particular given distribution *a priori* may be nearly equivalent to complete ignorance *a priori*". This is one of those places in SMSI where significant changes were made between the first and second editions. In [3] (pp. 68–69), Fisher noted that the fiducial approximate and Jeffreys' posterior means in fact agree to fourth (N^4) order if allowance is made for the effects of the non-normality of the binomial distribution.

Of course, given that the second and third methods closely approximate Fisher's fiducial approximation (approximate because the variate is discrete), he could scarcely argue (as Venn did in the second edition of his *Logic of Chance*) that the answer given by the rule of succession was absurd on its face.

4.3. Bayesian Prediction: A Conundrum

As a final example of Fisher's discussion of predictive inference, let us turn to a curious passage on pp. 116–117 of SMSI. Returning to the fourth and last of the mathematical consequences of Bayes' postulate listed earlier, Fisher observes:

≪It may be noticed that the last factor in the expression developed above [predicting (c,d) from (a,b) assuming a uniform prior],

$$\frac{(c+d)!}{c!\,d!}$$

stands only for the binomial coefficients forming the last line, or base, of Fermat's arithmetical triangle; but

$$\sum_{c=0}^{c+d} \frac{(c+d)!}{c!\,d!} p^c q^d$$

is not the only polynomial in p,q the value of which is constantly equal to unity.≫

Fisher then gives the following interesting diagram by way of illustration (reproduced courtesy of University of Adelaide Library, Rare Books and Manuscripts) and comments:

≪If, in fact, the triangle is extended to any chosen boundary, as for example in the diagram, the thirteen totals outside the boundary are the coefficients $\omega(c,d)$ of a polynomial

$$\sum \omega(c,d) p^c q^d = p^4 + 4p^6 q + 18p^6 q^2 + \ldots$$

of which the value is unity for all values of p.

Then, based on previous experience of a successes out of $a+b$, we may infer the probability of reaching the terminal value (c,d) to be

$$\frac{(a+b+1)!}{a!\,b!} \cdot \frac{(a+c)!(b+d)!}{(a+b+c+d+1)!} \cdot \omega(c,d),$$

if the subsequent trial were made with these endpoints.≫

What is going on here? In ordinary coin-tossing, stopping after n-tosses is a simple stopping rule, which permits us to partition the space of outcomes into those sequences of heads and tails that result in k heads, $0 \le k \le n$. However, more generally, suppose the stopping rule is to stop when, as in Figure 1, one exits the area delineated by the dark line. Then, one can again decompose the sequence of possible outcomes into those resulting in c heads and d tails. The probability of any one such sequence is, as before,

$$\frac{(a+b+1)!}{a!b!} \cdot \frac{(a+c)!(b+d)!}{(a+b+c+d+1)!}$$

but now we have to multiple this single-path probability by the number of such paths $w(c,d)$ to obtain the probability of having c heads and d tails when we stop. Since we have to terminate in some (c,d) pair, the sum of their probabilities is one. In effect, Fisher is calculating the consequences of a sequential stopping rule assuming Bayes' postulate.

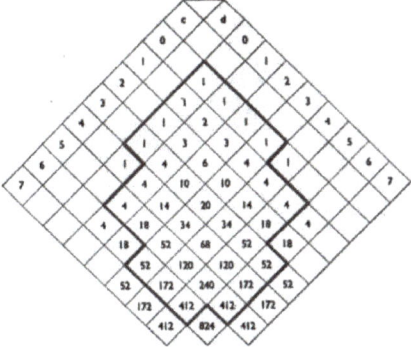

Figure 1. The arithmetical triangle extended.

4.4. Fiducial Prediction

Of course, having first considered Bayesian prediction, it was only to be expected Fisher would also discuss prediction from the fiducial standpoint as well (Section 5.3). Suppose that for $j = 1, 2$ that X_j is the sum of N_j independent exponentials (parameter θ). Then, $X_j \sim \Gamma(N_j, \theta)$, and one has the "pivotal" quantity:

$$\chi^2_{2N_j} = 2\theta X_j.$$

The fiducial distribution of θ given X_1 is $\Gamma(N_1, X_1)$; that is, it has density:

$$\frac{x_1^{N_1}}{(N_1 - 1)!} \theta^{N_1 - 1} e^{-x_1 \theta} d\theta.$$

The predictive distribution of X_2 given X_1 therefore has density:

$$f(X_2|X_1) = \frac{N_1 + N_2 - 1)!}{(N_1 - 1)!(N_2 - 1)!} \cdot \frac{X_1^{N_1} X_2^{N_2 - 1}}{(X_1 + X_2)^{N_1 + N_2}}.$$

Suppose that θ has an initial distribution given the *Jeffreys' prior* $d\theta/\theta$. Then, as Fisher notes, the posterior distribution of θ given X_1 *is the same as the fiducial distribution* given by Fisher. Therefore, once again, the Fisher and Jeffreys' approaches arrive at similar (here, in fact, identical) conclusions. However, Fisher had much earlier [12] rejected the use of this *improper* Jeffreys' prior.

Predictive fiducial inference in the case of the normal distribution had been considered much earlier in [34] and was discussed by him at length in [2] (Section 5.4).

4.5. Are Fiducial Probabilities Verifiable?

At the end of his discussion of fiducial prediction for exponential variates, Fisher remarks [2] (p. 118):

≪Such fiducial probability statements about future observations are verifiable by subsequent observations to any degree of position required.≫

Fisher's comment is readily understood in the case of this particular example, because the ratio $(N_1 X_2)/(N_2 X_1)$ has an F distribution on N_2, N_1 degrees of freedom. For example, if $N_1 = N_2 = 10$, then $P(X_2/X_1 < 2) = 0.8552$, and the prediction that the observed value of X_2 will be less than twice the observed value of X_1 exactly 86% of the time can be directly verified.

Fisher's more sweeping statement, however, that fiducial probabilities are in general verifiable, made both elsewhere in [2] (pp. 62 and 70) and on other occasions (as in his correspondence with Tukey; see [22] (pp. 221–231)), led to considerable mystification. When asked by Tukey about the case of the Behrens–Fisher distribution, Fisher justified his claim by pointing to the distributions of the two t-statistics entering into the calculation [22] (p. 230)); this suggests that what Fisher had in mind was the use of pivotals as building blocks in arriving at the final probability, but hardly answered Tukey's question.

5. The Fiducial Argument Revisited

The fiducial argument as presented in SMSI both differs from and expands on Fisher's earlier prewar views in several important ways and is related to his changing views about Bayes.

5.1. The Logical Status of the Fiducial Distribution

In 1930, Fisher wrote that fiducial and Bayesian posterior distributions differed not only numerically, but also in their logical meaning. A quarter of a century later, Fisher no longer believed this, writing in [2] (p. 59):

≪It is essential to introduce the absence of knowledge *a priori* as a distinctive datum in order to demonstrate completely the applicability of the fiducial method of reasoning to the particular and real experimental cases for which it was developed. This last point I failed to perceive when, in 1930, I first put forward the fiducial argument for calculating probabilities. For a time this led me to think that there was a difference in logical content between probability statements derived by different methods of reasoning. They are in reality no grounds for any such distinction. This contradicts Fisher's later statement in [2] (p. 105) that he had simply failed to make clear the need for this "distinctive datum" in his 1930 paper (scarcely credible given essentially the same language appears in two later papers, [12,34]), and then compounded this by going on to criticize Neyman and Pearson for not perceiving its necessity at the time.≫

This shift did not occur for some time: in two papers written shortly after, Fisher reprised the fiducial argument, using the examples of estimating the variance σ^2 by s^2 ([12]) and μ using t ([34]) in the case of the normal distribution, in terms virtually identical to those in his 1930 paper. (In his 1935 paper, Fisher added the caveat that "the statistics used contain the whole of the relevant information which the sample provides." This reflected what he regarded as an important distinction between Neyman's confidence intervals and a fiducial distribution.) Indeed, as late as 1940, in an exchange of letters with the French mathematician Maurice René Fréchet (1878–1973), Fisher sent Fréchet a copy of his 1930 paper as representative of his current views on the fiducial argument and specifically reiterated that "statements of fiducial probability have a logical content different from the more familiar statements of inverse probability" ([22] p. 121).

5.2. Is T Fixed?

The logical status of the fiducial distribution was closely related to the observed value of the statistic T. Initially, Fisher did not view T as fixed. In his 1940 correspondence with Fréchet, Fisher was very clear on this:

≪For the population of cases relative to which a fiducial probability is defined, the value of any relevant statistic T is not regarded as fixed. This I have deliberately exerted myself to make clear since my first writings on the subject...

I shall be glad to give you all possible support in dissuading mathematicians from thinking that they can obtain a true probability statement logically equivalent to one of the kind aimed at by Bayes' theorem, yet without using the approximate basis of this theorem. Believe me, I have never attempted anything so foolish. The inferences which can be drawn without the aid of Bayes' axiom seem to me of great importance, and quite precisely defined, but are certainly not statements of the distribution of a parameter θ over its possible values in a population defined by random samples selected to give a fixed estimate T.≫ [22] (p. 124)

Later, Fisher changed his mind on this. Initially, he apparently thought the matter so obvious he did not think through the matter carefully, writing in his 1955 paper that first accepting a symbolic statement such as:

$$Pr\{(\bar{x} - ts) < \mu < (\bar{x} + ts)\} = \alpha,$$

but then, after substituting observed values of $\bar{x}$ and s, proceeding to reject the resulting numerical statement, say:

$$Pr\{92.99 < \mu < 93.01\} = 95 \text{ per cent.},$$

was to violate "the principles of deductive logic", denying ≪the syllogistic process of making a substitution in the major premise of terms which the minor premise establishes as equivalent≫. This was an embarrassing gaffe: it is trivial to think of counterexamples where such a substitution is invalid. For example, if $a < b < c < d$ and X, Y are independent random variables such that

$$P(X = a) = P(X = d) = P(Y = b) = P(Y = c) = \frac{1}{2},$$

then

$$P(Y > X) = \frac{1}{2}, \quad \text{but} \quad P(Y > a) = 1 \quad \text{and} \quad P(Y > d) = 0.$$

Neyman of course pounced on this in a response, and it is surely significant that Fisher never again used this argument. Instead, he conceded (in SMSI, p. 57) there was a legitimate issue:

≪The applicability of the probability distribution to the particular unknown value of θ sought by an experimenter, without knowledge *a priori*, on the basis of the particular value of T given by his experiment, has been disputed, and certainly deserves to be examined.≫

In doing so, Fisher came up with an interesting and much more defensible (if ultimately still flawed) defense.

5.3. Recognizable Subsets

This invoked the concept of a *recognizable subset*. The simple illustrative example invoked by Fisher was that of tossing a die, and the question was the justification for identifying the relative frequency of, say, an ace in a long sequence of trials with the probability of obtaining an ace in a single toss. Fisher argued [2] (p. 35):

≪Before the limiting ratio of the whole set can be accepted as applicable to particular throw, a second condition must be satisfied, namely that no [subset having a different limiting ratio] can be *recognized*. This is a necessary and sufficient condition for the applicability of the limiting ratio of the entire aggregate of possible future throws as the probability of any one particular throw. On this condition we may think of a particular throw, or succession of throws, as a *random* sample from the aggregate, which is in this sense subjectively homogeneous and without recognizable stratification (There is an obvious and close connection here with Richard von Mises's concept of a *kollektiv*, with its twin assumptions of the existence of a limiting frequency (Fisher's first condition) and its invariance under place selection (Fisher's second condition). Note however the two concepts serve very different purposes. von Mises denied the existence of single-case probabilities; the *kollektiv* was designed instead to give a formal mathematical definition of random sequence, and the absence of place selections with differing limiting frequencies was what characterized randomness. For Fisher in contrast the absence of recognizable subsets permitted the identification of the class frequency with the probability of the individual. For Fisher's views on infinite populations and continuous variates as convenient fictions, see [2] (pp. 35 and 114). He was never a frequentist in the von Mises sense).≫

Somewhat surprisingly, Fisher went on to write [2] (p. 33):

≪This fundamental requirement for the applicability to individual cases of the concept of classical probability shows clearly the role of subjective ignorance, as well as that of objective knowledge in a typical probability statement.≫

The knowledge here is that of a "well-defined" population having a known limiting frequency ratio; the ignorance is our "inability to discriminate any of the different sub-aggregates having different limiting frequency ratios". There are obvious connections here to de Finetti's concept of exchangeability (see [35] for an interesting discussion) and more generally subjective Bayesianism (so much so that Fisher appears to have regretted his wording here, and in the third edition of SMSI, "subjective" and "objective" were changed to "well-specified" and "specific").

5.4. Recognizability and Fiducial Inference

In general, whether recognizable subsets exist in any specific setting is ultimately a matter of judgment, but in the specific setting of statistical estimation, their existence is a matter of mathematical fact. For the basic example of the t and the probability of the inequality:

$$\mu < \bar{x} - \frac{1}{\sqrt{N}} ts,$$

Fisher asserted [2] (p. 84):

≪The reference set for which this probability statement holds is that of the values of $\mu, \bar{x}$, and s corresponding to the same sample, for all samples of a given size of all normal populations. Since $\bar{x}$ and s are jointly Sufficient for estimation, and knowledge of μ and σ *a priori* is absent, there is no possibility of recognizing any sub-set of cases, within the general set, for which any different value of the probability should hold. The unknown parameter μ has therefore a frequency distribution *a posteriori* defined by Student's distribution (Contrast this with Fisher's discussion on pp. 61–62, where he advances a different, equally lapidary justification ("in the absence of a prior distribution of population values there is no meaning to be attached to the demand for calculating the results of random sampling among populations, and it is just this absence which completes the demonstration"). Savage [36] (p. 476), quoting at length from the passage containing this statement, refers to it as illustrating "Fisher's dogged blindness about it all").≫

No proof of this bold statement on Fisher's part was ever given however; presumably, he thought it so obvious as not to require one. Unfortunately for this approach, however, and somewhat surprisingly, recognizable subsets with variable frequencies do exist in the case of the t; see [37–39]. One cannot make the Bayesian omelet without breaking the Bayesian eggs.

5.5. Fiducial Reprise

Fisher's original fiducial construct gave a general method for constructing confidence intervals in the continuous one-dimensional case. This indeed had a logical content entirely different from that of a Bayesian prior. However, over time, the fiducial distribution—initially so carefully yoked to a pure sampling theory interpretation—began to suffer from mission creep. In 1934, after Neyman had advanced a more general theory of confidence intervals for the multi-parameter case, Fisher added the requirement that the statistics employed in a fiducial argument had to be exhaustive, even though this property did not enter into the logic of his 1930 paper. Later, Fisher attempted to extend the fiducial argument to the multi-parameter case, but could only accomplish this by approaches in which the original sampling interpretation had to be jettisoned.

At the same time—despite his protestations to Fréchet in 1940—Fisher began to write more and more in conditional rather than unconditional terms, culminating in his appeal in 1956 to the absence of recognizable subsets. However, this meant fiducial distributions could not co-exist with Bayesian posterior distributions, since the existence of a Bayesian prior meant recognizable sets with differing relative frequencies might exist. The basic requirement for using the fiducial argument it turned out was an absence of knowledge about the underlying parameters. However, why the absence of knowledge ensured the absence of such sets was never explained—just asserted—and in the case of the t, in fact, turned out to be false.

5.6. Aftermath

From 1956, when SMSI appeared, until his death in 1962, Fisher's book generated considerable controversy. One instance of particular interest is Lindley's review of the book in the journal *Heredity* [40], because it sheds some further light on Fisher's attitude towards Bayes.

George Barnard later reported [41] (p. 184) that Fisher "had been much upset" by Lindley's review, as well he might: Lindley, who was by then a prominent Bayesian, was highly critical of the book, beginning with what he described as a mathematical criticism, which "demonstrates that an error has been made", followed by "other criticisms which are far more matters of opinion". Fisher of course would not have cared about the criticisms that were matters of opinion, but the charge of an outright mathematical error was quite another matter. Fisher had asserted that the concept of probability involved in the fiducial argument was "entirely identical with the classical probability of the early writers, such as Bayes" (SMSI, p. 54). Lindley gave an example to show that this could not be the case.

Here is Lindley's example, interesting precisely because it is so simple. Consider the family of probability densities in x:

$$\frac{\theta^2}{\theta+1}(x+1)e^{-x\theta} \quad (x \geq 0, \theta > 0).$$

Given one or more observations drawn from such a population, their sum is a sufficient statistic for θ. In particular, given two independent observations x, y, let θ_{xy} denote the result of using x to generate a fiducial distribution for θ, which one then in turn uses as a Bayesian prior to generate a posterior distribution for θ using the other observation y. Then, a trite calculation [42] shows that the result depends on the order in which the two observations are used: $\theta_{xy} \neq \theta_{yx}$, and both differ from the fiducial distribution ψ generated by using x and y simultaneously. If fiducial probabilities were ordinary probabilities, how could this be? Why should the order matter?

Fisher clearly found Lindley's example vexing. In both a letter to George Barnard dated 27 February 1958 (quoted in [41]) and a paper two years later, Fisher interpreted Lindley as criticizing a particular example in [2] (Chapter V, Section 6) "Observations of two kinds", the version of Bayes' calculation "more suitable for the 20th century" discussed earlier, in which a continuous time is combined with a discrete count (while in Lindley's examples, the two successive observations were of the same type). This was to completely miss Lindley's point, which was not the particular example or the method endorsed by Fisher of combining the two observations, but the much more general issue of the lack of consistency if probabilities generated by the fiducial argument were really probabilities in the classical sense. Fisher added in his letter to Barnard:

≪In fact, the more I consider it, the more clearly it would appear that I have been doing almost exactly what Bayes had done in the 18th century. As Lindley purports to be a protagonist of Bayes, it seems that his misunderstanding and confusion goes deeper than anyone could imagine.≫

As Barnard notes, in his response to Fisher, his own confusion was "hardly less than Lindley's" and "led to an acrimonious encounter between Fisher and myself at a conference later that year".

Here, Fisher has in some ways come almost full circle, in effect defending his example because of its agreement with Bayes' own approach. Of the many other contemporaneous critiques of Fisher's defense of fiducial inference in SMSI and other papers from this period, three of particular note are [43–45].

6. Conclusions

From relatively early on in his career, Fisher was a dedicated opponent of "inverse probability". This was certainly a matter of principle with him, but his vehemence may also have been spurred on by what he regarded as Karl Pearson's springing on him an unwarranted criticism of an important part of his famous 1915 paper on the correlation coefficient. However, it is difficult to replace something by nothing, and Fisher would certainly have recognized the unsatisfactory state of affairs arising from abandoning the Bayesian position without having an adequate alternative ready at hand.

Fisher later thought he had discovered such an alternative in 1930 in the guise of fiducial inference. However, the happy accident that in the case of a single parameter, fiducial percentiles can be spliced together to form a distribution function in the purely mathematical sense (that is, a function increasing from 0 to 1) led him to regard this construct as a viable and principled probabilistic replacement for a Bayesian posterior distribution. His inability to extend this construction to the multi-parameter setting in a way that won general acceptance never caused him to waver from this view. See [46] for an outstanding discussion of the complexities that arise in the case of the *Behrens–Fisher problem*, estimating a difference of means drawn from two normal populations having possibly different variances.

However, Fisher's evolving view of fiducial inference, one which downplayed its sampling theory origins, led him to view the Bayesian and fiducial approaches as viable alternatives based on qualitatively different states of knowledge, and this appears to have led him to view Bayes' original paper with increasing appreciation (Bayes' "mathematical contributions to the *Philosophical Transactions* show him to have been in the first rank of independent thinkers, very well qualified to attempt the really revolutionary task opened out by his posthumous paper" (SMSI, p. 8)). His last book on statistical methods and scientific inference, although controversial in many of its pronouncements, makes for fascinating reading, in part because of its discussion of prediction, contrasting and developing the Bayesian and fiducial approaches.

Note: There were three editions of *Statistical Methods and Scientific Inference*: 1956, 1959, and 1973 [2–5]. Page references in this paper are to the 3rd edition, as being the most available (in the 1990 Bennett reprint). Where there are differences between editions in a passage being cited, this is noted. Sometimes the changes between editions consisted of additional material being inserted (as in Fisher's discussion of the rule of succession in the 2nd edition or Todhunter in the 3rd); and sometimes subtle changes that can only be easily identified by a change to a lighter typeface (for an interesting example of the latter, see [28] p. 380). In the 3rd, posthumous edition the changes were based on material Fisher "had entered in his interleaved copy of the book for this purpose sometime before his death" (p. v).

Funding: This research received funding from the European Research Council (ERC) under the European Union's Horizon 2020 research and innovation programme under grant agreement No 817257.

Data Availability Statement: Not applicable.

Conflicts of Interest: The authors declare no conflict of interest.

References

1. Fienberg, S.E. When did Bayesian inference become "Bayesian"? *Bayesian Anal.* **2006**, *1*, 1–40. [CrossRef]
2. Fisher, R.A. *Statistical Methods and Scientific Inference*; Oliver & Boyd: Edinburgh, UK, 1956.
3. Fisher, R.A. *Statistical Methods and Scientific Inference*, 2nd ed.; Oliver and Boyd: Edinburgh, UK, 1959.
4. Fisher, R.A. *Statistical Methods and Scientific Inference*, 3rd (posthumous) ed.; Collier Macmillan: London, UK, 1973.
5. Fisher, R.A. *Statistical Methods and Scientific Inference*, 3rd ed.; Reprinted in Statistical Methods, Experimental Design, and Scientific Inference; Bennett, J.H., Ed.; Oxford University Press: Oxford, UK, 1990.
6. Bayes, T. An essay towards solving a problem in the Doctrine of Chances. *Philos. Trans. R. Soc. Lond.* **1763**, *53*, 370–418. [CrossRef]
7. Fisher, R.A. On an absolute criterion for fitting frequency curves. *Messenger Math.* **1912**, *41*, 155–160.
8. Fisher, R.A. Frequency distribution of the values of the correlation coefficient in samples from an indefinitely large population. *Biometrika* **1915**, *10*, 507–521. [CrossRef]
9. Fisher, R.A. On the "probable error" of a coefficient of correlation deduced from a small sample. *Metron* **1921**, *1*, 3–32.
10. Fisher, R.A. On the mathematical foundations of theoretical statistics. *Philos. Trans. R. Soc. A* **1922**, *222*, 309–368.
11. Fisher, R.A. Uncertain inference. *Proc. Am. Acad. Arts Sci.* **1936**, *71*, 245–258. [CrossRef]
12. Fisher, R.A. The concepts of inverse probability and fiducial probability referring to unknown parameters. *Proc. R. Soc. A* **1933**, *139*, 343–348.
13. Aldrich, J.R.A. Fisher and the making of maximum likelihood 1912–1922. *Stat. Sci.* **1997**, *12*, 162–176. [CrossRef]
14. Edwards, A.W.F. What did Fisher mean by "inverse probability" in 1912–1922? *Stat. Sci.* **1997**, *12*, 177–184. [CrossRef]
15. von Kries, J. *Die Principien der Wahrscheinlichkeitsrechnung, Eine Logische Untersuchung*; Mohr: Tübingen, Germany, 1886.
16. von Kries, J. *Die Principien der Wahrscheinlichkeitsrechnung, Eine Logische Untersuchung*, 2nd ed.; Mohr: Tübingen, Germany, 1927.
17. Keynes, J.M. *A Treatise on Probability*; Macmillan: London, UK, 1921.
18. Soper, H.E.; Young, A.W.; Cave, B.M.; Lee, A.; Pearson, K. On the distribution of the correlation coefficient in small samples. Appendix II to the papers of "Student" and R. A. Fisher. A cooperative study. *Biometrika* **1917**, *11*, 328–413. [CrossRef]
19. Edwards, A.W.F. R.A. Fisher on Karl Pearson. *Notes Rec. R. Soc. Lond.* **1994**, *48*, 97–106.
20. Coolidge, J.L. *An Introduction To Mathematical Probability*; Clarendon Press: Oxford, UK, 1925.
21. Fisher, R.A. Inverse probability. *Proc. Camb. Philos. Soc.* **1930**, *26*, 528–535. [CrossRef]
22. Bennett, J.H. *Statistical Inference and Analysis: Selected Correspondence of R. A. Fisher*; Clarendon Press: Oxford, UK, 1990.
23. Neyman, J. On the two different aspects of the representative method: The method of stratified sampling and the method of purposive selection. *J. R. Stat. Soc.* **1934**, *97*, 558–625. [CrossRef]
24. Zabell, S.L. R.A. Fisher and fiducial argument. *Stat. Sci.* **1992**, *7*, 369–387. [CrossRef]
25. Box, J.F. *R. A. Fisher: The Life of a Scientist*; Wiley: New York, NY, USA, 1978.
26. Edwards, A.W.F. Commentary on the arguments of Thomas Bayes. *Scand. J. Stat.* **1978**, *5*, 116–118.
27. Stigler, S.M. Thomas Bayes' Bayesian inference. *J. R. Stat. Soc. Ser. A Gen.* **1982**, *145*, 250–258. [CrossRef]
28. Zabell, S.L. Philosophy of inductive logic: The Bayesian perspective. In *The Development of Modern Logic*; Haaparanta, L., Ed.; Oxford University Press: Oxford, UK, 2009.
29. Pearson, K. The fundamental problem of practical statistics. *Biometrika* **1920**, *13*, 1–16. [CrossRef]
30. Edgeworth, F.Y. Molecular statistics. *J. R. Stat. Soc.* **1921**, *84*, 71–89. [CrossRef]
31. Burnside, W. On Bayes' formula. *Biometrika* **1924**, *16*, 189. [CrossRef]
32. Pearson, K. Note on the "fundamental problem of practical statistics". *Biometrika* **1921**, *13*, 300–301.
33. Pearson, K. Note on Bayes' theorem. *Biometrika* **1924**, *16*, 190–193. [CrossRef]
34. Fisher, R.A. The fiducial argument in statistical inference. *Ann. Eugen.* **1935**, *6*, 391–398. [CrossRef]
35. Lindley, D.V.; Novick, M.R. The role of exchangeability in inference. *Ann. Stat.* **1981**, *9*, 45–58. [CrossRef]
36. Savage, L.J. On rereading R. A. Fisher. *Ann. Stat.* **1976**, *4*, 441–500. [CrossRef]
37. Buehler, R.J. Some validity criteria for statistical inferences. *Ann. Math. Stat.* **1959**, *30*, 845–863. [CrossRef]
38. Buehler, R.J.; Feddersen, A.P. Note on a conditional property of student's t. *Ann. Math. Stat.* **1963**, *34*, 1098–1100. [CrossRef]

39. Brown, L. The conditional level of Student's *t* test. *Ann. Math. Stat.* **1967**, *38*, 1068–1071. [CrossRef]
40. Lindley, D.V. Review: Statistical Methods and Scientific Inference. *Heredity* **1957**, *11*, 280–283. [CrossRef]
41. Barnard, G.A.R.A. Fisher—A true Bayesian? *Int. Stat. Rev.* **1987**, *55*, 183–189. [CrossRef]
42. Lindley, D.V. Fiducial distributions and Bayes theorem. *J. R. Stat. Soc. B* **1958**, *20*, 102–107. [CrossRef]
43. Pitman, E.J.P. Statistics and science. *J. Am. Stat. Assoc.* **1957**, *52*, 322–330. [CrossRef]
44. Tukey, J.W. Some examples with fiducial relevance. *Ann. Math. Stat.* **1957**, *28*, 687–695. [CrossRef]
45. Dempster, A.P. Further examples of inconsistencies in the fiducial argument. *Ann. Math. Stat.* **1963**, *34*, 884–891. [CrossRef]
46. Wallace, D.L. The Behrens-Fisher and Fieller-Creasy problems. In *R. A. Fisher: An Appreciation*; Lecture Notes in Statistics; Fienberg, S.E., Hinkley, D.V., Eds.; Springer: Berlin/Heidelberg, Germany, 1980; Volume 1; pp. 119–147.

MDPI
St. Alban-Anlage 66
4052 Basel
Switzerland
Tel. +41 61 683 77 34
Fax +41 61 302 89 18
www.mdpi.com

Mathematics Editorial Office
E-mail: mathematics@mdpi.com
www.mdpi.com/journal/mathematics

www.ingramcontent.com/pod-product-compliance
Lightning Source LLC
LaVergne TN
LVHW070155120526
838202LV00013BA/1145